Data Mining and Statistical Analysis Using SQL

ROBERT P. TRUEBLOOD AND JOHN N. LOVETT, JR.

apress™

Data Mining and Statistical Analysis Using SQL
Copyright ©2001 by Robert P. Trueblood and John N. Lovett, Jr.

ISBN (pbk): 1-893115-54-2

Printed and bound in the United States of America 12345678910

Editorial Directors: Dan Appleman, Gary Cornell, Jason Gilmore, Karen Watterson

Technical Editor: Alexander Werner

Project Manager, Copy Editor, and Production Editor: Anne Friedman

Compositor: Susan Glinert

Cartoonist: John N. Lovett, Jr.

Artist: Tony Jonick

Proofreaders: Carol Burbo, Doris Wong

Indexer: Carol Burbo

Cover Designer: Karl Miyajima

Marketing Manager: Stephanie Rodriguez

Distributed to the book trade in the United States by Springer-Verlag New York, Inc., 175 Fifth Avenue, New York, NY, 10010

and outside the United States by Springer-Verlag GmbH & Co. KG, Tiergartenstr. 17, 69112 Heidelberg, Germany

In the United States, phone 1-800-SPRINGER; orders@springer-ny.com; http://www.springer-ny.com

Outside the United States, contact orders@springer.de; http://www.springer.de; fax +49 6221 345229

For information on translations, please contact Apress directly at 901 Grayson Street, Suite 204, Berkeley, CA, 94710 Phone: 510-549-5938; Fax: 510-549-5939; info@apress.com; http://www.apress.com

We wish to dedicate this book to our wives, Janie Lovett and Sue Trueblood, and to Chris and Jennifer Trueblood.

About the Authors

Robert P. Trueblood (on the left) has a B.S. in Mathematics from Auburn University, an M.S. in Computer Science from the University of Tennessee, and a Ph.D. in Computer Science and Applications from Virginia Polytechnic Institute and State University. While at Auburn, he met and later married Sue Stephens. After receiving his Ph.D. he taught computer science at the university level from 1979–1996. In 1991, he became an active consultant for QuantiTech, Inc., and in 1996, he joined the company full-time. While at QuantiTech, he has developed several specialized applications involving unique engineering techniques. These applications make extensive use of databases and are written in Visual Basic. He has published articles and VB tips in Pinnacle's *Visual Basic Developer.*

    John N. Lovett, Jr. (on the right) has a B.A. in Mathematics from Hendrix College, and an M.S. in Operations Research and a Ph.D. degree in Industrial Engineering from the University of Arkansas. He taught at two universities between 1976 and 1991. John has consulted privately with over 25 organizations. In 1997 he was asked to help QuantiTech, Inc., a small engineering firm in Huntsville, Alabama. There he met Robert Trueblood, and they began collaborating on several projects. The balance of his time is spent developing Falls Mill, an operating 1873 water-powered mill he and his wife Janie own in Franklin County, Tennessee, and its museum (see http://www.fallsmill.com), and helping in the restoration of over 20 other old mills around the country.

Contents at a Glance

Dedication .... *iii*

About the Authors .... *iv*

Introduction .... *xi*

Acknowledgments .... *xvi*

Chapter 1   Basic Statistical Principles
             and Diagnostic Tree *1*

Chapter 2   Measures of Central Tendency and Dispersion 9

Chapter 3   Goodness of Fit *41*

Chapter 4   Additional Tests of Hypothesis 85

Chapter 5   Curve Fitting *119*

Chapter 6   Control Charting *181*

Chapter 7   Analysis of Experimental Designs 229

Chapter 8   Time Series Analysis 277

Appendix A   Overview of Relational Database Structure
             and SQL 337

Appendix B   Statistical Tables 359

Appendix C   Tables of Statistical Distributions
             and Their Characteristics 373

Appendix D   Visual Basic Routines *381*

Bibliography 397

Index *399*

Contents

Dedication.. *iii*

About the Authors... *iv*

Introduction.. *xi*

Acknowledgments.. *xvi*

Chapter 1   Basic Statistical Principles
            and Diagnostic Tree... *1*

Categories of Data... *2*
Sampling Methods.. *2*
Diagnostic Tree... *5*
SQL Data Extraction Examples... *7*

Chapter 2   Measures of Central Tendency
            and Dispersion... *9*

Measures of Central Tendency... *10*
    Mean *10*
    Median *12*
    Mode *16*
    Geometric Mean *17*
    Weighted Mean.. *20*
Measures of Dispersion *22*
    Histogram Construction... *22*
    Range *32*
    Standard Deviation *33*
Conclusion.. *40*

Chapter 3   Goodness of Fit 41

Tests of Hypothesis 43
Goodness of Fit Test 46
Fitting a Normal Distribution to Observed Data 47
Fitting a Poisson Distribution to Observed Data 62
Fitting an Exponential Distribution to Observed Data 68
Conclusion 71
T-SQL Source Code 72
Make_Intervals 73
Combine_Intervals 74
Compare_Observed_And_Expected 80
Procedure Calls 82

Chapter 4   Additional Tests of Hypothesis 85

Comparing a Single Mean to a Specified Value 88
Comparing Means and Variances of Two Samples 94
Comparisons of More Than Two Samples 101
Conclusion 104
T-SQL Source Code 105
Calculate_T_Statistic 105
Calculate_Z_Statistic 107
Compare_Means_2_Samples 108
Contingency_Test 113
Procedure Calls 117

Chapter 5   Curve Fitting 119

Linear Regression in Two Variables 121
Linear Correlation in Two Variables 127
Polynomial Regression in Two Variables 130
Other Nonlinear Regression Models 136
Linear Regression in More Than Two Variables 141
Conclusion 147
T-SQL Source Code 147
Linear_Regression_2_Variables 148
Gaussian_Elimination 150
Array_2D 158
Polynomial_Regression 160
Exponential_Model 169
Multiple_Linear_Regression 172
Procedure Calls 179

Chapter 6   Control Charting *181*

Common and Special Causes of Variation *183*
Dissecting the Control Chart *193*
Control Charts for Sample Range and Mean Values *195*
Control Chart for Fraction Nonconforming *206*
Control Chart for Number of Nonconformities *213*
Conclusion *215*
  T-SQL Source Code *216*
    Sample_Range_and_Mean_Charts *216*
    Standard_P_Chart *219*
    Stabilized_P_Chart *222*
    C_Chart *224*
    Procedure Calls *227*

Chapter 7   Analysis of Experimental Designs *229*

One-Way ANOVA *231*
Two-Way ANOVA *238*
ANOVA Involving Three Factors *245*
Conclusion *260*
T-SQL Source Code *261*
    ANOVA *261*
    Procedure Calls *275*

Chapter 8   Time Series Analysis *277*

Simple Moving Average *278*
Single Exponential Smoothing *286*
Double Exponential Smoothing *292*
Incorporating Seasonal Influences *300*
Criteria for Selecting the Most Appropriate
    Forecasting Technique *308*
Conclusion *311*
T-SQL Source Code *311*
    Simple Moving Average *312*
    Weighted Moving Average *315*
    Single Exponential Smoothing *318*
    Double Exponential Smoothing...................................... *322*
    Seasonal Adjustment *327*
    Procedure Calls *332*

Appendix A Overview of Relational Database Structure and SQL *337*

Appendix B Statistical Tables *359*

Appendix C Tables of Statistical Distributions and Their Characteristics *373*

Appendix D Visual Basic Routines *381*

Bibliography *397*

Index *399*

Introduction

Where to Start

WITH THE EXPLOSION IN computer technology during the past 30 years, there has been an accompanying proliferation of data. In many cases, however, companies have stacks of printed data or electronic databases that are of little use to them. This is often because those who could benefit the most from this information lack experience using the tools and techniques that could help them extract knowledge from the numbers. Now, more and more databases are also placed on the Internet. In fact, at no other time in history has more information been accessible to so many people. But how can it help us?

Access to this information is important because data collected in the past may be used to predict future trends or to characterize a population from which a sample has been extracted. For example, a sample poll among 1,000 voters is often used to predict the outcome of an election. The latest computer-age term for this exercise is "data mining." Statistics is the science (and art) of mining or analyzing such data. Like lawyers and physicians, statisticians have traditionally insulated themselves from the masses by using jargon and Greek symbols, thereby helping to ensure the need for their services. Although certain aspects of the science are mathematically complex, many of the practical applications of statistics may be put to use without an advanced degree, or any degree at all. It is the purpose of this book to guide the user through these practical techniques, without delving deeply into the theoretical foundations that verify the validity of the results (we'll leave these in the ivory towers).

When working with computer databases, it is helpful to understand how to "query" the database to extract needed information. For example, in a large collection of data on heights and weights of a population of people, it may be useful to know the average height and weight, the spread of heights and weights about this average, and the height and weight that represent the upper 5 percent of the population. These types of questions can be answered by applying standard statistical techniques and using Structured Query Language (SQL). In order to know the questions to ask regarding the database, and effectively extract the desired information, it's necessary to understand some basic principles of statistics as well as the mechanics of SQL.

Who Can Benefit From This Book

Anyone who collects data and must extract information from the collected data can benefit from this book. If you're faced with the task of analyzing Web logs, extracting customer behavior patterns, or looking for ways to explore relationships among data elements, then you've found the right source. This book is designed and written for professionals who may possess minimal statistical knowledge, but are looking for a book that can provide specific "how to" information on accomplishing statistical calculations. Programmers, data administrators, Web administrators, managers, economists, marketers, engineers, scientists, and, yes, statisticians, can find this book of great value when the need arises for applying data analyses. In addition, students of statistics, computer science, engineering, and business management should find this book more useful in explaining the tools and techniques of data mining than their traditional textbooks.

This book also contains SQL statements in Microsoft Access 2000 and routines that accomplish the statistical calculations for the examples presented within each chapter. Although most of these statements are specific to our examples, they may easily be modified to handle any database of similar format. We have provided queries and routines for use with both Access and SQL Server. Knowledge of SQL isn't required to understand the statistics or follow the examples. We provide the SQL statements because most of the time the data that needs to be analyzed is maintained in a database, and most database management systems support a dialogue of SQL for extracting information from the database. Even though you may have limited knowledge of SQL, you'll be able to follow the statistics and examples. However, since SQL is very popular and relatively easy to read and understand, you might find the SQL statements a bonus for your work.

In an effort to provide more generic algorithms for solving the statistical problems presented in this book, we have also placed at the end of Chapters 3-8 generalized Transaction SQL (T-SQL) routines for Microsoft SQL Server 2000. These combine the individual steps of the Access SQL statements into a comprehensive solution package.

....

Robert's Rubies
Post Cold War Recycling

Having a good understanding of SQL can be very useful in manipulating data even in an unusual way. I recently had the opportunity to work on a project involving the demilitarization of obsolete or damaged military munitions. Various amounts of these munitions were stored at military bases across the country and around the world. However, only certain military bases had the capacity to dismantle selected types of munitions. Thus, many of the munitions had to be shipped from one base to another to be dismantled. The problem involved finding the most cost-effective shipping plan that would give the maximum amount of munitions dismantled under a fixed annual budget. Recognizing this as a linear programming problem, we quickly discovered that the problem involved over 40,000 variables and over 50,000 constraint equations. Linear programming is concerned with finding an optimal solution to a linear objective function subject to a set of linear constraint equations (usually a set of linear inequalities). The real problem came into full bloom when we asked ourselves who was going to write all these equations. Knowing that the data for the problem was stored in a relational database, it quickly came to me that I could generate the equations via a set of SQL queries. I used Visual Basic to formulate the SQL queries and to apply the queries to the database. The SQL query results were then used to formulate the linear equations that were written out to a text file. An off-the-shelf application package, called

LINGO by Lindo Systems, Inc., was used to solve the linear programming problem. However, that was not the end. We still had to take the solution and store the results into the database. Again using Visual Basic to generate SQL update queries, we were able to save the data in the database. The model is now being used by the government to determine the most cost-effective methods to recycle materials recovered from old munitions.

The Structure of the Book

The purpose of this book is to present to the programmer, database manager, analyst, or the non-statistician the ways in which statistical techniques and SQL can be employed to answer useful questions about a collection of data. Each chapter includes the following:

- Basic statistical tools and techniques that apply to a data mining situation

- The formulas needed to perform the required analysis

- A numerical example using a real-world database to illustrate the analysis

- Data visualization and presentation options (graphs, charts, tables)

- SQL procedures for extracting the desired results

The book is organized as follows:

Chapter 1: Basic Statistical Principles and Diagnostic Tree. This chapter provides a brief discussion of the basic principles of statistics and SQL, and presents a diagnostic tree that guides the user to a problem solution.

Chapter 2: Measures of Central Tendency and Dispersion. This chapter presents methods for determining basic measures of characterization for the data, such as mean and standard deviation.

Chapter 3: Goodness of Fit. This chapter focuses on how to fit a statistical distribution to the data set and test for goodness of fit. It includes the procedure for structuring a test of hypothesis and reaching a conclusion.

Chapter 4: Additional Tests of Hypothesis. This chapter discusses tests of normality, tests between means of two data sets, tests between variances, and tests of contingency.

Chapter 5: Curve Fitting. This chapter presents techniques for fitting lines and curves to data sets and testing for the "best" fit (regression and correlation methods).

Chapter 6: Control Charts. This chapter presents methods for plotting and analyzing data by using attribute or variable control charts and determining processes capability. The chapter also discusses the important notion of random variation in processes versus assignable causes.

Chapter 7: Analysis of Experimental Designs. This chapter features some basic techniques for analyzing data collected from experimental designs by analysis of variance (ANOVA), using the standard F-test, and a brief discussion of Taguchi methods.

Chapter 8: Time Series Analysis. This chapter deals with methods of forecasting trends, such as moving averages and exponential smoothing.

Appendixes: Four appendixes are included to supplement the chapter information. These include a brief overview of relational databases and SQL (Appendix A), a set of all statistical tables referenced in the book (Appendix B), extensive tables of information on various statistical distributions (Appendix C), and Visual Basic routines for performing statistical calculations (Appendix D).

What's on the Web?

This book is accompanied by a download from the Web site `http://www.apress.com` that contains the databases, SQL statements, and routines used in all the examples. The download also contains all the statistical tables given in Appendix B.

> **NOTE** *You will need to answer questions pertaining to this book in order to success-fully download the code.*

Invitation to the Reader

We have worked diligently to simplify the statistical principles and techniques presented to you, the reader, in this book. It is our hope that we have succeeded in producing an informative and readable narrative, supplemented with relevant examples and illustrations. While each chapter to some degree stands alone, we encourage you to read through the book sequentially and benefit from a "building block" approach. Don't be intimidated by the statistical symbols and jargon. All the tools you'll need to be a successful data miner can be simplified to a group of basic steps involving arithmetic calculations and computer database operations (SQL queries). Thanks for investigating this book. We hope it will help you.

Acknowledgments

WE WISH TO THANK our wives, Janie Lovett and Sue Trueblood, for their patience and encouragement during the collaboration and writing of this book. Also we are indebted to Karen Watterson for her initial request to produce this book for Apress, and for her subsequent advice and support. The help of editors Anne Friedman and Alexander Werner and managing editor Grace Wong, as well as those who assisted them, has been greatly appreciated. We also wish to thank Sheila Brown, president and CEO of QuantiTech, Inc., for allowing us to collaborate and use her company's facilities during the writing of this book. Finally, we thank in advance you, the reader and purchaser of this book, for your confidence in our ability to produce a readable and practical guide to data mining.

CHAPTER 1

Basic Statistical Principles and Diagnostic Tree

Where to Look?

Categories of Data

NO ONE IS BORN A DATA MINER. In order to grow expertise as a data miner or as an information analyst, you need to obtain certain basic knowledge. Then you need data to mine, as well as a way to measure the important characteristics of a process or phenomenon, so you can employ the appropriate statistical tools. Measuring doesn't necessarily mean using a ruler, calipers, or a scale. It can also be simply a "yes" or "no" decision. Statisticians and data miners typically categorize data as follows:

> *Variables data* represent actual measured quantities, such as weights, dimensions, temperatures, proportions, and the like. The measurements have units associated with them (for example, inches, pounds, degrees Fahrenheit, and centimeters). Because variables data may take on any value within a certain range (subject to the precision of the measuring instrument), these observations are sometimes said to be *continuous*.

> *Attributes data*, on the other hand, represent the classification of measurements into one of two categories (such as "defective" or "nondefective") or the number of occurrences of some phenomenon (such as the number of airplanes that arrive at an airport each hour). In most cases, these types of observations can only assume integer values, so attributes data are said to be *discrete*.

Collections of data are initially classified into either variables or attributes because the statistical tools used to analyze the data vary slightly depending on the category. It's important for you, as a data miner, to classify your observations appropriately.

Sampling Methods

Data values are collected by a variety of methods for a variety of reasons. Sometimes you'll want to inspect every possible item for a particular quality characteristic, such as the weights of boxes coming off a packaging line. This is known as 100 percent inspection. The alternative is sampling inspection, where information about the overall process is inferred by a relatively small sample collected from the process. Sampling inspection is often more cost effective than 100 percent inspection and still provides a high degree of accuracy.

For sampling inspection to be statistically sound, every candidate for measurement or inspection should have the same opportunity to be selected for the sample as every other candidate. Statisticians call such a sample a "random" sample. If this equal opportunity isn't assured, the statistical results may be biased or inaccurate. In many practical applications, however, assuring this random selection can be difficult.

Two approaches are commonly used for collecting sample data. In the first situation, a group of items might be selected randomly from a much larger group that has a finite or limited size. The larger group is known as the **population** or **universe**. Usually, the *sample* should usually contain at least 30 items from this population. If you calculate an average value from the 30 sample measurements, you'd expect the population average to be about the same if the sample is truly representative of the population. An example of this would be sampling 30 baskets from a corporation's large holiday order of 1,000 custom baskets. This concept is illustrated in Figure 1-1.

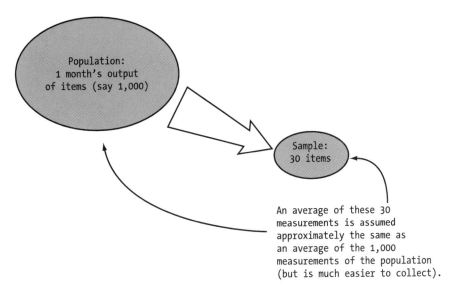

Figure 1-1. Sampling from a finite population

In the second situation, you view the process or population as continuous or infinite in nature, with no fixed stopping point. Sampling in this case might consist of collecting samples daily or hourly, each of relatively small size (4 or 5 observations per sample). One advantage of this approach is that you're able to detect changes in the process over time. Consider the daily fluctuations in a company's stock values. A single daily sample may not reveal much information, but the trend of stock values over a year's time might be detectable (see Figure 1-2).

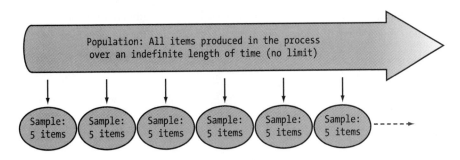

Figure 1-2. Sampling from a continuous (infinite) population.

In the case where a set of sample data represents one population and is straightforward, such as the monthly sales figures for a certain product over the past five years, the observations may be divided between two or more groups. For example, you might have sales figures by season (say, one group representing "winter" sales and one group representing "summer" sales). In the former case, several statistical tools may be invoked to answer questions about the data. These include measures of central tendency and dispersion, such as the average sales per month and range of sales (smallest to largest). It may also be helpful to group the data in bar chart format to reveal the shape and spread of the distribution of sales. This might indicate the shape of a known statistical distribution. We illustrate several examples in Chapters 2 and 3.

In the latter situation, where two or more data sets are collected, the analyst may wish to know whether the sets represent the same or distinct populations. In other words, are the winter sales figures significantly greater than the summer figures to warrant a campaign to increase summer sales? As time progresses, and monthly sales figures continue to be recorded, are trends revealed? Is the process stable (random) or predictable? Is it possible to fit a curve to the data to predict sales into the future? These types of questions can be answered by various statistical methods and data queries. These methods and the Structured Query Language (SQL) commands required to extract the desired information from the database are presented in subsequent chapters.

The Sampling Paradox

It is a common misconception that the size of a sample should remain a fixed percentage of the population size to maintain the same degree of accuracy for estimation purposes. For example, let's suppose we want to predict the result of a presidential election in one state. Maybe the voter base in that state is 10 million people. If we take a poll among 1,000 voters selected randomly, perhaps we could be 95% confident that the poll results would accurately reflect the preference of everyone who votes. This represents a 0.01% sample (1,000 out of 10,000,000). Now if we extend the poll to the entire United States voting base (let's assume it's 200 million just for simplicity), we might think we still need a 0.01 percent sample for the same degree of confidence in our prediction. This means, for 200 million voters, we must interview 20,000 people. Well, that's a lot of phone calls. Surprisingly, however, it turns out we can still achieve about the same degree of confidence in our results if we only interview 1,000 voters again! This saves time and money. The conclusion is, the confidence level of our prediction may be maintained if we stay with a fixed sample size, rather than let it increase proportionally with the population size. This is the Sampling Paradox.

Diagnostic Tree

Sometimes it's difficult to know what to do with the data or what questions to ask. We've developed a simple diagnostic tree to guide you through the maze of the various statistical measures and analysis techniques available. The tree is not intended to encompass all methodologies. It presents those most helpful and most commonly employed in statistical work. The tree is presented in Figure 1-3 in the form of a flow-chart, with references to the appropriate chapter at each decision point. Note that the tree begins with a check to determine if there is enough data to perform most statistical analyses. As a rule of thumb, there should be at least 30 observations, and preferably more for certain types of analyses, such as control charting.

In most applications, it will be necessary for you to calculate one or more of the basic measures of central tendency and dispersion discussed in Chapter 2. This relationship is indicated by dashed lines on the flowchart. Another guideline to note is that, in general, the techniques on the left portion of the flowchart involve static sets of sample data (one or more) from populations of variables or attributes (see Figure 1-1). The third column of flow symbols usually involve continuous (dynamic) data collected over time, again of variables or attributes form (see Figure 1-2). The typical data for the fourth column is either in tabular form (for analysis of variance), or in graphic format, such as in a two-dimensional coordinate system. In these situations, one of the variables depends on another (or several others) to generate its values. These concepts become clear in later chapters.

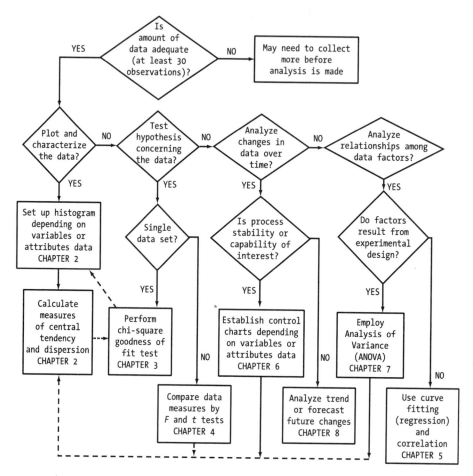

Figure 1-3. Diagnostic tree

John's Jewels
Let Statistics Work for You

Much of my teaching and consulting work was in the fields of quality assurance and quality management during the Great Quality Movement of the 1980s and 1990s. Organizations clamored to find people to teach their employees topics like Statistical Process Control (SPC) and Total Quality Management (TQM), and we professor types were always in need of supplementing our meager university incomes. I was frequently faced with the somewhat daunting task of teaching statistical principles to men and women with no background in the field whatever. In most situations, their supervisors required the implementation of SPC tools in the workplace, and the workers closest to the process were acknowledged to be the experts in their own areas, and so became the students of SPC. I started out, as we have in this book, with the types of data (variables and attributes), built up to histogram construction, measures, and usually wound up with control charting. One day I was invited into a plant where I was teaching workers SPC techniques. A man was eager to show me his control charts for his process. He operated a line of enormous die presses, stamping out parts for later assembly. He had begun applying a control chart we had studied to track the dimensions of the parts he was producing. One dimension had a tight tolerance that was critical to the later assembly process. He showed me how he selected his samples and measured this dimension with a micrometer. He then recorded the average dimension for his sample on his control chart. He walked to one of the machines that was dismantled and pulled the control chart. He pointed to the pattern of data and asked me enthusiastically, "See what this chart showed me?" The points on his chart varied randomly for a time, then began to trail up and up. "My parts were going out of tolerance," he said, "so I immediately stopped the machine and called in my supervisor. He agreed that we needed to tear down the press and find out why it was beginning to produce defective parts. We did, and found a shaft bearing going out. Maintenance is fixing it now. Back in the old days, I might have stamped 5,000 of these parts before the folks down the line found out they wouldn't fit and got word back to me!" This fellow made my day. I was finally seeing the practical benefit of statistics! I hope you will, too, as you read on.

SQL Data Extraction Examples

Throughout the chapters of this book, you will find sets of queries that perform statistical calculations for the examples presented. The queries are written in the relational database query language known as SQL (Structured Query Language). Please do not confuse the name SQL for Structured Query Language with software products known as SQL Servers, written by Microsoft, Sybase, or others. Microsoft's SQL Server is a relational database management system that uses an extended dialogue of the Structured Query Language. If perhaps you need a quick refresher on SQL, Appendix A provides an overview of the fundamentals of relational database structure and SQL. A very important feature of SQL that we use extensively is its ability to view the results of a query as another table of data in its own right. The beauty of this is that we can write a query that is based on a previous query. Thus, we can build from one query to the next to form a sequence of stair steps that take us from a set of raw data values to our desired statistical value.

   Most of the time a statistical data set is simply a collection of numbers. The collection often contains repeated values. Actually, repeated values can violate First Normal Form, but not the relational algebra. That is, when viewed as a database table, there can be duplicate (non-unique) rows. In order to work with this data in a tabularized form, we augment it with an additional column that can serve as a unique identifier. A simple, but very effective, augmentation is to number the data values. We make extensive use of this technique in our SQL examples.

As you progress through the chapters, you encounter query names, table names, and column names that look something like "Query 3_4" or "Table 3_1" rather than "Query 3-4" or "Table 3-1." In SQL the hyphen (dash) is the subtraction operator. Thus, we have to use the underscore (_) to identify queries, tables, and columns in SQL. However, when referring to a figure or table appearing in the text, we continue to use the "hyphen" notation, such as "Table 3-1." This is not a plot to confuse you. It simply means that "Table 3-1" in the text will appear as "Table 3_1" when referenced as a query.

Now grab your pick, shovel, and mule, and we'll begin to mine some data. Don't forget the hardtack and jerky.

CHAPTER 2

Measures of Central Tendency and Dispersion

Field of Dispersion

PROBABLY THE SIMPLEST and most descriptive data measures are averages and spreads. Because there are several available, they fall under the general categories of measures of central tendency and dispersion, respectively. A measure of central tendency is a value about which the other values in the data tend to cluster. The majority of the observed results usually crowd closely around this value, but a few deviate more and more to each side. This deviation may be characterized as an absolute deviation, or spread, or as an average deviation around the value of central tendency. Both measures are useful to us under different circumstances.

If the database is large enough (say at least 30 observations), it is common practice to organize and pictorially represent the data as a *histogram*. Then we can determine the appropriate measures we need to characterize the data. This sequence is shown in the diagnostic tree in Figure 1-3, as the left-most branch. We now discuss how to accomplish this.

Measures of Central Tendency

There are several measures of central tendency that qualify as "averages." These include the mean, median, mode, geometric mean, and weighted mean. Each is discussed and illustrated in the following sections.

Mean

The most common measure of central tendency is called the *mean*. Some statisticians refer to it as the "arithmetic mean," because it's simply the average of a set of observations. To obtain the mean, you just add all the values in the data and divide by the total number of values. This is an easy task, as illustrated by the example below.

Assume the following seven values are numbers of employees in a large company who opted for early retirement each month over a seven-month period:

44 39 45 48 40 45 47

The mean, commonly denoted $\bar{x}$, we calculate as follows:

$$\bar{x} = \frac{44+39+45+48+40+45+47}{7} = \frac{308}{7} = 44$$

Thus, the average number of employees per month opting for early retirement was 44 during the seven-month period. This exercise hardly drew sweat.

SQL/Query

Data like that given in the example above, are not often readily available. For example, the Employee database might contain a table called Retirees that lists each retired employee along with the date of retirement and other relevant information. Table 2-1 illustrates this.

Table 2-1. Table of Retirees

EMPLOYEE NUMBER	EMPLOYEE NAME	DATE HIRED	DATE RETIRED
1001	Adams	05/01/1988	12/10/1999
1004	Blake	07/01/1990	10/15/1999
1012	Clark	09/15/1979	11/30/1998
1023	Deans	01/15/1982	11/30/1999
1052	Evans	11/01/1985	12/31/1997
...	...	...	...

The question before you, the data miner, is "How do we calculate the mean for a number of retirees leaving monthly?" To answer this question, we first extract those records from the Retirees table that are within the seven-month period, then tally the number of retirees in each month. The following SQL query accomplishes this task:

Query 2_1:

```
SELECT MONTH([Date Retired]) AS [Month Retired],
COUNT([Employee Number]) AS [Number Retired]
FROM Retirees
WHERE [Date Retired] BETWEEN #06/01/1999# AND #12/31/1999#
GROUP BY Month([Date Retired]);
```

The Results of Query 2_1 are illustrated in Table 2-2.

Table 2-2. Results of Query 2_1

MONTH RETIRED	NUMBER RETIRED
6	44
7	39
...	...
12	47

11

Next, we need to calculate the mean. We do this by writing another SQL query that determines the average (mean) from the Number Retired column. The following query accomplishes this task:

Query 2_2:

```
SELECT AVG([Number Retired]) AS [Mean]
FROM Query 2_1;
```

The result is, of course, 44.

Median

In some situations, a more appropriate measure of central tendency may be warranted. For example, suppose in the above situation that the third data value was 10 instead of 45. If a new mean is calculated, its value is 39. Due to the significantly lower value of 10 compared to the other data, this mean tends to be smaller than might be appropriate to characterize the data. Therefore, another measure of central tendency called the *median* could be used. The median is simply the middle observation in a sequenced (or "arrayed") set of data. In other words, half the observations fall below the median, and half above. The new set of data in sequence is shown below.

    10 39 40 44 45 47 48

The median of this data, denoted by $\tilde{x}$, is the number right in the middle, namely $\tilde{x} = 44$.

This median value is more representative of the average number of employees retiring early than the mean value of 39, due to the "outlier" of 10. Incidentally, the median of the original seven values is 45. Yes, you're probably wondering how to find the median if the number of data values is even instead of odd, and technically there is no "middle" value. In this case, the common practice is to find the *two* data values that lie right in the middle and calculate their mean, or average, as the median. Let's see how we can accomplish this in SQL.

SQL/Query

Using SQL to find the median can be challenging because SQL does not have an operator or built-in function for calculating the median. Furthermore, SQL does not provide an operation or function for selecting the rows from the middle of a table. However, we can accomplish the calculation by sorting the values and by appending the sorted values to an empty table. In addition, the table must have an extra field whose data type is Counter (or autonumber).

> **NOTE**   *Depending on your use of various database management software packages, the data type autonumber might be known as Counter or Identity. In Microsoft Access 2000, Table Design View uses the data type autonumber whereas the SQL Create Table data definition language in Access 2000 uses the data type Counter. The data type Counter is also used in SQL statements via programming languages like Visual Basic. Microsoft SQL Server uses the data type Integer and Identity is given as the field attribute.*

First, the empty table called MedianTable is created as illustrated in Query 2_3. The field Seq acts as a counter that numbers the records as they are appended to the table. The data type of Val is shown as single, but should be set to match the data type of the set of values for which the median is to be determined.

Query 2_3:

```
CREATE TABLE MedianTable
(Seq Counter (1, 1),
Val as Single);
```

Next, we populate the table with the sorted values. Query 2_4 shows how this can be done using the retirement data that is extracted via Query 2_1. As each retirement number is added to the MedianTable, the value of Seq is incremented by one. When the last record is added, Seq is the count of records. Table 2-3 illustrates the results of Query 2_4.

Query 2_4:

```
INSERT INTO MedianTable ([Val])
SELECT [Query 2_1].[Number Retired])
FROM [Query 2_1]
ORDER BY [Query 2_1].[Number Retired]);
```

Table 2-3. Results of Query 2_4

SEQ	VAL
1	10
2	39
3	40
4	44
5	45
6	47
7	48

The next step is to find the middle record(s) so that we can extract the median value. As noted earlier, the value of the median depends on the total number of values. If there is an odd number of values, the median is simply the middle number. If there is an even number of values, the median is the average of the two middle values. Here lies the challenge. How do we do this in a single SQL query?

At first glance, we could simply add 1 to the record count and divide the sum by 2 and select the record whose Seq number is equal to the quotient. Looking at the maximum value for Seq in Table 2.3 we obtain the value 7. Adding one to this value and dividing by 2 yields the value 4. By selecting the row (record) whose Seq value is 4, we obtain the corresponding value 44 as the median. Query 2_5 gives the SQL for accomplishing this. In the query, the function Max returns the largest value of Seq, and the function Int rounds the value of the subquery to the nearest smallest integer. Query 2_5 works great if there is always an odd number of records. However, as luck would have it, we often have an even number of records.

Query 2_5:

```
SELECT Val As Median
FROM MedianTable
    WHERE MedianTable.Seq =
    SELECT Int((Max([Seq])+1)/2)
    FROM MedianTable);
```

To handle the case of an even number of records, we calculate the average of the middle two values. One way to accomplish this is to repeat the Where clause condition and change Max([Seq])+1 to Max([Seq])+3. This is shown in Query 2_6. Query 2_6 works great as long as the number of records is even. The question now arises, "Can we write a single query that works with either an even or odd number of values?" Yes, but we have to use a trick.

Query 2_6:

```
SELECT Avg([Val]) As Median
FROM MedianTable
WHERE MedianTable.Seq =
    (SELECT Int((Max([Seq])+1)/2)
    FROM MedianTable);
or MedianTable.Seq =
    (SELECT Int((Max([Seq])+3)/2)
    FROM MedianTable);
```

The trick is to rewrite the Where clause of Query 2_6 in such a way that if Max([seq]) is odd, the same middle record is selected twice. To accomplish this we take advantage of the Int function when dealing with negative numbers. For example, Int(7.5) is 7 and -Int(-6.5) is also 7. Using this fact, we rewrite the second Where clause which yields Query 2_7. Query 2_7 gives the correct median regardless of the number of records in the table. Thus, to find the median of a set of values, we do Query 2_3, Query 2_4, and Query 2_7.

Query 2_7:

```
SELECT Avg([Val]) As Median
FROM MedianTable
WHERE MedianTable.Seq =
    (SELECT Int((Max([Seq])+1)/2)
    FROM MedianTable);
or MedianTable.Seq =
    (SELECT -Int((-Max([Seq])-1)/2)
    FROM MedianTable);
```

NOTE *Some implementations of SQL provide the Ceil and Floor functions rather than the Int function. In those cases we replace Int((Max([Seq]+1)/2) with Floor((Max([seq])+1)/2, and we replace –Int((-Max([Seq])-1)/2) with Ceil((Max([Seq])+1)/2).*

If you need to recalculate the median, you should delete the table MedianTable created by Query 2_3 and rerun Query 2_3 before rerunning Query 2_4. The reason is that it resets the counter to one; otherwise, the counter continues to increment from its last value.

John's Jewels
Mean or Median?

When my students asked me when a median was a better measure of "average" than a mean, I offered the exam score example. It goes something like this: Now suppose I graded your exams, and the scores generally ranged from the 70s through the 90s, but one poor devil made a 15%. Now if you asked me the average score, I could quote the mean, but it would be somewhat biased, or artificially low, due to the "outlier" score of 15. However, if I ranked the scores from lowest to highest, then found the one right in the middle, this would better represent the "average," don't you think? Of course, a numerical example always helps.

Mode

There is yet another measure of central tendency that has its place in certain situations. This is the most frequently occurring value in the data set, called the *mode*. Again, it's easiest to find the mode if the data are sequenced. The mode of the original employee data above is 45, because it occurs twice, whereas each of the other values occurs only once. Incidentally, sometimes the data may have the same number of occurrences for several numbers, and in that case the data has multiple modal values. Data with two modes are said to be "bimodal".

SQL/Query

To obtain the mode via SQL, we need two queries. The first query groups and counts the values. The second query selects the value with the highest count. The first query is illustrated as Query 2_8. The selection of the most frequent value is accomplished by Query 2_9.

Query 2_8:

```
SELECT [Query 2_1].[Number Retired],
Count([Query 2_1].[Number Retired]) AS FreqCount
FROM [Query 2_1]
GROUP BY [Query 2_1].[Number Retired];
```

Query 2_9:

```
SELECT [Query 2_8].[Number Retired]
FROM [Query 2_8]
WHERE (((([Query 2_8].FreqCount)=
(SELECT Max(FreqCount) FROM [Query 2_8]))));
```

To summarize, the employee retirement data has mean 44, median 45, and mode 45. If these numbers are all equal, the data are said to be *symmetric*. We'll say more about this later in this chapter.

Geometric Mean

Special circumstances require other measures of central tendency. One you may some-times encounter is the "rate of change" situation, where a sequence of data values represents a growth or decline. In this case, each observation bears an approximately constant ratio to the preceding one. An average rate of change is a factor that, when multiplied successively by each observation, yields the final data value in the sequence. For example, let's suppose we are interested in finding the average increase in our salary over the past five years. Maybe we want to compare it to the rate of inflation to see if the salary has kept up. The annual salary figures are, let's suppose, the ones shown below.

$38,400 $41,500 $42,500 $47,000 $49,300

First, we could calculate the ratios for each pair of figures, proceeding through the years:

$$41,500/38,400 = 1.0807$$
$$42,500/41,500 = 1.0241$$
$$47,000/42,500 = 1.1059$$
$$49,300/47,000 = 1.0489$$

This shows that from year 1 to 2, our salary increased 8.07% ([1.0807 - 1] × 100%). From year 2 to 3 the increase was 2.41%, from 3 to 4 it was 10.59%, and from 4 to 5, 4.89%. Logic would tell us to add these four results and divide by four to obtain the average per year increase. But this would be incorrect. The arithmetic mean always overestimates the average rate of change, because it is an additive rather than a multi-plicative relationship. The correct method is to take the fourth root of the product of the four ratios to obtain what is known as the *geometric mean* $\bar{x}_g$, as follows:

$$\bar{x}_g = \sqrt[4]{(1.0807)(1.0241)(1.1059)(1.0489)} = 1.0645 \text{ or } 6.45\% \text{ increase}$$

If we apply this percent increase to the first year's salary, then apply it progressively to each result obtained, we hit \$49,300 at the end. An equivalent calculation is simply $\$38,400(1.0645)^4 = \$49,300$. Therefore, 6.45% is our average salary increase per year.

Have you already figured out a shortcut to this result? All we need to do is note the number of data values we have, subtract one to give the root (in this case 4), divide the last number by the first in the sequence (49,300/38,400), and take the fourth root of the quotient. The result is again 1.0645 or 6.45% increase. Much quicker! Now, let's take a look at how we can do this with SQL queries.

SQL/Query

Now suppose the above annual salary figures are stored in a field called Salary that is contained in a database table named Income. Further suppose that the table Income contains a field called Year that represents the year in which the salary was earned. Table 2-4 is the Income table for the five-year period beginning with 1995.

Table 2-4. The Income Table

YEAR	SALARY
1995	38,400
1996	41,500
1997	42,500
1998	47,000
1999	49,300

To calculate the geometric mean over the five-year period of 1995–1999, we must extract the 1995 salary and the 1999 salary and place them in the same row of another table, say GM. Table 2-5 illustrates the desired results.

Table 2-5. Contents of the Table GM

FIRST_YEAR	FIRST_SALARY	LAST_YEAR	LAST_SALARY
1995	38,400	1999	49,300

First, we execute Query 2_10 to create the table GM. Notice how the year fields are of type Integer rather than Text or Date so that we can use them in our calculations later. Next we insert the first year and its salary into the table GM by processing Query 2_11. We then update the GM table with the last year and salary by running Query 2_12. The result is the table GM shown in Table 2-5.

Query 2_10:

```
CREATE TABLE GM
(First_Year      Integer,
 First_Salary    Single,
 Last_Year       Integer,

 Last_Salary     Single);
```

Query 2_11:

```
INSERT INTO GM (First_Year, First_Salary)
(SELECT Year, Salary
 FROM Income
 WHERE Year = 1995);
```

Query 2_12:

```
UPDATE GM, Income
SET GM.Last_Year = Income.Year,
GM.Last_Salary = Income.Salary
WHERE GM.First_Year=1995 AND Income.Year=1999;
```

> **NOTE**   *You may combine Query 2_11 and Query 2_12 into one query as shown below.*

```
INSERT INTO GM ( First_Year, First_Salary, Last_Year, Last_Salary )
SELECT I1.Year, I1.Salary, I2.Year, I2.Salary
FROM Income AS I1, Income AS I2
WHERE (((I1.Year)=1995)) and (((I2.Year)=1999));
```

> **NOTE**   *However, if the Income table happens to contain a large number of records, the implied cross product (i.e., missing join condition in the WHERE clause) may produce a significant performance degradation. Thus, the two separate queries, Query 2_11 and Query 2_12, might be more desirable.*

Next, we calculate the geometric mean, GeoMean, by running Query 2_13. Notice how the year values were used to obtain the value for the root. The result for GeoMean is 1.0645.

Query 2_13:

```
SELECT (Last_Salary/First_Salary) ^ (1.0/(Last_Year - First_Year))
    AS GeoMean
FROM GM;
```

The geometric mean may also reflect a rate of decrease. Suppose the incidence of a certain disease in a population of animals is decreasing each year, as shown by the following numbers of cases recorded in a local veterinary hospital:

54   48   43   33   29   23   14

Applying the same process as before, take the sixth root of 14/54 to obtain 0.7634. The rate of decrease is, in this case, found as 1 - 0.7634 = 0.2366 or 23.66% per year. If the geometric mean is greater than one, then the sequence is increasing; otherwise, the sequence is decreasing.

Weighted Mean

Occasionally, data observations may be assigned different degrees of importance, or weights. When it's necessary to calculate a mean, these weights must be considered. The result is known as the *weighted mean*. This statistic is best illustrated by an example.

Suppose a business has determined that over the last five years it spent 18% of its expense budget on wages and salaries, 12% on benefits, 36% on taxes, 21% on purchases and supplies, 5% on utilities, and 8% on miscellaneous expenses. If during this same period, wages and salaries increased 34%, benefits increased 27%, taxes increased 16%, cost of purchases and supplies increased 12%, and utilities increased 7%, what was the average (combined) percentage increase in the cost of these categories (ignoring miscellaneous expenses)? The weighted mean $\bar{x}_w$ is found by summing the products of the cost category percentage and its increase, and dividing the result by the sum of the cost category percentages, as follows:

$$\bar{x}_w = \frac{(0.18)(0.34)+(0.12)(0.27)+(0.36)(0.16)+(0.21)(0.12)+(0.05)(0.07)}{0.18+0.12+0.36+0.21+0.05}$$

$$= 0.1955 \text{ or } 19.55\%$$

We can determine the weighted mean through an SQL query in the following way:

SQL/Query

Suppose that the expense budget percentages are contained in the database table called Expense as shown in Table 2-6, and that the inflation rates are in the database table called Increase as shown in Table 2-7. Notice that both tables contain the Category column.

Table 2-6. Contents of Expense Table

CATEGORY	SPENT
Wages	0.18
Benefits	0.12
Taxes	0.36
Purchases	0.21
Utilities	0.05
Miscellaneous	0.08

Table 2-7. Contents of Increase Table

CATEGORY	INCREASE
Wages	0.34
Benefits	0.27
Taxes	0.16
Purchases	0.12
Utilities	0.07

We can calculate the weighted mean by executing Query 2_14. Observe how the Sum function was used twice—once in the numerator and again in the denominator. To obtain the percentage, we multiply by 100.

Query 2_14:

```
SELECT (Sum(Spent*Increased)/Sum(Spent))*100 AS Weighted_Mean
FROM Expense, Increase
WHERE Expense.Category = Increase.Category;
```

The result is again 19.55% and it shows the average increase in the total costs of running the business.

John's Jewels
If Only More Than Half Were Above the Median

Once I was talking with a high school counselor. He was upset, he said, because "about half of our senior students score below the median on their final exams." "Yes, that would make sense," I thought to myself. "If we could only get more students above the median, I think we would be in a better placement position," he concluded. I hated to tell him he would never be able to have more than half his students above their median score!

Measures of Dispersion

The spread or dispersion of a distribution of data is an important measure in many applications. If most of the data values cluster tightly around the measure of central tendency, such as the mean, then the dispersion is not great. If the process generating the data does not change significantly, then future observations could be expected to vary little from the mean. If, on the other hand, the spread is wide, this sometimes indicates a lack of stability in the process, and observations often vary considerably to either side of the mean. The concept of dispersion is best illustrated by constructing a *histogram*, usually in the form of a bar chart, for the data. Such a chart offers a visual interpretation of dispersion and is also useful for applying some more advanced statistical tools.

Histogram Construction

You see histograms daily in newspapers, magazines, and on television. A histogram is simply a pictorial representation (or "snapshot") of a collection of observed data showing how the observations tend to cluster around an average and then taper away from this average on each side. Histograms are particularly useful in forming a clear image of the true character of the data from a representative sample of the population.

To illustrate the construction of a histogram, consider the data shown in Table 2-8 below. These numbers represent a company's stock trading prices per share (in dollars), taken each working day over a 15-week period. The time of day the observation was collected was chosen at random. The values are rounded to the nearest dollar for convenience, although it's possible to record them to the nearest cent if desired. Since the stock prices may assume any possible value between zero and some upper limit, the observations may be considered continuous, or variables, data. The data are presented in the table in the order in which they were collected.

There are two options for constructing a histogram for these variables data values. Since many of the values in the table repeat themselves, a histogram may be set up as shown in Figure 2-1. The horizontal scale at the bottom indicates all possible values in the data from smallest to largest. In this case, the smallest value is 4, and largest is 29. On a vertical scale to the left, numbers 0, 1, 2, 3, …, represent the frequencies of occurrence of specific values in the data. (In the figure, only the even numbers are shown to save space.) This scale should be high enough to accommodate the data value that occurs most often (that's right, it's the *mode*). In the table, it is observed that the value 20 (dollars) is the mode, occurring 16 times. Now that the two scales are established, go through the data values one by one, marking with a dot (or "X") where they fall relative to the horizontal scale. Each occurrence of a value gets its own dot, so there are 75 dots on the final histogram. The dots create vertical lines as you count repeated occurrences of the same values. When all values are exhausted, the histogram is complete. The height of each vertical line of dots (which can be converted to a bar as shown in Figure 2-1) represents the number of times that particular value occurred in the data (its *frequency*), and can be read directly from the vertical scale.

Table 2-8. Stock Price Data

WEEK	DAY OF WEEK				
	Monday	Tuesday	Wednesday	Thursday	Friday
1	21	29	26	29	29
2	29	24	23	20	23
3	20	29	20	23	22
4	21	21	20	22	21
5	17	20	19	20	16
6	13	22	20	20	22
7	17	19	19	20	18
8	20	18	17	17	18
9	22	20	19	17	19
10	20	18	19	18	18
11	17	15	25	20	25
12	4	22	20	16	19
13	14	17	20	20	12
14	17	18	19	19	22
15	19	19	19	18	19

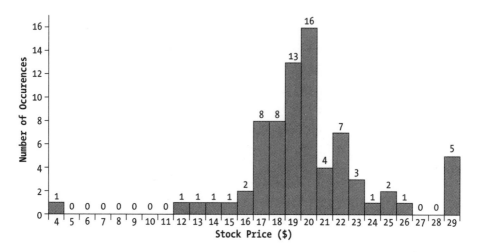

Figure 2-1. Histogram (bar chart) of stock price data

A second approach in histogram construction is used when very few individual values repeat themselves in the data set. If this had been the case in the above data, our vertical bars would all be short, and there would be no discernible histogram shape. In this situation, it is customary to group data in intervals rather than count individual values. Figure 2-2 shows this type of histogram for the same data with an interval width of three. Note how the intervals are denoted below each bar. The interval [4 - 7) means all values greater than or equal to 4 and less than 7 are inside the interval. If a value falls on the right end of an interval, such as the value 7, it is counted in the next interval to the right. The interval width is 7 minus 4, which is 3. Since our data values are integers, the interval [4 – 7) contains the values 4, 5, and 6. We use this generalized notation [4 - 7) so that if we had the data value 6.8 it would reside within the interval. The number of intervals chosen is arbitrary. However, if too few or too many are selected, the shape of the histogram may not be representative of the data. An old rule of thumb is to divide five into the total number of data values to yield a recommended number of intervals. This works pretty well until the number of observations exceeds 50 or so, as it does in this example. In that case, the number of intervals may be too large. It is also helpful to avoid decimal fractions in the interval width, and if possible, make the width a whole number. Finally, the interval width should be kept constant across the width of the histogram.

A visual inspection of the histogram indicates that the mean most likely falls between 19 and 22, in the highest bar, since the shape is relatively symmetric (or "mirrored") around this bar. The actual mean value of the data is 19.85 or about 20. The data are dispersed between 4 and 29. We can determine the extent of either the absolute dispersion or the average dispersion, by employing two new statistical concepts presented below. Before we do, however, let's use SQL to generate the histogram.

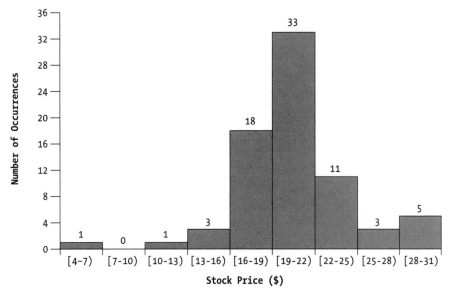

Figure 2-2. Histogram (bar chart) for the stock price data, with intervals

SQL/Query

SQL by itself does not have the capability to create graphical charts. However, many database management systems do provide this capability. We illustrate this by using Microsoft Chart via Microsoft Access 2000 and SQL queries to create the histogram shown in Figure 2-1. First, we notice that the data given in Table 2-8 is horizontal; that is, each row represents a week's worth of stock prices. This layout is relatively common for reports and is visually appealing. However, it's not in an appropriate format for SQL data manipulation operations. So, let's rotate it to a vertical layout as illustrated in Table 2-9. We'll name this table Vertical Price. Notice in the table that the days of the week appear sorted in alphabetical order.

Table 2-9. Sample Data from the Vertical Price Table

WEEK	DAY	PRICE
1	Friday	29
1	Monday	21
1	Thursday	29
1	Tuesday	29
1	Wednesday	26
2	Friday	23
2	Monday	29
…	…	…

To accomplish the rotation, we use two SQL queries. The first query, Query 2_15 shown below, uses five Select statements to extract the data for each day of the week and uses the SQL Union All operator to combine the data into vertical lists. The second query, Query 2_16, takes the results of Query 2_15 and places them into the table called Vertical Price. When Query 2_16 is executed, the SELECT...INTO... query replaces the previous contents of the table with the query results. Notice in Table 2-9 that the days are sorted alphabetically.

Query 2_15:

```
SELECT Week, "Monday" AS [Day], Monday AS [Price] FROM StockPrice
UNION ALL
SELECT Week, "Tuesday" AS [Day], Tuesday AS [Price] FROM StockPrice
UNION ALL
SELECT Week, "Wednesday" AS [Day], Wednesday AS [Price] FROM StockPrice
UNION ALL
SELECT Week, "Thursday" AS [Day], Thursday AS [Price] FROM StockPrice
UNION ALL
SELECT Week, "Friday" AS [Day], Friday AS [Price] FROM StockPrice;
```

Query 2_16:

```
SELECT Week, Day, Price
INTO [Vertical Price]
FROM [Query 2_15]
ORDER BY Week, Day;
```

Our next task is to count the occurrences of each price. We do this by grouping the price data and applying the Count function to each group to determine the number of occurrences. The results are used to determine the height of the histogram bars. Query 2_17 achieves this task, and produces the table named Bar Heights that is illustrated in Table 2-10.

Query 2_17:

```
SELECT Price, Count(Price) AS [Number Of Occurrences]
INTO [Bar Heights]
FROM [Vertical Price]
GROUP BY Price;
```

Table 2-10. Bar Heights Table

PRICE	NUMBER OF OCCURRENCES
4	1
12	1
13	1
14	1
15	1
16	2
17	8
18	8
19	13
20	16
21	4
22	7
23	3
24	1
25	2
26	1
29	5

The data given in Table 2-10 provides the height for each histogram bar. If we create a bar chart directly from the Bar Heights table, all the bars appear side-by-side and there are no spaces between bar 4 and bar 12 or between bar 26 and bar 29 as shown in Figure 2-1. To obtain the spacing requires that some additional records be included in the plot data. That is, we need to add a bar with zero height for the nine stock prices 5 through 11, 27 and 28. We could do nine INSERT queries to add the needed records to the Bar Heights table. However, our objective in this book is to provide a more general-ized technique that can be adapted to any set of data. Unfortunately, SQL does not offer a means for repeating a query for a specified number of times (such as in For loops, Do loops, and so on). As a result, we have to resort to writing a Visual Basic function that employs a For…Next loop. The Visual Basic function needs to do several things. First, we need to know the range for our loop (the beginning value and the ending value). For the stock prices histogram the loop begins with 4 and ends with 29, and these are also the minimum and maximum values of the stock prices in the Bar Heights table. Second, since the missing records (that is, 5-11, 27, and 28) could be any subset of records in between the minimum and maximum, our procedure should handle this situation auto-matically. We can do this by creating a new table that has a zero value for the number of occurrences for each stock price between 4 and 29, inclusively, and then update the number of occurrences for those records with a stock price listed in the Bar Heights table. The function is called Gen_Histogram_Data and is accomplished as follows:

```
Public Function Gen_Histogram_Data()

    Dim i       As Long    ' Loop index
    Dim CntInc  As Long    ' Counter increment
    Dim CntBgn  As Long    ' Counter beginning value
    Dim CntEnd  As Long    ' Counter ending value
    Dim rs      As New ADODB.Recordset
    Dim Q       As String  ' A query


    ' Build query to obtain beginning and ending values
    Q = "SELECT Min(Price) AS [BeginVal], Max(Price) AS [EndVal] "
    Q = Q & "FROM [Bar Heights]; "

    ' Process the query
    rs.Open Q, Application.CurrentProject.Connection

    ' Set counter parameters
    CntBgn = rs("BeginVal")
    CntInc = 1
    CntEnd = rs("EndVal")

    ' Close out query
    rs.Close
    Set rs = Nothing



    ' Turn off records modification confirmation message so that
    ' it does not appear while creating the Histogram Data table.
    DoCmd.SetWarnings False

    ' Execute the query to remove old table
    DoCmd. RunSQL "DROP TABLE [Histogram Data];"

    ' Establish table and set the initial value and increment for the counter
    DoCmd.RunSQL "CREATE TABLE [Histogram Data] " & _
            "(Price Counter(" & Str(CntBgn) & ", " & Str(CntInc) & "), " & _
            "[Number Of Occurrences] Long);"
```

```
    ' Append the desired number of zero records
    For i = CntBgn To CntEnd Step CntInc
       DoCmd. RunSQL "INSERT INTO [Histogram Data] " & _
             "( [Number Of Occurrences] )VALUES (0);"
    Next i


    ' Update the table with the sampled bar heights data
    DoCmd. "UPDATE [Histogram Data] " & _
          "INNER JOIN [Bar Heights] ON [Histogram Data].Price=[Bar Heights].[Price] " & _
          "SET [Histogram Data].[Number Of Occurrences] = [Bar Heights].[Number Of Occur-
rences];"


    ' Turn on confirmation message
    DoCmd.SetWarnings True


End Function
```

The Gen_Histogram_Data function begins by declaring all variables. The variables CntBgn and CntEnd are used to denote the price range. In our example, the range is 4 to 29, inclusively. The variables rs and Q are used to obtain the beginning and ending values of Price from the Bar Heights table. We use ADO to execute a query to obtain the minimum value (4) and maximum value (29). This was necessary in order to assign the values to the variables CntBgn and CntEnd, respectively. Next, the existing Histogram Data table is destroyed by executing a query to drop the old Histogram Data table. The Histogram Data table is then recreated. This creation sets up the Counter by specifying the counter's initial value (CntBgn) and increment (CntInc) and by resetting the counter back to the initial value 4. A For…Next loop is used to execute a query to insert a row into the Histogram Data table. Each time a row is appended, the value of the counter (Price) is automatically incremented by CntInc, and a zero value is assigned to Number of Occurrences field. After executing the loop, the Histogram Data table contains a row for each price value in the range with an occurrence value of zero. For our example, there are 26 rows. Finally, a query is executed to update the Histogram Data table by replacing the zero value for Number of Occurrences in the Histogram Data table with the corresponding value in the Bar Heights table by matching the Price values between the two tables.

NOTE *The function Gen_Histogram_Data uses ADO for Access 2000. However, if you are using Access 97, you may need to change the function to the code shown next.*

```
Dim dbs as Database
Dim rs as Recordset
Dim Q as string
Q = "SELECT Min(Price) AS [BeginVal], Max(Price) AS [EndVal] "
Q = Q & "FROM [Bar Heights]; "Set dbs = CurrentDb
Set rs = dbs.OpenRecordset(Q)
CntBgn = rs("BeginVal")
CntInc = 1
CntEnd = rs("EndVal")
rs.Close
Set dbs = Nothing
```

TIP   *One way to run a Visual Basic function is to simply make a single line macro as follows:*

```
RunCode Gen_Histogram_Data ()
```

Save the macro and give it a name such as "Run_Gen_Histogram_Data." To run (execute) the macro, select the Tools menu and choose Macro. From the Macro list select "Run Macro..." Select the macro named "Run_Gen_Histogram_Data" and click OK. Once you click OK, Access will run the macro that in turn invokes the Visual Basic function Gen_Histogram_Data.

Now that we have the data for drawing the histogram, we create a chart on a Microsoft Access 2000 form. Use the following steps to create the histogram via the Chart Wizard:

TIP   *Some installations of Access do not have the Chart Wizard installed. To install the Chart Wizard, rerun Microsoft Access or Microsoft Office Setup program, click Add/Remove, and then the click the Advanced Wizards check box.*

1. Select Form and click New. The New Form window will appear.

2. Select Chart Wizard and select Histogram Data for the object's data table. Click OK and the Chart Wizard window appears.

3. Click the double arrowhead button to choose both fields. Click Next.

4. For the chart type select the vertical bar chart. Click Next.

5. To lay out the data on the chart, if not already shown, drag Price to the *x*-axis and drag Number of Occurrences to the *y*-axis. Double click the *y*-axis data to bring up the Summarize window. Choose None and click OK. Click Next.

6. Click Finish and the chart appears. The histogram should look similar to the one in Figure 2-1.

7. The size of the form might need to be enlarged and the axis's font size and scale might need to be adjusted. The following directions explain how to make these adjustments:

8. Right click on the form and select Properties. Under the Format tab set Size Mode to Stretch. To resize the form move the cursor to the edge of the form and drag.

9. Click once on the chart to select it. Drag the resize handle (that is, the little black squares located around the borders of the chart area) to resize the chart.

10. Double click on the chart to invoke the Chart Editor. Right click on the axis and select Format Axis. Click the Font tab to change the font type and font size. Click the Scale tab to change the axis limits and tick mark units. Click OK to apply the changes.

The flowchart given in Figure 2-3 recaps the steps and processes performed in creating the histogram shown in Figure 2-1. The rectangles denote tables or the results of a query or process. Beside each arrow is the query or process used. The annotation, denoted by the dashed line to the right of each rectangle, identifies the figures or tables that illustrate the example data.

Now, suppose we wish to create the histogram shown in Figure 2-2 in which the data are grouped into intervals. We can do this in a relatively straightforward manner if we use the Histogram Data table produced by the Gen_Histogram_Data function. All we need is Query 2_18 and Query 2_19. Query 2_18 groups the prices into intervals and determines the lower and upper values of the interval. Query 2_19 determines the label text to be displayed beneath each bar of the histogram (see Figure 2-2). If we wish to establish a different interval width, say a width of 5 rather than of 3, all we need to do is change the number 3 in Query 2_18 to 5.

Query 2_18:

```
SELECT [Histogram Data].Price, Fix((([Price]-1)/3) AS [Interval],
(Fix((([Price]-1)/3))*3+1 AS LowEnd,
(Fix((([Price]-1)/3)+1)*3+1 AS HiEnd,
[Histogram Data].[Number Of Occurrences]
FROM [Histogram Data];
```

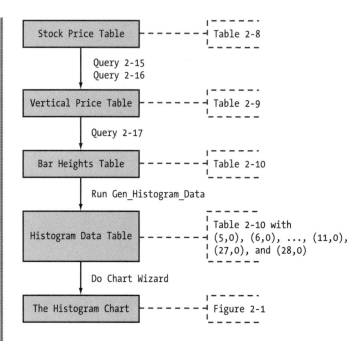

Figure 2-3. Processing Steps

Query 2_19:

```
SELECT "[" & [LowEnd] & " - " & [HiEnd] & ")" AS [Price Interval],
Sum([Query 2_18].[Number Of Occurrences]) AS [SumOfNumber Of Occurrences]
FROM [Query 2_18]
GROUP BY "[" & [LowEnd] & " - " & [HiEnd] & ")", [Query 2_18].LowEnd
ORDER BY [Query 2_18].LowEnd;
```

With the histogram thus created, we are ready to move on to our discussion of the important measures of dispersion.

Range

The *range* is one of the simplest of all statistical measures to calculate. It's the difference between the largest and smallest values in a sample and is a measure of the absolute dispersion or spread of a data set. In the stock price data above, the range is the difference between 29 and 4, or 25, if the data are taken as one large group. The range plays an important role in control charting (see Chapter 6) as well as other applications.

SQL/Query

The range can be calculated with a single SQL query. Query 2_20 shows how the Min and Max function can be used together in calculating the range.

Query 2_20:

```
SELECT Max(Price) - Min(Price) AS Range FROM [Vertical Price];
```

Standard Deviation

The *standard deviation* is one of the most important and useful statistical measures. Basically, it's the average dispersion of a set of data. Looking again at the histogram of stock price data in Figure 2-2, and remembering that the mean is about 20, it appears that most of the data are found within only a few units of this mean. In other words, there is a close clustering around 20, with fewer and fewer values less than 16 or greater than 25. We could judge that most of the clustering appears to be within about four units to either side of the mean. The true sample standard deviation (which is an estimate of the standard deviation of the population of stock price data from which this sample was taken) is 3.93 (dollars). The process for calculating this number is outlined by the following steps:

1. Sum all the data values (perhaps this was already done to calculate the mean). Using the 75 stock price data values, the sum is 1,489. This is commonly denoted by Σx, where x is used to represent a single data value, and Σ (the Greek letter *sigma*) is the universal statistical symbol for the sum of a group of data values.

2. Sum all the squares of the data values. This is denoted by Σx^2 and is equal to 30,705.

3. Count the number of data values and call this n. Here $n = 75$.

4. Calculate $\dfrac{(n)\left(\sum x^2\right)-\left(\sum x\right)^2}{(n)(n-1)}$

This is called the *variance* estimate and is denoted by s^2. Taking the square root of it, we have the standard deviation (denoted by s), as follows:

$$s = \sqrt{\frac{(75)(30,705) - (1,489)^2}{(75)(75-1)}}$$

$$= 3.93 \text{ (dollars)}$$

This result is pretty close to the visual conclusion of four dollars on average to either side of the mean.

SQL/Query

Some implementations of SQL offer a built-in function for calculating the standard deviation. For example, Microsoft Access 2000 offers the function called StDev(). Query 2_21 shows how the standard deviation can be determined using the StDev() function. However, some implementations of SQL do not offer this function. Query 2_22 shows how to compute the standard deviation using the above formula.

Query 2_21:

```
SELECT StDev([Price]) AS S
FROM [Vertical Price];
```

Query 2_22:

```
SELECT Sqr((Count([Price])*(Sum([Price]^2)) -
    (Sum([Price])^2))/(Count([Price])*(Count([Price])-1))) AS S
FROM [Vertical Price];
```

NOTE *Some implementations of SQL have the square root function as SQRT() rather than SQR().*

Before we look at another case, pay attention to the shape of the histogram in Figure 2-2. It starts low on the left, rises to a high point near the middle of its range, then tapers off again. If a smooth line is drawn through the tops of the bars, it almost looks like a bell shape. This is very important, and provides an introduction to the concepts presented in the next chapter. Now let's move on to another example that illustrates several statistical techniques presented in this chapter.

A Moment of Kurtosis

Statisticians have developed a language all their own. We talked about the mean and variance as measures of central tendency and dispersion. The mean is sometimes referred to as the "first moment about the origin." A *moment-generating function* can be defined to derive other moments. The variance is called the "second moment about the mean." The moment-generating function may be used to produce the "third moment about the mean." It is a measure of the skewness of a distribution of data. A data distribution is skewed to the right if it rises to a high point and then tapers gradually off to the right. It is skewed to the left if the opposite is true. It is symmetric if it tapers the same to the left and to the right. Now a distribution has a characteristic peakedness to it also. It may be very pointed and narrow, like the Matterhorn, or broad and low, like Mount Etna. The peakedness of a distribution is called *kurtosis* (sounds like a liver disease, doesn't it?). The "fourth moment about the mean" is a measure of kurtosis.

An Example Illustrating Data Organization and Measures of Central Tendency and Dispersion

Limousine Service at an Airport

The owners of a limousine service, whose headquarters are near the International Airport in Atlanta, are currently charging customers by the mile for transportation to the airport. They are primarily contacted by phone or through their Web site on the Internet. They think it would be easier to simply charge a flat fee for any pick-up within a reasonable distance of the airport and advertise that rate on the Web site as an incentive to potential customers to use their service over competitors' services. In order to have some idea of how to set this fee, a random sample of 100 pick-up records from the last quarter was collected. The (one-way) distances the limousines were dispatched (in miles) and the locations of the calls were known. The distances are shown in Table 2-11, in the row order in which the records were pulled.

Table 2-11. Distances (in Miles One-way) Limousines Were Dispatched from Airport to Customers in a Sample of 100 Records

21	8	36	54	7	14	43	7	3	24
22	41	56	2	10	39	48	2	43	14
36	29	49	8	42	35	23	35	41	28
56	30	22	9	46	15	58	28	21	56
37	40	14	55	41	36	23	55	53	52
69	35	51	62	41	15	65	15	29	26
60	29	22	70	44	42	16	48	77	44
29	22	28	41	48	3	54	55	42	68
30	29	11	46	17	69	68	4	76	28
31	32	47	56	21	55	36	52	48	83

The locations of the calls were plotted on an *x-y* coordinate system at the appropriate distances from the limousine service headquarters. The results of this plot, along with the location of the airport limousine service, are shown in Figure 2-4. The limousine service headquarters is in the center of the plot at (90,90) and is denoted by a small gray circle. Although the points are plotted to scale, the coordinate values are completely arbitrary.

The first step is to construct a histogram of distances. This is easier to accomplish if the data are sorted from smallest to largest value. For 100 data values, it is customary to have between 10 and 15 histogram intervals. In this case, it is convenient to use 7 as the interval width, starting with 0 and ending with 84. Thus the 12 resulting intervals encompass all values. We again use the convention that the left endpoint of an interval is inclusive and the right endpoint is exclusive to determine into which interval a border-line value is placed. For example, the first interval ranges from 0 to (but not including) 7, and the second from 7 to 14. A value of 7 would therefore be counted in the second interval. To construct the histogram, the number (or frequency) of occurrences of values in the data is counted for each interval. The heights of the histogram bars reflect these counts, as shown in Figure 2-5. The sum of the heights, or frequencies, of course equals 100.

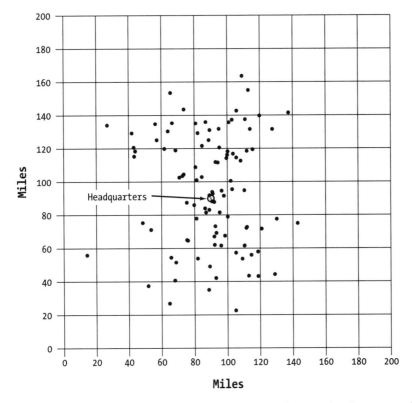

Figure 2-4. Locations of customer pick-up points relative to headquarters of airport limousine service

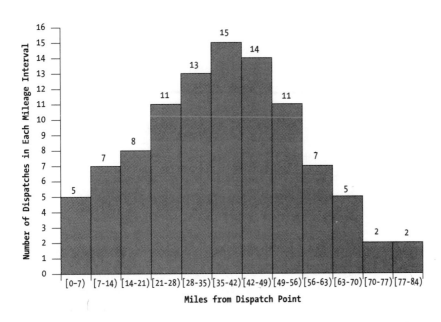

Figure 2-5. Histogram of limousine dispatch distances (in miles one-way from airport headquarters)

The mean of these 100 values is found by adding them and dividing by 100. The result is $\bar{x} = 36.56$, or about 37 miles, which falls in the tallest bar of the histogram. Assuming the sample is representative of the population of distances from all calls, the average call is 37 miles (one-way) from the limousine headquarters at the airport. The standard deviation s is calculated from the formula previously presented and is equal to 19.25 miles. This represents the average dispersion of values around the mean, so the average spread of the distribution of distances is between $36.56 - 19.25 = 17.31 \cong 17$ miles and $36.56 + 19.25 = 55.81 \cong 56$ miles. Of course some will fall outside this range.

By sorting the data from smallest to largest values, the overall range is found to be $R = 83 - 2 = 81$. The median value falls directly in the middle of the sorted distances and, in this case, is equal to the average of the two middle values (both of which are 36), or $\tilde{x} = 36$. This is very close to the mean. By observing the shape of the histogram, the data appear to be almost symmetric around the mean. This means that if the histogram is folded along a vertical line through the mean, the two halves would almost coincide, or mirror each other. Again, a smooth line drawn through the tops of the bars would look bell-shaped. We need to keep that in mind for the next chapter's discussion.

Now to the problem at hand. If the distribution of distances is perfectly bell-shaped (which it isn't, but it's close), the distances occur randomly, and we may use the standard deviation value of 19.25 to advantage. Borrowing from a control chart concept (see Chapter 6), we can draw a circle centered at the limousine headquarters with radius equal to the mean distance plus one standard deviation, or $36.56 + 19.25 = 55.81$ miles. The area of this circle encompasses approximately 68.26% of the customer locations calling for pick-up. The mean plus two standard deviations, or $36.56 + 2(19.25) = 75.06$ miles, generates a circle encompassing 95.44% of the expected customer locations. The mean plus three standard deviations (94.31 miles) encompasses 99.74% of the locations. These percentages assume the distances are randomly distributed and follow the bell curve (more about this and the source of the mysterious percentages in Chapter 3; for now please accept them on faith). Table 2-12 is a summary of these results, showing the actual and expected number of customer locations within each circle. Figure 2-6 is a plot of the customer locations and circles.

Table 2-12. Expected and Actual Customer Locations Within Each Circle

APPROXIMATE DISTANCE FROM HEADQUARTERS (RADIUS OF CIRCLE IN MILES)	EXPECTED NUMBER OF CALLS WITHIN CIRCLE OF THIS RADIUS	ACTUAL NUMBER OF CALLS OBSERVED IN SAMPLE THAT FALL WITHIN CIRCLE OF THIS RADIUS
$\bar{x} + 1s = 56$	68	68
$\bar{x} + 2s = 75$	95	98
$\bar{x} + 3s = 94$	100	100

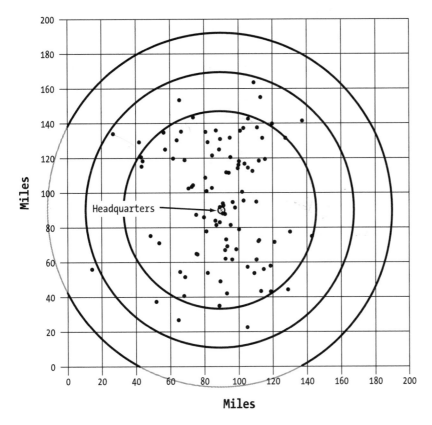

Figure 2-6. One, two, and three standard deviation (above the mean) circular contours centered at the headquarters of the limousine service

Suppose one of the limousine service owners looks at these concentric circles, or contours, and asks, "What if we charged $20 to pick up anyone within the smallest circle, $35 to go out into the next one, and $50 to range into the largest circle?" The other owner says this sounds reasonable, but the split rate is too complicated. Why wouldn't the original flat rate idea apply? A statistician, waiting to pay for his ride to the airport, overhears their conversation and says, "Just use a weighted mean to find the flat rate." In other words, remember that 68.26% of the customers are expected to be within the small circle and would pay the $20 rate. An additional 27.18% (99.44% - 68.26%) would pay the $35 rate, and an additional 4.30% (99.74% - 95.44%) would pay the $50 rate. The weighted mean gives the desired flat rate, as follows:

$$\bar{x}_w = (0.6826)(\$20) + (0.2718)(\$35) + (0.0430)(\$50) = \$25.32$$

The owners decide to round this to a $25 flat rate. For any customers who might call for pick-up outside the 94-mile "largest" circle, a surcharge is added. Wasn't that fun?

Robert's Rubies
Density Contours

John and I were faced with a problem similar to the limousine service example not long ago. There was an old World War II ordnance site in California used for training and maneuvers. Much of the ordnance fired for practice and training of the troops was still buried throughout the site, so the task included assessing the extent of this ordnance (some of which was still live) by field sampling and investigation. Our involvement was to take the field data and determine whether or not we could verify historical target and impact areas from the sampled finds. We developed a statistical model to calculate the ordnance densities in the sampled areas, find the point in the field about which the ordnance finds clustered, then determine the standard deviation of distances of these ordnance locations from the central point. We used SQL to query the database of locations. Once the measures of central tendency and dispersion were known for the data, we created what we called a "density contour." This was a circle centered at the central point, with radius equal to the sum of the mean and standard deviation of the distances of all the ordnance locations from the central point. Since the histogram of locations was approximately bell-shaped (or normally distributed as we see in Chapter 3), we could assume that about 68% of the ordnance finds would fall within this first contour. We then created contours with radii equal to the mean plus two, then three, standard deviations. As in the limousine example, these contours included about 95.4% and 99.7%, respectively, of the ordnance finds, assuming they were distributed randomly throughout the area. What we found seemed to verify the historical target areas, and this information was used to aid in clean-up decisions at the site.

Conclusion

Understanding the measures of central tendency and dispersion in this chapter is fundamental for all that follows. Although some of the measures presented are encountered infrequently or only in special applications, the mean and standard deviation are used extensively in virtually every statistical problem. As we proceed into the chapters ahead, we constantly refer back to these basic concepts and their methods of calculation.

CHAPTER 3
Goodness of Fit

Poor Fit

THE AIRPORT LIMOUSINE EXAMPLE in Chapter 2 illustrates how important the knowledge of random distribution of data and the shape of the histogram can be to drawing statistically sound conclusions. Although we didn't state it in the example, the assumption that the data followed a known statistical distribution was inherent in the calculations of the three contour radii based on the standard deviation of the data values. In this chapter, we present the technique of fitting a statistical model to a set of observed data and testing for "goodness of fit."

Once we represent a set of observed data pictorially by a histogram, and we calculate measures of central tendency and dispersion, our interest may shift to the shape of the distribution of data. By constructing the histogram, the data moved from a formless collection of numbers to a visual representation that has meaning. Frequently in statistical studies the focus is to attempt to associate the observed data values with a known statistical model. The result may reveal important information about the population from which the data were sampled. In fact, if it can be shown that a known distribution "fits" the data, we can make predictions regarding the population as a whole.

Three steps are necessary to attempt to fit a known statistical distribution to a set of observed data, assuming a histogram has already been set up. These are as follows:

1. Classify the data as variables or attributes (see Chapter 1) and study the data characteristics to determine a candidate statistical model.

2. Observe the shape of the histogram to further direct the search for a model that may fit the data.

3. After deciding on a candidate model, check to see how well the observed data fits the model by conducting a "goodness of fit" test.

The classification of the data into either variables or attributes narrows your choices for a candidate statistical model, thereby making a selection somewhat easier. Statistical distributions fall into one of these two categories, and the sections that follow present the more commonly encountered distributions of each type. Distributions have unique graphs or plots (although some are very similar), so it's helpful to compare such shapes to the histogram of the observed data. Finally, once a candidate model has been selected, we follow a procedure to determine how well the model characterizes, or "overlays," the histogram of the observed data. The degree, or goodness, of fit is tested most commonly by what is known as a chi-square test (the Greek symbol used is χ^2). The test is so named because the calculation used to compare the observed to the assumed model closely follows the ideal χ^2 statistical distribution, if the observed data and the candidate model do not differ significantly. The degree to which they deviate determines the outcome of the test.

Tests of Hypothesis

Such a test falls under a category of statistical testing known as "hypothesis testing." In the diagnostic tree in Figure 1-3, note that the second branch addresses hypothesis testing. To satisfy the pure-blood statistician who might by chance be reading this, a brief discussion of tests of hypothesis is presented. Basically, a test of hypothesis begins with a statement of the hypothesis to be tested. Makes sense so far, right? The hypothesis might be that a statistic such as the mean, calculated from a set of observed sample data, differs little from the mean of the population from which the sample was collected. The problem becomes how to define "differs little." Fortunately, folks smarter than we are years ago developed a general solution to this dilemma, and thereby acquired tenure at their respective institutions. The degree of deviation of the observed from the population (or theoretical) statistic is compared to a tabulated value based on the theoretical distribution appropriate to the test. If the calculated value exceeds the table value, the conclusion reached is that the observed statistic differs so significantly from the table's that the hypothesis test fails, and the observed data therefore does not behave the way we thought. Well, it's a bit more complicated than that.

For one thing, you must establish a degree of significance before the test is conducted. This is usually stated as a percent or decimal fraction, such as 5% (which is probably the most commonly used value). This significance level is easier to understand if viewed pictorially. If we look back at the histogram for the limousine service data in Figure 2-5, also shown in Figure 3-1, we notice that the bulk of the observations cluster around the center of the bar chart, near the mean value. The histogram is tallest in this area, representing the higher number of data values. As we move to the right or to the left of center, the heights of the bars decrease, indicating fewer and fewer observations the further we deviate from the mean. Now, we draw a smooth line near the tops of the bars on the histogram as illustrated in Figure 3-1. This curve looks much like the curve in Figure 3-2 that represents a candidate statistical model.

Of course, the percentage of area under the curve of the candidate model is 100% and represents the entire population. If we want to select a region representing only 5% of this area, one option would be to pick a point in the "upper tail" of the curve to the right of which is 5% of the area. This is illustrated in Figure 3-2. Thus, any calculation or observation that falls within this relatively small area of the distribution is considered significantly different than most of the rest of the observations. The table we mentioned earlier (available in statistics texts and in our Appendix B) is used to determine the value of the distribution to the right of which is this 5% area.

Statisticians call this type of test a "one-tailed" test, since only one tail of the distribution has been employed. However, we may in some situations want to run a "two-tailed" test. In these cases, the total significance area is usually divided equally between both tails of the curve. If we stay with 5%, that means that 2.5% falls in each tail, as shown in Figure 3-3. The test is a bit more involved, so for now let's stick with the one-tailed variety.

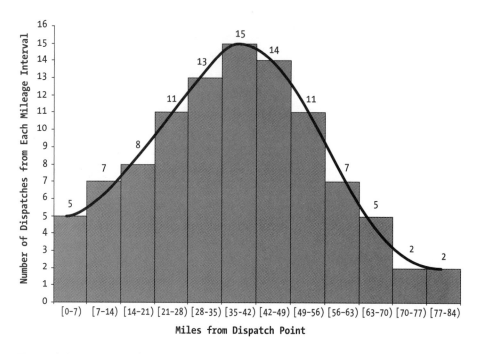

Figure 3-1. Histogram from Figure 2-5 with the bar height curve added

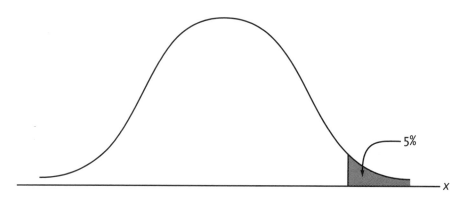

Figure 3-2. A typical one-tailed significance level for a test of hypothesis

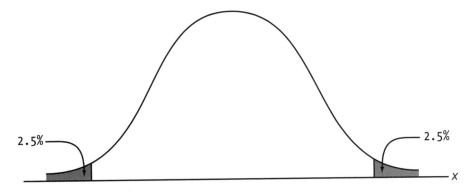

Figure 3-3. A typical two-tailed significance level for a test of hypothesis

There is one more parameter that must be used to extract the appropriate value from the table. This is the number of **degrees of freedom** (you know it's important if it's in bold italics). This is one of the most difficult concepts in statistics to explain to anyone. Perhaps it is easiest to present it as follows: Recall from Chapter 2 that the calculation of the standard deviation of a collection of sample values was

$$s = \sqrt{\frac{(n)\left(\sum x^2\right) - \left(\sum x\right)^2}{(n)(n-1)}}$$

An equivalent form of this expression is

$$s = \sqrt{\frac{\sum (x - \bar{x})^2}{n-1}}$$

where $\sum (x - \bar{x})^2$ is calculated by first finding the mean $\bar{x}$ of all n data values, then subtracting it from each individual data value (represented by x), squaring the results one at a time, and adding all the n squared results together. The $n-1$ in the denominator of the expression is the number of degrees of freedom for the statistic.

Suppose we want to calculate the standard deviation of four values, say 3, 5, 6, and 10. The mean $\bar{x}$ of these values is 6. To calculate the standard deviation, we subtract the mean from each of the four values, as follows:

$$3 - 6 = -3 \qquad 5 - 6 = -1 \qquad 6 - 6 = 0 \qquad 10 - 6 = 4$$

Now square each of these results:

$$(-3)^2 = 9 \qquad (-1)^2 = 1 \qquad (0)^1 = 0 \qquad (4)^2 = 16$$

Then sum the four values, yielding 26. Divide by $n-1=3$, take the square root of the result, and the standard deviation is equal to 2.94.

Now let's explain the degrees of freedom. Notice that the four values found by subtracting the mean from each observation sum to zero. Therefore, you can take any three of these differences and obtain the fourth with this knowledge. For example, knowing the first three differences are -3, -1, and 0, and the sum of all four is 0, the fourth difference has to be 4. Since any three of the four differences yield the fourth, there are three "free" variables, hence three degrees of freedom. For the sample standard deviation calculation, the number of degrees of freedom is always equal to one less than the number of observations (i.e., $n-1$). Another way of stating this is that, since the sample statistic $\bar{x}$ was used to estimate the standard deviation, and its value was obtained from the sample data itself, this fact reduced the number of degrees of freedom by one. Clear as mud? Good, now we can proceed from shaky ground to the quicksand.

John's Jewels
The Degrees of Freedom of Flightless Birds

In Chapter 7 we revisit the concept of degrees of freedom when we discuss experimental design analyses. At that point we state that the number of degrees of freedom is basically the number of independent variables in a problem less the number of constraints. For example, consider a bird in an aviary. If the bird can fly, it can move in any direction of length, width, and height of the aviary, so it has three degrees of freedom in which to travel. However, think about a flightless bird in the aviary. It can basically only move across the floor, in two directions (length and width). It is constrained by its inability to fly. Therefore, it has only two degrees of freedom (three directions less one constraint). Maybe this makes the concept a bit easier to understand.

Goodness of Fit Test

Generally speaking, tests of hypotheses have similar characteristics. However, for specific testing, there are slight variations, and the goodness of fit test is a typical example. After the candidate distribution model is selected for the goodness of fit test, it is necessary to use this model and one or more parameters estimated from the observed data (such as the mean and standard deviation) to calculate values predicted by the model. These values are compared to the observed values in intervals of the histogram. The comparison involves calculating a statistic that is the basis for the chi-square test. This will (perhaps) become clear when we present the following examples.

Fitting a Normal Distribution to Observed Data

The histogram of limousine service data presented in Figure 3-1 has a shape that we very commonly encounter in real-world situations, where the values are measured quantities (variables data). If we trace a smooth curve through the tops of the bars on the histogram, the resulting shape looks like a bell, so it is known as a bell curve. Statisticians prefer a more academic term, so the curve is said to represent a *normal* distribution. The normal statistical distribution is basic to an understanding of data mining and analysis. It is a type of *continuous* distribution, which means that the data whose histogram is bell-shaped assume continuous, or variables, values. In the case of the limousine service data, the observations were measured in miles from a point. The precision of these measurements was limited only to the method of measurement and the needs of the analyst. In this example, the results were simply rounded to the nearest mile, but in general they could have been measured to tenths or hundredths of a mile, or even more precise values. For the purpose of the analysis, this was not necessary, but the point is that the values could potentially assume any number within the observed range between 2 and 83 miles. This is why the data are termed *continuous*.

A continuous statistical distribution like the normal is represented by what is known as a *density* function. This is simply an algebraic expression that defines the shape of the distribution, and may be used to calculate areas under the normal curve between points on the horizontal axis. These points represent the various values that the range of the distribution might be expected to assume. In the case of an ideal normal distribution, the values range, in general, over the entire set of real numbers (both positive and negative). For most practical data sets, the numbers range from zero to (positive) infinity. If we integrate the normal density function between any two desired values within its range, it is possible to determine the area under the curve in that interval. This is especially useful in testing for goodness of fit. Fortunately, it is not necessary to actually integrate the horrible looking functional expression to obtain these areas (in fact, it is impossible to integrate it exactly!). Tables have been developed to accomplish this. For those stalwart few who wish to see the actual normal density function expression, cast your eyes below.

$$\frac{1}{\sqrt{2\pi}\sigma}e^{-(1/2)[(x-\mu)/\sigma]^2}$$

In the normal density function, μ denotes the mean, σ denotes the standard deviation, and x is the variable (e.g., miles traveled). Often μ and σ are unknown and we estimate them from the sample data by $\bar{x}$ and s, respectively. Of course $\pi = 3.14159$.

The procedure for testing to see if the limousine data actually fit a normal distribution is a little involved, so please bear with us. The first step is to calculate the mean and standard deviation of the original data. These are $\bar{x} = 36.56$ miles and $s = 19.25$ miles. Both these parameters are necessary to determine expected numbers of observations (or expected frequencies) to compare to the actual observations. The method by which this is accomplished is to calculate the area in each bar of the histogram. If the top of that bar was really determined by the curve of the (ideal) normal distribution, then multiply the result by the total number of observations (in this case 100). The results are compared to the observed frequencies interval by interval to determine the extent of deviation. However, there is a little rule that you should follow: If in any interval (or bar) of the histogram, the number of data values *expected* is less than 5, this interval is combined with one or more neighboring intervals until the total values are at least 5. The same values are then combined for the observed frequencies. This rule will be demonstrated when we calculate the expected frequencies a little later. Table 3-1 summarizes the data from the histogram in Figure 3-1.

Table 3-1. Number of Limousine Pick-Up Points Falling Within Each Mileage Interval from the Dispatching Location at the Airport

HISTOGRAM INTERVAL (MILES)	OBSERVED FREQUENCY (O)
[0 - 7)	5
[7 - 14)	7
[14 - 21)	8
[21 - 28)	11
[28 - 35)	13
[35 - 42)	15
[42 - 49)	14
[49 - 56)	11
[56 - 63)	7
[63 – 70)	5
[70 – 77)	2
[77 – 84)	2

SQL/Query

The data shown in Table 3-1 can be generated by executing a set of SQL queries. In our first query we create an initial version of Table 3-1 by forming the intervals of width 7 and by giving each interval an identification number. Query 3_1 accomplishes the task. Just a reminder: Notice that the table in Query 3_1 is named Table 3_1 rather than Table 3-1. This is because most implementations of SQL do not allow a hyphen in a table or field name. The result of Query 3_1 is illustrated in Table 3-2.

Query 3_1:

```
SELECT Fix(([Miles])/7) AS [Interval],
(Fix(([Miles])/7))*7 AS LowEnd,
(Fix(([Miles])/7+1))*7 AS HiEnd,
Count([Limo Miles].Miles) AS CountOfMiles
INTO [Table 3_1]
FROM [Limo Miles]
GROUP BY Fix(([Miles])/7), (Fix(([Miles])/7))*7, (Fix(([Miles])/7+1))*7
ORDER BY Fix(([Miles])/7);T
```

Table 3-2. Result of Query 3_1

INTERVAL	LOWEND	HIEND	COUNTOFMILES
0	0	7	5
1	7	14	7
2	14	21	8
3	21	28	11
4	28	35	13
5	35	42	15
6	42	49	14
7	49	56	11
8	56	63	7
9	63	70	5
10	70	77	2
11	77	84	2

To calculate the frequencies we would expect if the data were exactly normally distributed, we must either employ the table of areas under the normal curve found in Appendix B or use the Visual Basic code in Appendix D. Most forms of this table give the area under the normal curve from negative infinity up to the desired value. The normal curve used in the table, however, is called a *standardized* normal distribution, so that it can be used for any normal curve, so long as a transformation of variables is accomplished. The standardized normal curve has its mean at zero and standard deviation equal to one. For our example, the transformation is accomplished described in the text that follows:

The first interval on the histogram is from 0 to (but not including) 7 miles. If we overlay a normal distribution onto the histogram, a portion of the curve actually falls in the negative range to the left of zero. In practice we know we can't have any distances less than zero, but we still need to accommodate this very small (theoretical) tail region when we determine the area under the normal curve up to 7 miles. This area is illustrated in Figure 3-4. To use the normal table to determine this area, we need to calculate the following expression, which represents the transformation of variables:

$$\frac{x - \bar{x}}{s}$$

where x is the right-hand endpoint of the histogram interval, $\bar{x}$ is the mean of the data, and s is the standard deviation. For the first interval, this expression is equal to

$$\frac{7 - 36.56}{19.25} = -1.536$$

This value is called the **normal variate**, and is used to find the desired area in the normal table. If we find -1.53 in the table and interpolate for the third decimal place (which usually isn't necessary), the corresponding area is 0.0623, or 6.23%. This means that 6.23% of the area under the normal curve of mileage is to the left of 7 miles. For 100 observations, this means 6.23 is an expected frequency of occurrence. As a comparison, in the actual data there were 5 customer calls less than 7 miles from the limousine headquarters, or 5% of the sample data.

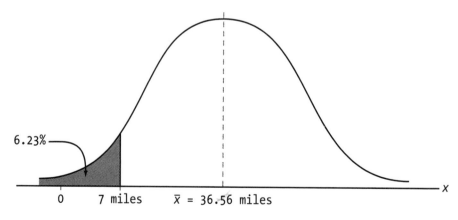

Figure 3-4. Theoretical normal distribution of limousine customer pick-ups, showing the area under the curve to the left of 7 miles

The next histogram interval is between 7 and 14, so we need to calculate the area under the normal curve between these two values. To do this using the normal table, we would first find the area to the left of 14 by using the procedure outlined above, then subtract from it the area to the left of 7 (already calculated). The difference would be the area between 7 and 14. Therefore, the following calculation is first performed:

$$\frac{14-36.56}{19.25} = -1.172$$

The area to the left of this value in the normal table is 0.1206, or 12.06%. By subtracting 6.23% from this area, we obtain the desired area between 7 and 14. This is 5.83%, or 5.83 occurrences for 100 values. Continuing in this manner, Table 3-3 is generated. The last expected frequency (1.782) may be found by subtracting the sum of all previous expected frequencies from 100 (in this case).

Table 3-3. Observed and Expected Frequencies of Occurrence of Call Distances (in Miles) Based on the Sample of 100 Records of the Limousine Service

HISTOGRAM INTERVAL (MILES)	OBSERVED FREQUENCY (O)	EXPECTED FREQUENCY (E)
< 7	5	6.227
[7 – 14)	7	5.833
[14 – 21)	8	8.895
[21 – 28)	11	11.861
[28 – 35)	13	13.956
[35 – 42)	15	14.369
[42 – 49)	14	12.945
[49 – 56)	11	10.289
[56 – 63)	7	7.153
[63 – 70)	5	4.353
[70 – 77)	2	2.337
≥ 77	2	1.782

SQL/Query

Now let us see how we can generate Table 3-3 using some SQL queries on Table 3-1 along with the standard normal table, StdNormal, described in Appendix B. First, we use Query 3_2 to calculate the sample mean and standard deviation. Next we execute Query 3_3 to create a seven column temporary work table called TempWork. The first column of TempWork is an interval identification number that is used as identify each interval in sequence. The second and third columns correspond to the information in Table 3-1. The fourth and fifth columns are the sample mean and standard deviation. The sixth column is the normal variate, $(x - \bar{x})/s$. The last column is the value from the StdNormal table for the normal variate. When we execute Query 3_3, it populates the first six columns of TempWork table and temporarily fills the last column with zeros. Notice that Query 3_3 extracts data from two different sources (i.e., Table 3_1 and Query 3_2) without a join condition. Whenever the join condition is omitted, a cross product is formed between the two sources. Since the result of Query 3_2 is a single row, we are in effect appending the sample mean and standard deviation to the selected rows from Table 3_1. Also notice the use of the function "Round(0, 3)" in place of "0" for the value of StdNorm. We did this so that the data type for StdNorm would be a real number rather than an integer. Table 3-4 illustrates the contents of TempWork after executing Query 3_3. It should be pointed out that the last value (2.4648) of the normal variate in the table is not used, so acts only as a place holder.

Query 3_2:

```
SELECT Avg(Miles) AS xBar, StDev(Miles) AS s, Count([Miles]) AS n
FROM [Limo Miles];
```

Query 3_3:

```
SELECT [Table 3_1].Interval,
[Table 3_1].LowEnd, [Table 3_1].HiEnd,
[Table 3_1].CountOfMiles,
[Query 3_2].xBar, [Query 3_2].s,
([HiEnd] - [xbar]) / [s] AS Nvariate,
Round(0, 3) AS StdNorm
INTO TempWork
FROM [Query 3_2], [Table 3_1];
```

Table 3-4. Initial Contents of TempWork

INTERVAL	LOWEND	HIEND	COUNTOFMILES	XBAR	S	NVARIATE	STDNORM
0	0	7	5	36.56	19.2471	-1.5358	0
1	7	14	7	36.56	19.2471	-1.1721	0
2	14	21	8	36.56	19.2471	-0.8084	0
3	21	28	11	36.56	19.2471	-0.4447	0
4	28	35	13	36.56	19.2471	-0.0811	0
5	35	42	15	36.56	19.2471	0.2826	0
6	42	49	14	36.56	19.2471	0.6463	0
7	49	56	11	36.56	19.2471	1.0100	0
8	56	63	7	36.56	19.2471	1.3737	0
9	63	70	5	36.56	19.2471	1.7374	0
10	70	77	2	36.56	19.2471	2.1011	0
11	77	84	2	36.56	19.2471	2.4648	0

Our next task is to update the StdNorm column in TempWork. Query 3_4 accomplishes this task. Notice the WHERE clause and the use of the Round function to truncate both values to three decimal places. This assures us of an equality match. Also the StdNormal table was generated to three decimal places.

Query 3_4:

```
UPDATE TempWork, StdNormal
SET TempWork.StdNorm = StdNormal.Area
WHERE Round(Nvariate,3) = Round(StdNormal.x,3);
```

Now that we have our normal values, we need to calculate the expected frequency value for each interval. As you may have already observed, the first interval is everything less than 7, and the last interval is everything greater than or equal to 77. All the intervals between the first and last have specific beginning and ending values. Thus, to calculate the expected value, we need to write three queries: one query for the first interval, one query for all intervals between the first and last, and one query for the last interval. Before we can write the queries, we need to know the lowest interval value (i.e., 7) and the highest interval value (i.e., 77). Query 3_5 accomplishes this task by extracting the interval number from the TempWork table.

Query 3_5:

```
SELECT Min(Interval) AS Lowest, Max(Interval) AS Highest
FROM TempWork;
```

Using the lowest interval value found by Query 3_5, we write Query 3_6a to obtain the expected frequency for the first interval. Notice how the expected frequency is multiplied by the total number of data values (in this specific example,100) to convert it to a number of observations, and then rounded to three decimal places.

Query 3_6a:

```
SELECT TempWork.Interval,
TempWork.LowEnd, TempWork.HiEnd,
TempWork.CountOfMiles AS O,
Round([StdNorm] * 100, 3) AS E
INTO [Table 3_3]
FROM TempWork, [Query 3_5]
WHERE TempWork.Interval = [Lowest];
```

Our next task is to calculate the expected frequency for all of the in-between intervals. Query 3_6b performs this task. Remember that the expected frequency is determined by subtracting the normal value of the preceding interval from the normal value of the current interval. We can match these two intervals by equating the number of the current interval to its preceding interval's number plus one. To do this we need to join the TempWork table with itself. To make sure that we only work with the in-between intervals, we select only those rows from TempWork with an interval number greater than the lowest but less than the highest.

Query 3_6b:

```
INSERT INTO [Table 3_3] ([Interval], LowEnd, HiEnd, O, E)
SELECT TempWork.Interval,
TempWork.LowEnd, TempWork.HiEnd,
TempWork.CountOfMiles AS O,
Round((([TempWork].[StdNorm] - TempWork_1.StdNorm) * 100, 3) AS E
FROM TempWork, TempWork AS TempWork_1, [Query 3_5]
WHERE ((TempWork.Interval = ([TempWork_1].[Interval] + 1)
And TempWork.Interval > [Lowest]
And TempWork.Interval < [highest]));
```

To determine the expected frequency for the last interval, we need to subtract the sum of all preceding expected frequencies (E) from the number of values, n. Query 3_6c and Query 3_6d perform this task. Notice in the query the use of the Round function with which we subtract the normal value from 1, and notice in the WHERE clause how we select the interval just preceding the highest.

Query 3_6c:

```
SELECT Sum([Table 3_3].E) AS Sum_Less_Last
FROM [Table 3_3];
```

Query 3_6d:

```
INSERT INTO [Table 3_3] ([Interval], LowEnd, HiEnd, O, E)
SELECT TempWork.Interval, TempWork.LowEnd, 999 AS HiEnd,
TempWork.CountOfMiles AS O,
[n] - [Sum_Less_Last] AS E
FROM TempWork, [Query 3_5], [Query 3_6c], [Query 3_2]
WHERE TempWork.Interval = [Highest];
```

The combined result of Query 3_6a through Query 3_6d is shown in Table 3-5.

Table 3-5. Result of Query 3_6a Through Query 3_6d

INTERVAL	LOWEND	HIEND	O	E
0	0	7	5	6.227
1	7	14	7	5.833
2	14	21	8	8.895
3	21	28	11	11.861
4	28	35	13	13.956
5	35	42	15	14.369
6	42	49	14	12.945
7	49	56	11	10.289
8	56	63	7	7.153
9	63	70	5	4.353
10	70	77	2	2.337
11	77	999	2	1.782

Since we need the number of expected frequencies in each interval greater than or equal to 5, we first combine all the adjacent intervals whose combined sum is less than 5. In Query 3_7, we save these combined intervals in a table called "CombInterval." The results of Query 3_7 are given in Table 3-6.

Query 3_7:

```
SELECT [Table 3_3].Interval,
[Table 3_3].LowEnd, [Table 3_3_1].HiEnd,
[Table 3_3].[O] + [Table 3_3_1].O AS O,
[Table 3_3].[E] + [Table 3_3_1].E AS E
INTO CombInterval
FROM [Table 3_3], [Table 3_3] AS [Table 3_3_1]
WHERE ((([Table 3_3].HiEnd) = [Table 3_3_1].[LowEnd])
AND (([Table 3_3].[E] + [Table 3_3_1].[E]) < 5));
```

Table 3-6. Result of Query 3_7

INTERVAL	LOWEND	HIEND	O	E
10	70	999	4	4.119

Those intervals (rows) that were combined in Query 3_7 are replaced with the rows in the CombInterval table. To do this we first tag for deletion those rows in Table 3_3 that participated in forming the combined intervals, by setting the interval identification number to –1 (Query 3_8). Afterward we execute Query 3_9 to delete the tagged rows (intervals) from the table. Then we run Query 3_10 to append the newly formed combined interval to Table 3_3. Repeat Query 3_7 through Query 3_10 until no more intervals can be combined.

Query 3_8:

```
UPDATE CombInterval, [Table 3_3]
SET [Table 3_3].[Interval] = -1
WHERE ((([Table 3_3].LowEnd) >= [CombInterval].[LowEnd])
AND (([Table 3_3].HiEnd) <= [CombInterval].[HiEnd]));
```

Query 3_9:

```
DELETE [Table 3_3].Interval
FROM [Table 3_3]
WHERE [Table 3_3].Interval = -1;
```

Query 3_10:

```
INSERT INTO [Table 3_3]
SELECT CombInterval.*
FROM CombInterval;
```

Next, we replace Query 3-7 with Query 3_7a and run repeatedly the query sequence Query 3_7a, Query 3_8, Query 3_9, and Query 3_10 until no more intervals are combined. Table 3-7 shows Table 3_3 after the modifications.

Query 3_7a:

```
SELECT [Table 3_3].Interval,
[Table 3_3].LowEnd, [Table 3_3_1].HiEnd,
[Table 3_3].[O] + [Table 3_3_1].O AS O,
[Table 3_3].[E] + [Table 3_3_1].E AS E
INTO CombInterval
FROM [Table 3_3], [Table 3_3] AS [Table 3_3_1]
WHERE (([Table 3_3].HiEnd = [Table 3_3_1].[LowEnd])
AND ([Table 3_3].E < 5) AND ([Table 3_3_1].E < 5));
```

Table 3-7. Table 3_3 After Combining Intervals

INTERVAL	LOWEND	HIEND	COUNTOFMILES (O)	EXPECTED FREQUENCY (E)
0	0	7	5	6.227
1	7	14	7	5.833
2	14	21	8	8.895
3	21	28	11	11.861
4	28	35	13	13.956
5	35	42	15	14.369
6	42	49	14	12.945
7	49	56	11	10.289
8	56	63	7	7.153
9	63	999	9	8.472

At this point we have combined all the adjacent intervals whose combined sum is under 5 or whose individual values are under 5. This leaves us with a table in which we still might have an interval with an expected frequency less than 5. If we combine this interval with its adjacent row, the result is an interval frequency greater than or equal to 5. Thus our next query extracts the row pairs that can be combined to yield intervals greater than 5. These combined intervals are placed in the table "RemInterval" as illustrated in Table 3-8. However, since our example does not have any intervals to combine, the entries in Table 3-8 are all zeros, but in reality the result of Query 3_11 is empty (no values).

Query 3_11:

```
SELECT [Table 3_3].Interval,
[Table 3_3].LowEnd, [Table 3_3].HiEnd,
[Table 3_3].O, [Table 3_3].E
INTO RemInterval
FROM [Table 3_3]
WHERE ((([Table 3_3].E) < 5));
```

Table 3-8. Table RemInterval

INTERVAL	LOWEND	HIEND	O	E
0	0	0	0	0

Combine this remaining interval with its adjacent interval and place the result in the table "CombInterval" which is shown in Table 3-9. Again, we show all zeros since our example produced an empty result.

Query 3_12:

```
SELECT [Table 3_1].LowEnd, RemInterval.HiEnd,
[Table 3_1].[O] + [RemInterval].[O] AS O,
[Table 3_1].[E] + [RemInterval].[E] AS E
INTO CombInterval
FROM RemInterval, [Table 3_1]
WHERE ((([Table 3_1].HiEnd) = [RemInterval].[LowEnd]));
```

Table 3-9. Table CombInterval

INTERVAL	LOWEND	HIEND	O	E
0	0	0	0	0

Repeat Query 3_8, Query 3_9, and Query 3_10 to replace those intervals, if any, that have been combined with the newly combined intervals. The result of Query 3_10 is given in Table 3-10. Notice in this case that the contents of Table 3-10 are identical to Table 3-7 because there are no more intervals to combine.

Table 3-10. Results of Query 3_10

INTERVAL	LOWEND	HIEND	COUNTOFMILES	E
0	0	7	5	6.227
1	7	14	7	5.833
2	14	21	8	8.895
3	21	28	11	11.861
4	28	35	13	13.956
5	35	42	15	14.369
6	42	49	14	12.945
7	49	56	11	10.289
8	56	63	7	7.153
9	63	999	9	8.472

Once these values are known, a chi-square (χ^2) statistic is calculated by comparing pairwise the observed (O) and expected (E) frequencies, as follows:

$$\chi^2 = \sum \frac{(O - E)^2}{E}$$

$$= \frac{(5 - 6.227)^2}{6.227} + \frac{(7 - 5.833)^2}{5.833} + \ldots + \frac{(9 - 8.472)^2}{8.472}$$

$$= 0.892$$

We find the number of degrees of freedom for the chi-square statistic in this case by counting the number of terms in the above sum (which is 10), subtracting 2 (representing the number of parameters needed from the data to calculate the expected frequencies—specifically the mean and standard deviation), then subtracting 1 (to account for the use of sample data). The result is of course 7 degrees of freedom. If we would like less than a 5% chance of reaching the wrong conclusion about the goodness of fit of the normal distribution, we would choose 0.05 as our significance level for the test, and either look up the χ^2 value in the appropriate statistical table in Appendix B, Table B-2, or use the Visual Basic code in Appendix D. The table value is equal to 14.067. Obviously, the calculated χ^2 value is much less than this table value, so our conclusion is that the distance data do appear to follow a normal distribution.

SQL/Query

Writing an SQL query to calculate the χ^2 can be accomplished in a relatively straightforward manner. Query 3_13 uses the Sum function to tally measured differences between the observed and expected frequencies. Either Query 3_14a or Query 3_14b can be used to obtain the χ^2 value. Query 3_14a looks up the χ^2 value from the chi-square statistics table. Query 3_14b uses a Visual Basic function to link to Excel to obtain the χ^2 value.

Query 3_13:

```
SELECT Sum((([o] - [e]) ^ 2) / [e]) AS ChiSq
FROM [Table 3_2];
```

Query 3_14a:

```
SELECT ChiSquare.ChiSq
FROM ChiSquare
WHERE ChiSquare.Percent = 0.05
AND ChiSquare.Degress_Of_Freedom = 7;
```

Query 3_14b:

```
SELECT Chi_Sq(0.05, 7) AS [Chi Sq]
FROM [Query 3_13];
```

NOTE *In Query 3_14b the FROM clause is not used. However, SQL requires a FROM clause in all queries.*

The results of the χ^2 test may be used to lend credence to the previous calculations in Chapter 2, and support the assumption of random distribution of distances. We can also use the normal distribution to answer questions about the distances. For example, what percentage of customers could be expected to call for limousine service say, more than 40 miles from the headquarters at the airport? This question is illustrated in Figure 3-5. The solution is to find the area under the normal curve to the right of 40 miles. To do this, we may find the area to the left of 40 and subtract the result from 100%, using the normal table. Thus

$$\frac{40-36.56}{19.25} = 0.1787$$

The area to the left of this value in the normal table is 0.5709, so the shaded area in Table 3-3 is equal to 1-0.5709, or 0.4291. Therefore, 42.9% of the customers could be expected to call for service more than 40 miles from the limousine headquarters.

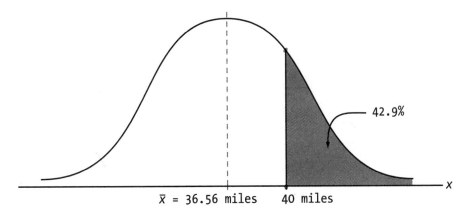

Figure 3-5. Theoretical normal distribution of limousine customer pick-ups, showing the area under the curve to the right of 40 miles

SQL/Query

An SQL query that can be used to obtain the answer to the "more than 40 miles" question is given in Query 3_15. Notice that there is no join condition, since Query 3_2 is a single row, and that the Round function is used to assure proper matching.

Query 3_15:

```
SELECT round((1.0 - [Area]) * 100, 1) AS Answer
FROM [Query 3_2], StdNormal
WHERE round([X], 3) = round((40.0 - [xBar]) / [s], 3);
```

....

John's Jewels
The Chance of an Accident Given the Opportunity

I recently had an opportunity to fit a statistical distribution to a set of data. We were working in the field of risk analysis and were trying to relate a certain accident risk to the opportunity for the accident to occur. An analogy would be the following: If you drove your car to work 1,000 times over a period of years, what is the likelihood that you would have an accident? Of course, the data required to develop such a model must come from past accident statistics. In our case, these data were scarce, since the types of accidents we were interested in analyzing were extremely rare. We ended up plotting the total accidents by years of occurrence, using a 7-year interval. The graph of the resulting data looked much like the curve of the gamma statistical distribution (see Appendix C), as accident occurrences became more infrequent in recent years. I tried the gamma as my model and ran a standard chi-square test on the results. They showed the gamma to be a pretty good fit for the data, so we were then able to use this information to predict the chance of a future accident.

....

Fitting a Poisson Distribution to Observed Data

A small retailer maintains a Web site for ordering online. The owner is particularly interested in the e-commerce sales of one high-profit item. The daily online sales of each unit of this item are tracked for 75 days, with the results recorded in Table 3-11. Each number represents the units per day in sales of this item.

Table 3-11. Number of Items Sold per Day for 75 Days

1	3	3	5	4	4	1	3	5	3	9	4	6	4	0
4	5	6	4	0	5	6	6	8	2	4	6	4	2	3
5	7	12	3	6	2	7	1	6	5	2	2	3	4	4
6	9	0	5	4	3	4	4	0	1	3	5	6	3	3
2	7	6	10	3	4	6	4	5	4	3	4	4	5	3

The owner desires to know if sales follow some random statistical distribution, based on the sample data, so that predictions may be made regarding future sales. First, a histogram is set up for the data. Since the values represent counts, the data are *discrete*, or based on *attributes*. Since so many individual values repeat themselves, a frequency of occurrence may be determined for each value. After the counts are completed, the histogram appears in Figure 3-6.

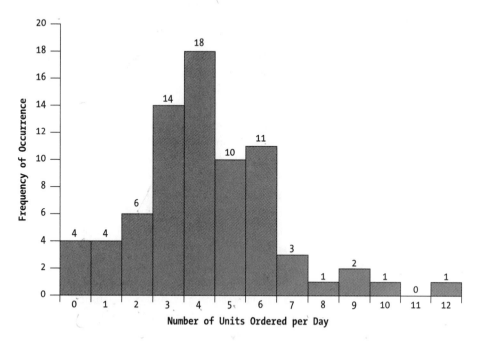

Figure 3-6. Histogram of daily sales

The shape of this histogram (rising fairly rapidly to a peak, then tapering gradually to the right) and the nature of the phenomenon being observed, point to a well-known statistical distribution called the *Poisson*. A Poisson distribution typically characterizes the occurrence of a discrete "event," such as ordering a retail item, when the number of orders may be counted. However, the "nonoccurrence" of the event makes no sense (i.e., the number of people who did not order on a given day is unknown and irrelevant). The Poisson also commonly characterizes arrival and departure phenomena, and the random spatial distribution of observations such as the number of trees per acre in a forest, or prairie dog colonies per square mile in a wildlife habitat.

Poisson Sets the Groundwork for Horse Kick Analysis

The great mathematician Simeon Denis Poisson was born in 1781 in Pithiviers, France. He was first forced to study medicine, but later became a student of the legendary mathematicians Laplace and Lagrange. During his lifetime, he excelled in applications of mathematics to astronomy, electricity and magnetism, elasticity, differential equations, potential theory, and mechanics. He published between 300 and 400 mathematical works. One of the most famous was entitled *Recherches sur la probabilite des jugements*... (1837), in which Poisson first introduced to the world the statistical distribution that still bears his name. Legend has it that one of the first practical applications of the Poisson distribution was by von Bortkewitch in 1898 (some sources say 1880). He actually employed the distribution to analyze the deaths due to horse kicks in the Prussian army. He found that, among 10 army corps over a 20-year period, 122 men had died from being kicked by their horses. When the number of such deaths per year is plotted against a theoretical Poisson distribution with mean 0.61 deaths per corps per year (the actual average), they coincide very closely. A chi-square goodness of fit test on the observed distribution shows great significance. This example has been used in many statistics texts over the years. Its relevance has diminished in a practical sense (there aren't as many horse corps as there once were), but its importance in the history of statistics, along with Poisson, is well assured. Incidentally, Poisson died in 1840 near Paris.

But do the sales data really follow a Poisson distribution? To find out, we first need to know the density function (usually called "mass" function for discrete distributions) for the Poisson. It is

$$\frac{e^{-\mu}\mu^x}{x!}$$

where μ is the mean of the distribution (usually estimated as $\bar{x}$ from the sample data), x is the count of the number of orders per day ($x = 1, 2, 3, \text{K}$), $e^{-\mu}$ is the exponential constant (= 2.71828) raised to the power -μ, and $x!$ is "x factorial" (defined as $1 \cdot 2 \cdot 3 \cdot 4 \cdot 5 \cdots x$, with 0! defined as equal to 1). Although the expression may look horrible, it's easy to calculate on a hand calculator. First, though, we need to know the mean of the 75 data values (the average number of orders per day). We can add all 75 values and divide by 75, or take advantage of the work we already expended to develop the histogram. Thus

$$\bar{x} = \frac{4 \cdot 0 + 4 \cdot 1 + 6 \cdot 2 + 14 \cdot 3 + 18 \cdot 4 + 10 \cdot 5 + 11 \cdot 6 + 3 \cdot 7 + 1 \cdot 8 + 2 \cdot 9 + 1 \cdot 10 + 1 \cdot 12}{75}$$

$$= 4.2 \text{ orders per day}$$

$\bar{x}$ is an estimate of μ. Although it's not a nice whole number, that's O.K. It's usually quoted as a decimal fraction anyway. (Incidentally, the standard deviation of the 75 values is 2.27 orders per day.) As a demonstration, let's calculate the Poisson expression for $x = 6$ orders per day, as follows:

$$\frac{e^{-4.2}(4.2)^6}{6!} = \frac{(0.0150)(5,489.0317)}{1 \cdot 2 \cdot 3 \cdot 4 \cdot 5 \cdot 6} = 0.1143$$

If this number is multiplied by 75, we obtain the number of days out of 75 we could expect 6 orders, if the distribution of orders per day is exactly Poisson with mean 4.2 orders per day. The result is about 8.5741, or nearly 9 days. If we repeat this procedure for $x = 0, 1, 2$, etc., multiplying each result by 75 as above, we generate Table 3-12. Now we have *expected* occurrences to compare to those actually *observed* in the sample. Notice that since no days were recorded in the sample when more than 12 orders were placed, the observed value is 0. However, the theoretical Poisson is an infinite distribution, so it requires a small estimate even beyond 12. This is equal to 75 less the sum of all the expected values from 0 to 12, or about 0.0323.

Table 3-12. Observed and Expected Daily Sales Assuming a Poisson Model

NO. OF ORDERS PER DAY (*X*)	NO. OF DAYS OBSERVED WITH *X* ORDERS	POISSON EXPRESSION FOR *X*	NO. OF DAYS EXPECTED WITH *X* ORDERS	CUMULATIVE POISSON VALUES FROM COLUMN 3
0	4	0.01500	1.1247	0.01500
1	4	0.06298	4.7236	0.07798
2	6	0.13226	9.9196	0.21024
3	14	0.18517	13.8874	0.39541
4	18	0.19442	14.5818	0.58983
5	10	0.16332	12.2487	0.75315
6	11	0.11432	8.5741	0.86747
7	3	0.06859	5.1444	0.93606
8	1	0.03601	2.7008	0.97207
9	2	0.01681	1.2604	0.98888
10	1	0.00706	0.5294	0.99594
11	0	0.00269	0.2021	0.99863
12	1	0.00094	0.0707	0.99957
> 12	0	0.00043	0.0323	1.00000

It is again recommended that contiguous expected values be combined to assure that each expected number of days with x orders is at least 5. The observed days are combined as well. This results in 7 comparisons of observed and expected values, as shown in Table 3-13. In this case, only the mean is needed to calculate the Poisson expression, so the number of degrees of freedom is equal to $7-1-1 = 5$. Finally, chi-square is calculated as before, from the combined intervals in Table 3-13:

$$\chi^2 = \frac{(8-5.85)^2}{5.85} + \frac{(6-9.92)^2}{9.92} + K + \frac{(8-9.94)^2}{9.94}$$

$$= 4.621$$

For a 5% significance level and 5 degrees of freedom, the tabulated $\chi^2 = 11.070$. Since the calculated value is less than the table value, the distribution of orders per day is judged to be Poisson.

Table 3-13. Table 3-12 After Combining Intervals

NO. OF ORDERS PER DAY (X)	NO. OF DAYS OBSERVED WITH X ORDERS	POISSON EXPRESSION FOR X	NO. OF DAYS EXPECTED WITH X ORDERS
<1	8	0.07798	5.8485
2	6	0.13226	9.9196
3	14	0.18517	13.8874
4	18	0.19442	14.5818
5	10	0.16332	12.2487
6	11	0.11432	8.5741
>7	8	0.13253	9.9401

With this information, we might want to ask the following questions:

- How many units of the item should be kept in daily inventory to meet 95% of the demand?

To answer this question, we can accumulate the Poisson values in the third column of Table 3-12 until we reach a total of 0.95. The cumulative Poisson values are given in the last column of Table 3-12. This occurs between 7 and 8 orders per day, so holding 8 items in daily inventory would satisfy at least 95% of the demand. We can also obtain the answer to this question by executing Query 3_16, which is shown on the next page.

SQL/Query

Query 3_16:

```
SELECT Min([Table 3_12].[Orders Per Day]) AS [Daily Inventory]
FROM [Table 3_12]
WHERE [Table 3_12].[Cumm Poisson] >= 0.95;
```

- What percent of daily sales fall between 2 and 6 units per day?

A similar approach may be used to answer this question. Add the Poisson values from 2 through 6 to yield 0.78949. Thus about 79% of daily sales fall between 2 and 6 units. This can also be obtained by executing Query 3_17.

SQL/Query

Query 3_17:

```
SELECT Sum([Table 3_12].[Poisson])*100 AS [Percent of Daily Sales]
FROM [Table 3_12]

WHERE [Table 3_12].[Orders Per Day] Between 2 And 6;
```

- Is there a justification for keeping more than 15 units per day in inventory?

To answer this question, we need to calculate the Poisson expression for $x = 13, 14$, and 15, which we don't have yet. These will be very small numbers, as shown below:

$$\frac{e^{-4.2}\,(4.2)^{13}}{13!} = 0.0003047$$

$$\frac{e^{-4.2}\,(4.2)^{14}}{14!} = 0.0000914$$

$$\frac{e^{-4.2}\,(4.2)^{15}}{15!} = 0.0000256$$

From Table 3-12, we know the chance that the daily orders are less than or equal to 12 is the sum of the Poisson expression from 0 through 12, or 0.99957. Adding the three values above brings this total to 0.99999, so the likelihood of orders of more than 15 units in a day is 1-0.99999 = 0.00001, or 0.001% (a very slim chance). We can obtain this result by running Query 3_18. The Poisson function is defined in Appendix D.

SQL/Query

Query 3_18:

```
SELECT
Round(1 - ([Cumm Poisson] + Poisson(13,4.2) +
Poisson(14,4.2) + Poisson(15,4.2)), 5) AS [More than 15 units]
FROM [Table 3_12]
WHERE [Table 3_12].[ Orders Per Day] = 12;
```

Fitting an Exponential Distribution to Observed Data

A service company maintains an Internet Web site to attract potential customers. The webmaster is presently tracking the elapsed time (in minutes) between successive "hits" on the site. Fifty observations during the day generated the data in Table 3-14.

Table 3-14. Fifty Random Observations of Elapsed Times (in Minutes) Between "Hits" on an Internet Web Site

17	20	10	9	23	13	12	19	18	24
12	14	6	9	13	6	7	10	13	7
16	18	8	13	3	32	9	7	10	11
13	7	18	7	10	4	27	19	16	8
7	10	5	14	15	10	9	6	7	15

A histogram may be set up for these observations, as shown in Figure 3-7. We've used a convenient interval width of 5 minutes, again with the left-hand endpoint inclusive and the right-hand endpoint exclusive for each interval.

We note from the histogram that there were two occurrences where the time between successive hits was less than 5 minutes. The most representative elapsed time, however, appears to be between 5 and 20 minutes. The mean is calculated by summing the 50 observations and dividing the result by 50. This yields $\bar{x} = 12.32$ minutes as the average time between "hits." Since the data are continuous (variables data), an appropriate model should be selected from among the continuous models available. At first glance, the data values seem to be exponentially decreasing. Consequently, an exponential distribution is chosen initially to characterize the data. The density expression

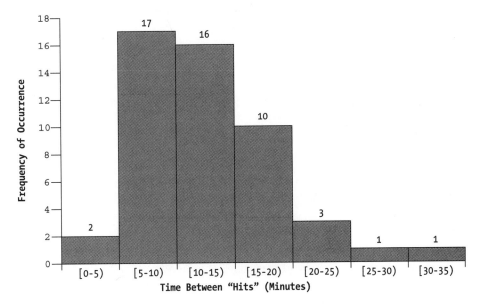

Figure 3-7. Histogram of times between hits from Table 3-14

for such a model is defined below in terms of the estimated mean $\overline{x}$ and the random variable x:

$$\frac{1}{\overline{x}} e^{-x/\overline{x}}, \quad x \geq 0$$

However, this is not the most convenient expression to yield the expected frequencies. In practice, we are required to integrate the above expression and multiply by 50 (the number of observations) to obtain the expected frequency in any particular interval of the histogram. Instead of that cumbersome approach, we can use the following expression instead:

$$50\left(1 - e^{-x/\overline{x}}\right)$$

This yields the cumulative expected frequency at any point (x value) on the histogram. The expression in parentheses is obtained by integrating

$$\frac{1}{\overline{x}} e^{-x/\overline{x}}$$

from 0 to x. It is known as the cumulative exponential distribution function.

To use it, we begin with the first interval from 0 to 5. Substituting $x = 5$ and $\bar{x} = 12.32$ into the cumulative distribution function, we obtain

$$50\left(1-e^{-5/12.32}\right)=16.68$$

This is the expected number of observations between 0 and 5 minutes. Next we calculate the expected number of observations between 0 and 10 minutes, as shown below:

$$50\left(1-e^{-10/12.32}\right)=27.79$$

By subtracting the expected number between 0 and 5 minutes from the expected number between 0 and 10 minutes, we obtain the expected number between 5 and 10 minutes, as follows:

$$27.79-16.68=11.11$$

Proceeding in this manner, we can generate Table 3-15.

Table 3-15. Observed and expected frequencies of times between successive "hits"

HISTOGRAM INTERVAL	OBSERVED FREQUENCY	EXPECTED FREQUENCY
[0 – 5)	2	16.68
[5 – 10)	17	11.11
[10 – 15)	16	7.41
[15 – 20)	10	4.94
[20 – 25)	3	3.29
[25 – 30)	1	2.19
[30 – 35)	1	1.46
≥ 35	0	2.92

Combining expected frequencies to insure each interval has a total of at least 5, we obtain Table 3-16. It also shows the χ^2 terms for each resulting interval.

Table 3-16. Calculation of Terms for the χ^2 Statistic

HISTOGRAM INTERVAL	OBSERVED FREQUENCY (O)	EXPECTED FREQUENCY (E)	$\dfrac{(O-E)^2}{E}$
$[0-5)$	2	16.68	12.920
$[5-10)$	17	11.11	3.123
$[10-15)$	16	7.41	9.958
$[15-25)$	13	8.23	2.765
≥ 25	2	6.57	3.179

The χ^2 statistic is the sum of the last column in Table 3-16, or about 31.94. The number of degrees of freedom is equal to 5 intervals (from Table 3-16) less one parameter (the mean) less one, or 3. For a 5% significance level and 3 degrees of freedom, the table χ^2 value is 7.815. Since the calculated value exceeds the table value, we conclude that the distribution of times between successive "hits" is significantly different from the exponential distribution. (That's statistician jargon to simply say "it didn't fit.") We therefore might want to try a different model for better characterization. It turns out that a gamma distribution of the form

$$0.00162 x^{3.10} e^{-0.33x}$$

is a reasonably good fit. This could have been foreseen from the original histogram, since the shape was not continually decreasing (a characteristic of the exponential distribution). Rather, it started low, rose to a maximum, then tapered down to the right.

Conclusion

There are many distributions, either discrete or continuous, that are candidates for sample data characterization. The nature of the data and shape of the histogram are clues that help guide the data miner toward a possible model. The χ^2 test is then helpful in comparing the observed data distribution to a theoretical model. In Appendix C we present a list of some of the more common distributions and their characteristics, in a table format for easier use. There are many models available. Please understand that in some cases the mechanics of calculating expected frequencies for certain models might be quite complicated, unless tables or cumulative distribution expressions are available. In addition, some of the distributions presented (such as χ^2, F, and Student's t) are more commonly associated with hypothesis testing than with characterizing data sets. Their density function expressions are pretty intimidating, as you can see from the table in Appendix C.

John's Jewels
The Godfather of Statistical Distributions

I'll tell you something you probably won't believe. There are dozens of statistical distributions, both discrete and continuous. However, there is one umbrella distribution that can be used to derive any of these known distributions. It is called the H function distribution. By substituting appropriate values for its parameters, we obtain the binomial, the Poisson, the normal, the exponential, and in fact any other statistical distribution, discrete or continuous. The procedure for accomplishing this, however, is quite complex, so the H function is little known and rarely encountered. It's like the godfather of the statistical family of distributions.

T-SQL Source Code

There are three T-SQL source code listings for performing the chi-square goodness of fit test for the normal distribution described in this chapter. The first listing is called Make_Intervals. Make_Intervals takes a table of values, groups the values based on the range (interval width), and counts the number of values within each interval. The second listing is called Combine_Intervals. Combine_Intervals takes the result of Make_Intervals, finds those intervals with a count value less than 5, and combines these smaller intervals with respect to their adjacent intervals until all intervals have a count greater than or equal to 5. The last listing is called Compare_Observed_And_Expected. Compare_Observed_And_Expected takes the results of Combine_Intervals and performs the statistical calculations necessary to compare the observed values to the expected values. Following the procedure listings is a section illustrating the call statements to each procedure with parameter specifications.

Make_Intervals

```
SET QUOTED_IDENTIFIER OFF
GO
SET ANSI_NULLS OFF
GO

CREATE PROCEDURE Make_Intervals
@Interval_Size smallint = 7,
@SourceTblName  VarChar(50) = '[Limo Miles]',
@SourceColName  VarChar(50) = '[Miles]',
@ResultTblName  VarChar(50) = '[Table 3_1]'
AS


/***********************************************************/
/*                                                         */
/*                   MAKE INTERVALS                        */
/*                                                         */
/*    This procedure takes a table of values, groups the   */
/* values into intervals of a specified width, and then    */
/* counts the number of values within each interval.       */
/*                                                         */
/* INPUTS:                                                 */
/*    SourceTblName - table containing sample data         */
/*    SourceColName - column containing sample data values */
/*    ResultTblName - table to receive intervals           */
/*                                                         */
/* NOTE:                                                   */
/* The result table has four columns:                     */
/*    Interval - interval ID number begins with zero       */
/*    LowEnd - lower bound of interval                      */
/*    HiEnd - upper bound of interval                       */
/*    ResultColName - (see INPUTS above)                    */
/*                                                         */
/***********************************************************/


/* Clear the result table */
EXEC('DELETE  FROM ' + @ResultTblName)
```

```
/* This query forms the intervals */
EXEC('INSERT INTO '+ @ResultTblName +
     'SELECT Floor((' + @SourceColName + ')/' +
           @Interval_Size + ') AS Interval, ' +
     '(Floor((' + @SourceColName + ')/' + @Interval_Size +
           '))*'+@Interval_Size + ' AS LowEnd, ' +
     '(Floor((' + @SourceColName + ')/' + @Interval_Size +
           '+1))*' + @Interval_Size + 'AS HiEnd, ' +
     'Count(' + @SourceTblName + '.' + @SourceColName +
           ') AS CountOf ' +
     'FROM ' + @SourceTblName + ' ' +
     'GROUP BY Floor((' + @SourceColName + ')/' +
           @Interval_Size + '), ' +
     '(Floor((' + @SourceColName + ')/' + @Interval_Size +
           '))*' + @Interval_Size + ', ' +
     '(Floor((' + @SourceColName + ')/' + @Interval_Size +
           '+1))*' + @Interval_Size + ' ' +
     'ORDER BY Floor((' + @SourceColName + ')/' +
           @Interval_Size + ')')


GO
SET QUOTED_IDENTIFIER OFF
GO
SET ANSI_NULLS ON
GO
```

Combine_Intervals

```
SET QUOTED_IDENTIFIER ON
GO
SET ANSI_NULLS ON
GO


CREATE PROCEDURE Combine_Intervals
@Src1TblName  VarChar(50) = 'Limo Miles',
@Src1ColName  VarChar(50) = 'Miles',
@Src2TblName  VarChar(50) = 'Table 3_1',
@RstTblName   VarChar(50) = 'Table 3_3'
AS
```

```
/*********************************************************/
/*                                                     */
/*              COMBINE INTERVALS                      */
/*                                                     */
/*    This procedure takes the result table from the   */
/* Make_Interval procedure and combines those intervals */
/* with an expected frequency less than 5 with their    */
/* adjacent intervals until all intervals have values   */
/* greater than or equal to 5.                         */
/*                                                     */
/* INPUTS:                                             */
/*    Src1TblName - table name for sample data         */
/*    Src1ColName - column name for sample data         */
/*    Src2TblName - result table name from Make_Interval */
/*    RstTblName - result table name from combined interval */
/*                                                     */
/*********************************************************/


/* Local Variables */
DECLARE @Q Varchar(500)          /* Query string */
DECLARE @Xbar Float              /* Sample mean */
DECLARE @SD Float                /* Sample Standard deviation */
DECLARE @N Int                   /* Number of sample data values */
DECLARE @ID Int                  /* Interval counter */
DECLARE @LO Int                  /* Lower bound of interval being combined */
DECLARE @HI Int                  /* Upper bound of interval being combined */
DECLARE @Lowest Int              /* Lower bound of all intervals */
DECLARE @Highest Int             /* Upper bound of all intervals */
DECLARE @SumO Int                /* Sum of the sum of observed vallues */
DECLARE @SumE Float              /* Sum of the expected frequencies */
DECLARE @Sum_Less_Last Float     /* Sum E's except last one */


/* SETUP FOR INTERVALS */

/* Obtain sample mean and standard deviation */
SET @Q = 'SELECT Avg(Convert(Float, [' + @Src1ColName + '])) AS Xbar, ' +
     'StDev([' + @Src1ColName + ']) AS SD, ' +
     'Count([' + @Src1ColName + ']) AS N ' +
     'INTO ##TempXbar_SD_N ' +
     'FROM [' + @Src1TblName + '] '
EXEC(@Q)
SELECT @Xbar = Xbar, @SD = SD, @N = N FROM ##TempXbar_SD_N
```

```
/* Establish intermediate work table */
CREATE TABLE ##TempWork (Interval Int,
    LowEnd Int, HiEnd Int, CountOf Int,
    Xbar Float, SD Float, NVariate Float, StdNorm Float null)

/* Populate the work table */
SET @Q = 'INSERT INTO ##TempWork (Interval, LowEnd, HiEnd, ' +
    'CountOf, Xbar, SD, NVariate, StdNorm) ' +
    'SELECT Interval, LowEnd, HiEnd, CountOf, ' +
    str(@Xbar,8,3) + ' AS Xbar, ' +
    str(@SD,8,3) + 'AS SD, ' +
    '(HiEnd - ' + str(@Xbar,8,3) + ')/' +
    str(@SD,8,3) + ' AS NVariate, 0.0 AS StdNorm ' +
    'FROM [' + @Src2TblName + ']'
EXEC(@Q)

/* Update work table with Standard Normal */
UPDATE ##TempWork
SET StdNorm =
    (Select Area FROM StdNormal
    WHERE Str(Nvariate,9,3) = Str(StdNormal.X,9,3))

/* Get lowest, highest interval values */
SELECT @Lowest = Min([Interval]),
    @Highest = Max([Interval])
    FROM ##TempWork

/* Establish table of observed (O) and */
/* expected frequencies (E) and insert */
/* the first interval */
SET @Q = 'SELECT ##TempWork.Interval As Interval, ' +
    '##TempWork.LowEnd AS LowEnd, ' +
    '##TempWork.HiEnd AS HiEnd, ' +
    '##TempWork.CountOf AS O, ' +
    '##TempWork.StdNorm * ' + str(@N,8,1) + ' AS E ' +
    'INTO ##TempTable3_3 ' +
    'FROM ##TempWork ' +
    'WHERE ##TempWork.Interval = ' + Convert(Varchar(20), @Lowest)
EXEC(@Q)
```

```
/* Calculate the expected frequency for */
/* all of the in-between intervals */
SET @Q = 'INSERT INTO ##TempTable3_3 (Interval, LowEnd, HiEnd, O, E) ' +
    'SELECT ##TempWork.Interval AS Interval, ' +
    '##TempWork.LowEnd AS LowEnd, ' +
    '##TempWork.HiEnd AS HiEnd, ' +
    '##TempWork.CountOf as O, ' +
    '(##TempWork.StdNorm - ##TempWork_1.StdNorm) * ' +
    str(@N,8,1) + ' AS E ' +
    'FROM ##TempWork, ##TempWork ##TempWork_1 ' +
    'WHERE ##TempWork.Interval = (##TempWork_1.Interval + 1) ' +
    'AND ##TempWork.Interval > ' + Convert(Varchar(20), @Lowest) + ' ' +
    'AND ##TempWork.Interval < ' + Convert(Varchar(20), @Highest)
EXEC(@Q)


/* Sum all the preceding expected frequences (E) */
/* before appending the last interval */
SELECT @Sum_Less_Last = Sum(E) FROM ##TempTable3_3


/* Determine expected frequency for last interval */
/* and then append the last interval */
SET @Q = 'INSERT INTO ##TempTable3_3 (Interval, LowEnd, HiEnd, O, E) ' +
    'SELECT ##TempWork.Interval AS Interval, ' +
    '##TempWork.LowEnd AS LowEnd, ' +
    '999 AS HiEnd, ' +
    '##TempWork.CountOf as O, ' +
    Convert(Varchar(20), @N) + ' - ' +
    str(@Sum_Less_Last,8,3) + ' AS E ' +
    'FROM ##TempWork ' +
    'WHERE ##TempWork.Interval = ' + Convert(Varchar(20), @Highest)
EXEC(@Q)


/* COMBINE INTERVALS */

/* If result table exists, then drop it */
IF exists (SELECT id FROM ..sysobjects
        WHERE name = @RstTblName)
Begin
    SET @Q = 'DROP TABLE [' + @RstTblName + ']'
    EXEC(@Q)
End
```

```
/* Create the result table */
SET @Q = 'CREATE TABLE [' + @RstTblName + '] (Interval Int, '+
     'LowEnd Int, HiEnd Int, O Float, E Float)'
EXEC(@Q)


/* Define the cursor and the local      */
/* variables to receive the row values */
DECLARE Cr1 INSENSITIVE SCROLL CURSOR
     FOR SELECT Interval, LowEnd, HiEnd, O, E
     FROM ##TempTable3_3
     ORDER BY Interval
DECLARE @Intv1 Int, @Low1 Int, @Hi1 Int, @O1 Float, @E1 Float


OPEN Cr1


/* Set cursor to first record */
FETCH NEXT FROM  Cr1
     INTO @Intv1, @Low1, @Hi1, @O1, @E1


/* Initialize */
SET @ID = 0


/* Go through the table and combine those  */
/* intervals that have a count less than 5 */


WHILE @@FETCH_Status = 0
    Begin
        If @E1 >= 5
        Begin
            /* Do not combine, just copy over  */
            /* the interval and move cursor    */
            /* ahead one row and get next row. */
            SET @Q = 'INSERT INTO [' + @RstTblName + '] ' +
                'VALUES(' + str(@ID,8,0) + ', ' +
                str(@Low1,8,0) + ', ' +
                str(@Hi1,8,0) + ', ' +
                str(@O1,12,5) + ', ' +
                str(@E1,12,5) + ')'
            EXEC(@Q)

            SELECT @ID = @ID + 1

            FETCH NEXT FROM Cr1
                INTO @Intv1, @Low1, @Hi1, @O1, @E1
        End
```

```
Else
Begin
    /* Combine the two or more intervals into   */
    /* one interval until we have @SumE >= 5,   */
    /* also remember low point of this interval */
    SELECT @LO = @Low1
    SELECT @SumO = 0.0
    SELECT @SumE = 0.0

    While (@@Fetch_Status = 0 and @SumE < 5)
    Begin
        SELECT @SumO = @SumO + @O1
        SELECT @SumE = @SumE + @E1
        SELECT @HI = @Hi1

        FETCH NEXT FROM Cr1
            INTO @Intv1, @Low1, @Hi1, @O1, @E1
    End

    /* Did we exit the loop */
    /* because End-Of-Table? */
    If @@Fetch_Status  <> 0 and @SumE < 5
    Begin
        /* Combine with the last row since */
        /* there are no more intervals to  */
        /* combine to get a value >= 5     */
        SET @Q = 'UPDATE [' + @RstTblName + '] ' +
        'SET O = O + ' + str(@SumO,12,5) + ', ' +
        'E = E + ' + str(@SumE,12,5) + ', ' +
        'HiEnd = ' + str(@HI,8,0) + ' ' +
        'WHERE HiEnd = ' + Convert(varchar(20), @LO)
        EXEC(@Q)
    End
```

```
                    Else
                    Begin
                        /* Save the combined interval */
                        SET @Q = 'INSERT INTO [' + @RstTblName + '] ' +
                        'VALUES(' + str(@ID,8,0) + ', ' +
                        str(@LO,8,0) + ', ' +
                        str(@HI,8,0) + ', ' +
                        str(@SumO,12,5) + ', ' +
                        str(@SumE,12,5) + ')'
                        EXEC(@Q)
                    End

            End

    End   /* End While loop */

    Close Cr1
    Deallocate Cr1

    /* Replace CombInterval with the RemInterval */
    Delete FROM [CombInterval]
    INSERT INTO [CombInterval]
        SELECT * FROM RemInterval

GO
SET QUOTED_IDENTIFIER OFF
GO
SET ANSI_NULLS ON

GO
```

Compare_Observed_And_Expected

```
SET QUOTED_IDENTIFIER OFF
GO
SET ANSI_NULLS ON
GO

CREATE PROCEDURE [Compare_Observed_And_Expected]
@Alpha Float = 0.05,
@v Int = 7,
@Src1TblName VarChar(50) = 'Table 3_3'
AS
```

```
/***********************************************************/
/*                                                       */
/*       COMPARE OBSERVED AND EXPECTED FREQUENCIES        */
/*                FOR A GOODNESS OF FIT                    */
/*              TO THE NORMAL DISTRIBUTION                */
/*                                                       */
/*    This procedure takes the result table from the     */
/* Combine_Intervals procedure and determines whether or */
/* not the data fits a normal distribution. This is      */
/* accomplished by performing a Chi-square test between the */
/* observed frequencies and the expected frequencies.    */
/*                                                       */
/* INPUTS:                                                */
/*   Src1TblName - result table from Combine_Intervals   */
/*                                                       */
/*   TABLES:                                              */
/*   TableChiSQ - Chi square values                      */
/*       Contents:                                       */
/*       CalcChiSq - float, calculated chi square        */
/*       TableChSq - float, table value of chi square    */
/*   ChiSquare -- the statistical table of Chi Squares.  */
/*       Contents:                                       */
/*   Alpha - significance level                          */
/*       v - degrees of freedom                          */
/*       ChiSq - chi square for Alpha and v              */
/*                                                       */
/***********************************************************/


/* Local Variables */
DECLARE @CalcChiSq float      /* Calculate Chi Square */
DECLARE @TableChiSq float      /* Table Chi Square value */

/* Calculate Chi Square */
EXEC('SELECT Sum(Power((([O]-[E]),2)/[E]) AS V ' +
     'INTO ##TmpChiSQCalcTable ' +
     'FROM [' + @Src1TblName + '] ')
SELECT @CalcChiSq = V
     FROM ##TmpChiSQCalcTable
```

```
/* Look up the Chi Square table value */
EXEC('SELECT ChiSquare.ChiSq AS TableChiSq ' +
     'INTO ##TmpChiSQTblTable ' +
     'FROM ChiSquare ' +
     'WHERE ((ChiSquare.[Percent]=' + @Alpha + ') ' +
     'AND (ChiSquare.Degress_Of_Freedom=' + @v + '))')
SELECT @TableChiSq = TableChiSq
     FROM ##TmpChiSQTblTable

/* Save the Chi Square values, */
/* but first clear the table */
DELETE [ChiSq For Data]
INSERT INTO [ChiSq For Data]
     SELECT @CalcChiSq, @TableChiSq

GO
SET QUOTED_IDENTIFIER OFF
GO
SET ANSI_NULLS ON
GO
```

Procedure Calls

Below are the call statements for calling the procedures Make_Intervals, Combine_Intervals, and Compare_Observed_And_Expected.

```
DECLARE @RC int
DECLARE @Interval_Size int
DECLARE @SourceTblName varchar(50)
DECLARE @SourceColName varchar(50)
DECLARE @ResultTblName varchar(50)
EXEC @RC = [CH3].[dbo].[Make_Intervals] @Interval_Size=7,
@SourceTblName='[Limo Miles]', @SourceColName='[Miles]',
@ResultTblName='[Table 3_1]'
```

```
DECLARE @RC int
DECLARE @Src1TblName varchar(50)
DECLARE @Src1ColName varchar(50)
DECLARE @Src2TblName varchar(50)
DECLARE @RstTblName varchar(50)
EXEC @RC = [CH3].[dbo].[Combine_Intervals] @Src1TblName ='Limo Miles',
@Src1ColName='Miles',
@Src2TblName ='Table 3_1', @RstTblName='Table 3_3'


DECLARE @RC int
DECLARE @Alpha float
DECLARE @v int
DECLARE @Src1TblName varchar(50)
EXEC @RC = [CH3].[dbo].[Compare_Observed_And_Expected]
@Alpha=0.05, @v=7, @Src1TblName='Table 3_3'
```

CHAPTER 4

Additional Tests of Hypothesis

Accept or Reject?

IN THE PREVIOUS CHAPTER, we presented methods for pictorially representing a set of data values and determining whether the data follows a known statistical distribution. We stated that testing the data fell under a general category of tests of hypothesis, of which the chi-square goodness of fit test was one example. In this chapter, we discuss a few more tests of hypothesis that occasionally are of benefit to the data miner. These tests involve comparing the parameters (such as mean and variance) of two or more samples, or comparing one sample's parameters to a known population measure. The procedure for conducting each test is much the same as that for the goodness of fit situation, so we review that for you and add a few more interesting details that will provide good party conversation, should you find yourself sipping cocktails with your local statistician.

Remember that a statistical hypothesis is a statement about the parameters of a statistical distribution of data. The value of a parameter is compared to another value to decide whether the statement concerning the parameter is true or false. For example, we might know that our past monthly gross sales have averaged $8,400 for a certain item. This particular year our monthly gross sales averaged $10,250. Is this enough of a change to be significant, or is it just the result of random fluctuations in sales that are inevitable month to month and year to year? We may be able to answer this question through a test of hypothesis. But first we need a hypothesis. Usually this is stated in a formal way as what is called a *null hypothesis*, and it is denoted H_0. In our example, the null hypothesis might be stated as follows: There is no significant difference between the past average monthly sales (say μ) and the average monthly sales of the current year (say $\bar{x}$). Symbolically,

$$H_0 : \mu = \bar{x}$$

If there *is* significant difference between the average monthly sales versus the current figure, the alternative hypothesis H_1 may be true instead. That is

$$H_1 : \mu \neq \bar{x}$$

In hypothesis testing, there are four possibilities that may occur as a result of your conclusion. In two of these, you will be correct, and in the other two, you will err. For example, let's suppose your test supports the acceptance of the null hypothesis, and you conclude the current average monthly sales are not significantly different than the expected average. Then either the parameter you're testing lies in the region of acceptance (in which case you made the right conclusion), or it lies in the region of rejection, or significance (in which case you made a false conclusion). This latter situation can actually happen, because statistics is like horse racing — nothing is 100% sure. If it does happen, the statisticians say you committed a *Type II error*. Sounds criminal, doesn't it?

But wait! What if the opposite (sort of) occurred in your test? Let's say your test result rejected the null hypothesis, and you concluded that current sales were in fact significantly different from the yearly monthly average. If the parameter lies in the rejection, or significance region, you made the right conclusion. But if it doesn't, you made a *Type I error*. Another way of stating these situations, probably no less confusing, is as follows: If you conclude something is true when it's really false, you commit a Type II error; when you conclude something is false when it's really true, you commit a Type I error. Maybe Table 4-1 will make things a bit more clear.

Table 4-1. Possible Outcomes Following a Test of Hypothesis on the Two Sales Figures

WHAT WE KNOW	WHAT WE'D LIKE TO KNOW	POSSIBLE ANSWERS	APPROPRIATE HYPOTHESIS	POSSIBLE CONCLUSIONS	THE ACTUAL SITUATION (IF YOU COULD KNOW IT)	RESULTS
Average past monthly sales are $8,400. This years' monthly sales are $10,250.	Are the average monthly sales figures this year significantly different from the past average monthly sales?	Yes, they are.	H_0: There is no significant difference in the two figures.	Accept H_0 (no significant difference in the two figures) and reject H_1.	No significant difference in the two figures.	You made the correct conclusion.
					There is significant difference in the two figures.	Type II error
		No, They're not.	H_1: There is significant difference in the two figures.	Reject H_0 and accept H_1 (significant difference in the two figures).	There is significant difference in the two figures.	You made the correct conclusion.
					No significant difference in the two figures.	Type I error

Obviously, the astute data miner wants to minimize the chances of committing these types of errors. In fact, it would be nice if you could be 100% sure you would never commit one. But remember, a statistical conclusion is never a certainty. That is why the significance level is set as low as is usually practically possible, say 5%. This means the chance of committing a Type I error is no greater than 5%. (Incidentally, the significance level or rejection region in hypothesis testing is commonly denoted by the Greek letter alpha (α) which may be stated as a decimal fraction or a percentage. Consequently, the acceptance region is $1 - \alpha$.)

John's Jewels
The Chaos of Boring a Hole

A classical illustration of the disastrous results of making repeated Type I errors is the lathe operator, who is boring a hole in a wheel hub for later mating with a shaft. The diameter of the holes varies ever so slightly, but randomly, as the hubs are completed, so long as the lathe and tooling are in good working order. Therefore, the process of boring the hole is stable. The first hole bored is, let's say, two thousandths of an inch above specified tolerance. The inclination of the operator is to reset the tooling and try to pull the next hole diameter back closer to the design specifications. The next time, the hole diameter comes out under tolerance, so the operator must have over-compensated. For the third hole, another adjustment is made, this time in an attempt to increase the hole diameter. The long-term result of this effort is to throw a stable process into chaos, and thereby produce more and more undesirable parts. If the operator only realized at the beginning that the hole diameter was going to vary slightly around (hopefully) the desired average, no adjusting would have been necessary. The operator was repeatedly committing Type I errors by concluding that a stable process was really unstable.

Comparing a Single Mean to a Specified Value

Let's try to clarify this discussion with an example. In this case, a local Internet provider has established a customer billing plan based on the assumption that the average residential customer is online 26.0 hours per month with a standard deviation of 5.8 hours (based on an analysis of the past two years' usage figures). During the past month, 30 new customers were signed up, and now their usage hours are available and are shown in Table 4-2.

Table 4-2. Times (in Hours) of Internet Usage Among 30 Customers

32.5	36.9	34.1	21.8	21.9	31.3
21.2	15.2	24.6	27.5	37.5	29.2
27.3	28.3	35.4	28.9	29.6	38.4
20.6	33.7	24.1	21.3	24.8	31.0
25.4	29.5	29.4	25.0	30.2	18.6

You may recall that in Chapter 2, we recommended 30 as a minimum number of observations for most statistical analyses. In this case, we are just barely meeting that rule of thumb, so we will probably be safe in estimating the mean and standard deviation for the population from this sample. In addition, it would be wise to look at a histogram of the data to assess its shape. The resulting histogram is shown in Figure 4-1.

SQL/Query

The histogram can be created by executing Queries 4_1, 4_2, and 4_3. Query 4_1 removes the previous histogram data table, and Query 4_2 recreates the table. The deletion and recreation is necessary to reset the automatic numbering of Interval back to 1. Query 4_3 tabulates the desired intervals and populates the histogram data table.

Query 4_1:

```
DROP TABLE Histogram_Data;
```

Query 4_2:

```
CREATE TABLE Histogram_Data
(Interval Counter(1,1), Freq_Count Long);
```

Query 4_3:

```
INSERT INTO Histogram_Data ( [Interval], Freq_Count )
SELECT ((Fix(([Hours])/5))*5) & "-" & ((Fix(([Hours])/5)+1)*5) AS [Interval],
Count([Internet Usage].Hours) AS CountOfHours
FROM [Internet Usage]
GROUP BY ((Fix(([Hours])/5))*5) & "-" & ((Fix(([Hours])/5)+1)*5);
```

You might want to test your skills at fitting a normal distribution to the data, as we demonstrated in Chapter 3. However, when you combine intervals to meet the rule of an expected frequency of five or more in each interval, you end up with only three intervals. This is not an adequate number on which to run a χ^2 test, because you will have zero degrees of freedom for the test. Therefore, the histogram would require at least six intervals (maybe more), and even then the rule may not be satisfied to the letter. This demonstrates why it is always better to have a little more data when possible. But we'll proceed with the example with the assumption that the data appear close enough to a normal distribution for our purposes of illustration.

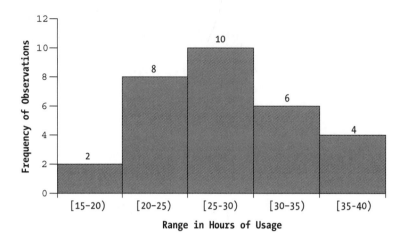

Figure 4-1. Histogram of the times (in hours) of Internet usage

The mean of the data is $\bar{x} = 27.8$ hours, and recall that the population standard deviation (which we'll call σ as before) is 5.8 hours. We can use a normal distribution test to check the significance (if any) of the sample mean from the "population" mean of $\mu = 26.0$ hours. Our null hypothesis H_0 states that there is no significant difference between the average usage hours of the one-month sample and the average of the previous database of users. In other words, $H_0: \mu = \bar{x}$. Now we set out to prove or disprove this hypothesis. First we set our significance level for the test, say $\alpha = 5\%$. The test in this application is two-tailed, so we have to divide the 5% between the two tails of the normal distribution, thereby leaving 2.5% in each tail. From the normal distribution table we used in Chapter 3 (and also given in Appendix B), we look up the variate value (called z) corresponding to an area of $(1 - \alpha) = 97.5\%$, or 0.975. This is found to be $z = 1.96$. The next step involves computing the statistic that will be compared to this table value. This is the normal statistic (or variate) expression presented in Chapter 3, except that the denominator standard deviation is an estimate from the sample values, so it is calculated as the sample standard deviation divided by the square root of the sample size n. This relationship is established in any statistics text, but it is probably not worth developing at this point. Let's just calculate the statistic we want below.

$$z = \frac{\bar{x} - \mu}{\sigma/\sqrt{n}} = \frac{27.8 - 26.0}{5.8/\sqrt{30}} = 1.70$$

Since 1.70 does not exceed the table value of 1.96, we accept the null hypothesis H_0 and conclude there is no significant difference between the latest 30 customers' average Internet usage and the average of the previous database. (Incidentally, sometimes the z calculation above is negative; in that case, we take its absolute value and work with the positive expression.)

SQL/Query

You can calculate the z statistic by running Query 4_4 and Query 4_5. Query 4_4 determines the sample mean and standard deviation. Notice the use of the Round function to round the calculated values to the same number of decimal places of the sample data given in Table 4.2. Query 4_5 computes the z statistic. The table [Population Stats] contains the mean, [Pop_Mean], of the population. For our example it is 26.0, which is the mean of the past two years' usage figures.

Query 4_4:

```
SELECT Count([Hours]) AS N,
Round(Avg([Hours]), 1) AS Mean,
Round(StDev([Hours]), 1) AS SD
FROM [Internet Usage];
```

> **NOTE**   *For Access 97 users, you may consider using the FORMAT function in place of the ROUND function. For example, replace Round(Avg([Hours]),1) with Format(Avg([Hours]), "#.0").*

Query 4_5:

```
SELECT ([Mean] - [Pop_Mean]) / ([SD] /Sqr([N])) AS Z
FROM [Population Stats], [Query 4_4];
```

But sometimes we don't have adequate population data to know a standard deviation (σ), so we can't use the normal test. Instead, we use a distribution with similar characteristics, called the *Student's t distribution*. Tables for this distribution are readily found in any statistics text and in Appendix B (also see Appendix C for more information about the t distribution). The procedure is similar to the previous example, except we compare the result to the table t value rather than the normal. We also have to know the number of degrees of freedom for the table value, which is equal to one less than the number of sample observations. Finally, we use the standard deviation for the sample (s) as an estimate for the unknown σ.

The Original "Student" of Statistics

Statistical theory and applications were developing dramatically after 1900. Many new distributions were presented and debated, and the advancement was rapid. In Dublin at the famous Guinness brewery, there was a chemist by the name of W. S. Gosset (1876 – 1937) who became interested in statistics. He was particularly concerned with the distribution of means of samples from a population with unknown variance. The distribution looked much like the well-known normal, but deviated somewhat in its peakedness and other characteristics. Gosset was able to develop a functional expression for this distribution, and became eager to publish his results, as was the custom. Unfortunately, however, the management of Guinness forbade any of its employees from publishing papers, so the legend goes. Gosset thought of a way to circumvent this problem. Since he fancied himself a student of statistics, he chose to present his findings to the world under the pseudonym "Student." Gosset's paper appeared in *Biometrika* in 1908 under the title *The Probable Error of a Mean*. Following the paper's appearance, Gosset collaborated with the great statisticians Sir Ronald A. Fisher and Egon S. Pearson until the early 1930s. To this day, Gosset's now famous *t* distribution is still commonly referred to as "Student's" distribution, or the "Student *t*" distribution.

To illustrate, let's take the same data as before, but just consider the first 20 observations (or first four columns of Table 4-2). Their mean $\bar{x}$ is about 27.1 and standard deviation s is about 5.6. The number of degrees of freedom is 19, so the table *t* value is found for 5% significance as 2.093 (see Appendix B for the table). The calculated statistic is

$$t = \frac{\bar{x} - \mu}{s/\sqrt{n}} = \frac{27.1 - 26.0}{5.6/\sqrt{20}} = 0.88$$

Again, this is less than the table value, so the conclusion is the same as before—accept $H_0: \mu = \bar{x}$ (i.e., the sample mean does not differ significantly from the population mean).

SQL/Query

We can also calculate the *t* statistic using SQL. Query 4_6 provides the mean and standard deviation for the first 20 values (i.e., first four columns of Table 4-2), and Query 4_7 performs the calculation.

Query 4_6:

```
SELECT Count([Internet Usage].Hours) AS N,
Round(Avg([Hours]),1) AS Mean,
Round(StDev([Hours]),1) AS SD
FROM [Internet Usage]
WHERE [Internet Usage].ID <= 20;
```

Query 4_7:

```
SELECT ([Mean]-[Pop_Mean])/([SD]/Sqr([N])) AS t
FROM [Query 4_6], [Population Stats];
```

To summarize, we have demonstrated procedures for comparing a single mean to a specified value (a population mean). If the population standard deviation is known or can be accurately estimated from a fairly large sample (say at least 30 observations), compute

$$z = \frac{\bar{x} - \mu}{\sigma/\sqrt{n}}$$

where z is the normal distribution variate, $\bar{x}$ is the mean of the sample data of interest, μ is the population mean (usually estimated from a much larger collection of data), σ is the population standard deviation (also estimated from the larger database if possible, or the sample), and n is the size of the smaller sample of interest. This value is then compared to the normal distribution variate from the table that corresponds to the appropriate significance level assumed, specifically 0.975 if the significance level is 5% (the table value in this case is always equal to 1.96).

If the population standard deviation is unknown and the sample is small (generally less than 30 observations), the procedure is similar except the t statistic is computed as follows:

$$t = \frac{\bar{x} - \mu}{s/\sqrt{n}}$$

where s is now the standard deviation estimated from the small sample. This value is compared to the table t value for the assumed significance level and a number of degrees of freedom equal to $n - 1$, where n is again the size of the small sample. This table t value varies with the number of degrees of freedom.

In either case, if the computed statistic is negative, just drop the negative sign. If the resulting value exceeds the table value, the null hypothesis is rejected, and the sample mean is judged to be significantly different than the population mean. Otherwise, the hypothesis is accepted, and the sample is said to be representative of the larger population of data.

Comparing Means and Variances of Two Samples

In the diagnostic tree of Figure 1-3, we showed tests of hypothesis along the second branch. When we had a single data set, we showed that a common test of hypothesis is the chi-square goodness of fit test presented in Chapter 3. There are other cases when the interest may involve comparing two sets of observations to see if they differ significantly, rather than comparing one data set to a population, as shown above. In this situation, a significance level is again chosen, commonly $\alpha = 5\%$ as before. Both the means and variances are computed for each sample. The variances are first compared for significant differences. If they do not differ significantly, another form of the t statistic is calculated to compare the means. If the variances do show significant differences, the t statistic may again be used to test the means, but its form is modified.

We've chosen to illustrate the procedure in the form of a diagram this time, as shown in Figure 4-2. A few comments about the diagram will be helpful. First, remember that the variance of a sample is just the square of the standard deviation, and is denoted by s^2. Since we are working with two samples now, it is also necessary to use the subscripts 1 and 2 to distinguish any parameter associated with either sample 1 or sample 2. In addition, you will notice in the diagram the use of the absolute value symbol, where $|x|$ is equal to x if x is positive, and is $-x$ if x is negative (in other words, the negative sign is just dropped if it is present).

To compare the variances of two samples, yet another weird statistical distribution is invoked. It is called the F distribution. It turns out that the ratio of the two sample variances (with the larger variance value always in the numerator) approximates an F distribution if they exhibit no significant difference. You may want to flip over to Appendix C to look at some additional information on the F distribution. The statistical tables for the F distribution are found in virtually all statistics texts and in Appendix B. There is usually one table for the 1% significance level, and one for the 5% level. Sometimes even more significance levels are represented. Since the degrees of freedom (denoted by the Greek letter ν in Figure 4-2) may differ between the samples (if they differ in size), the F tables accommodate this. Note in Figure 4-2 the statements of the null and alternative hypotheses.

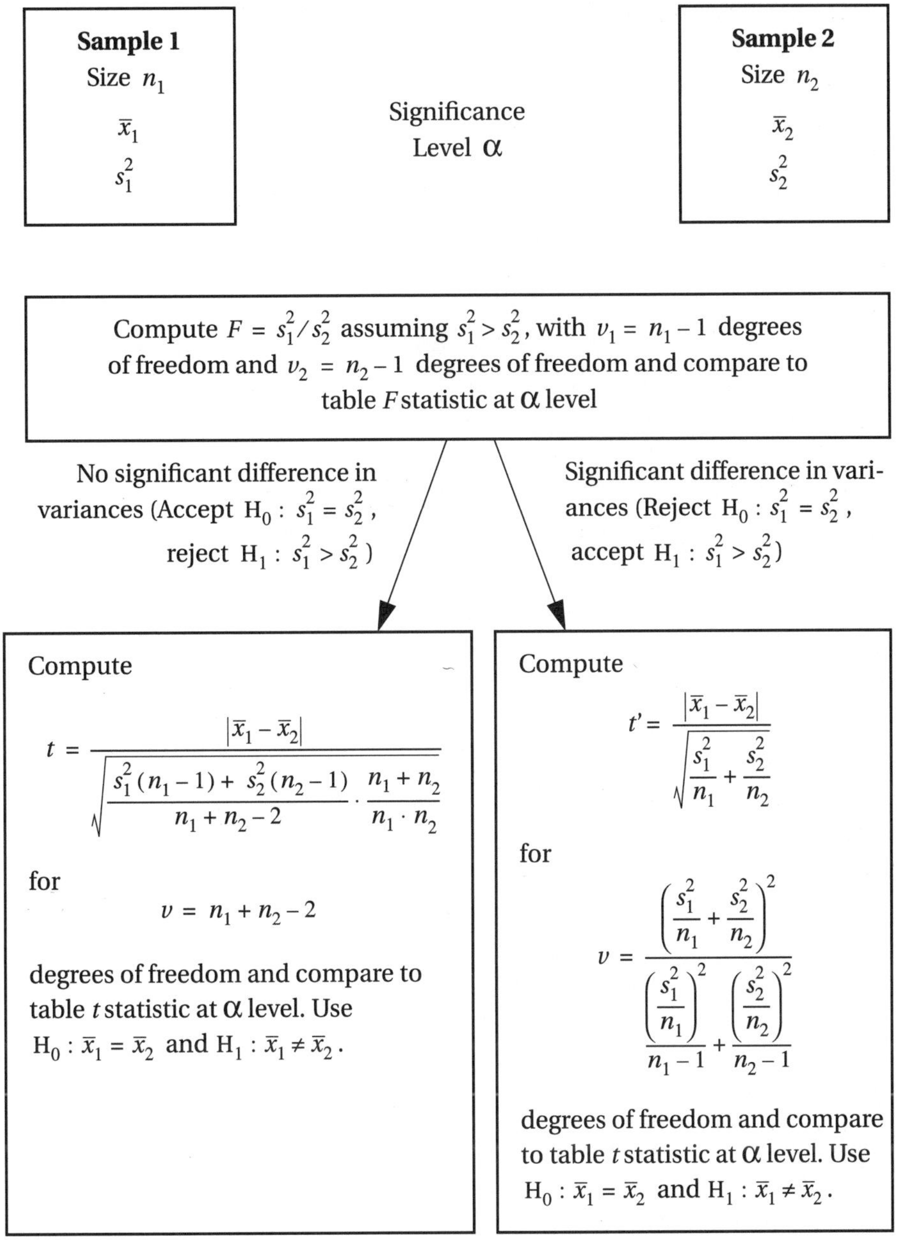

Sample 1 Size n_1 $\overline{x}_1$ s_1^2

Significance Level α

Sample 2 Size n_2 $\overline{x}_2$ s_2^2

Compute $F = s_1^2 / s_2^2$ assuming $s_1^2 > s_2^2$, with $v_1 = n_1 - 1$ degrees of freedom and $v_2 = n_2 - 1$ degrees of freedom and compare to table F statistic at α level

No significant difference in variances (Accept $H_0 : s_1^2 = s_2^2$, reject $H_1 : s_1^2 > s_2^2$)

Significant difference in variances (Reject $H_0 : s_1^2 = s_2^2$, accept $H_1 : s_1^2 > s_2^2$)

Compute

$$t = \frac{|\overline{x}_1 - \overline{x}_2|}{\sqrt{\dfrac{s_1^2(n_1-1) + s_2^2(n_2-1)}{n_1 + n_2 - 2} \cdot \dfrac{n_1 + n_2}{n_1 \cdot n_2}}}$$

for

$$v = n_1 + n_2 - 2$$

degrees of freedom and compare to table t statistic at α level. Use $H_0 : \overline{x}_1 = \overline{x}_2$ and $H_1 : \overline{x}_1 \neq \overline{x}_2$.

Compute

$$t' = \frac{|\overline{x}_1 - \overline{x}_2|}{\sqrt{\dfrac{s_1^2}{n_1} + \dfrac{s_2^2}{n_2}}}$$

for

$$v = \frac{\left(\dfrac{s_1^2}{n_1} + \dfrac{s_2^2}{n_2}\right)^2}{\dfrac{\left(\dfrac{s_1^2}{n_1}\right)^2}{n_1 - 1} + \dfrac{\left(\dfrac{s_2^2}{n_2}\right)^2}{n_2 - 1}}$$

degrees of freedom and compare to table t statistic at α level. Use $H_0 : \overline{x}_1 = \overline{x}_2$ and $H_1 : \overline{x}_1 \neq \overline{x}_2$.

Figure 4-2. General procedure for comparing means of two samples taken from populations with unknown variances

We can now illustrate the use of Figure 4-2 with a couple of examples. We'll keep the number of data values fairly small just to hold down the scale of the calculations involved, so you can check the numbers if you want to. In the first example, two high school coaches are looking over some prospects for their football and basketball teams, with an interest toward developing exercise programs for the players. The weights (in pounds) of all prospective players for the basketball team are as follows:

158 127 132 169 106 188 111 143 170 125

The weights of the prospective players for the football team are as follows:

249 236 198 221 204 216 218 225

The coaches wonder if the difference in weights between the two groups is significant enough to warrant distinct exercise plans. We can answer this. Let's call the basketball data Sample 1 and the football data Sample 2. We'll test the difference at the 5% significance level. Thus, the null and alternative hypotheses may be written

$$H_0 : \bar{x}_1 = \bar{x}_2 \text{ and } H_1 : \bar{x}_1 \neq \bar{x}_2$$

where $\bar{x}_1$ and $\bar{x}_2$ are the two sample means. This is a two-tailed test.

The mean and variance of the two samples are calculated in the usual manner. However, remember to calculate the *variances* instead of the standard deviations, for use in the F ratio test. The results are as follows:

$$\bar{x}_1 = 142.9 \text{ lb.} \quad s_1^2 = 749.9 \text{ lb.}^2 \quad \bar{x}_2 = 220.9 \text{ lb.} \quad s_2^2 = 268.1 \text{ lb.}^2$$

Following the steps in Figure 4-2, we first calculate the F ratio, remembering that the larger variance value goes in the numerator. For this preliminary test, we are using $H_0 : s_1^2 = s_2^2$ and $H_1 : s_1^2 > s_2^2$, a one-tailed test. We have

$$s_1^2 / s_2^2 = 749.9 / 268.1 = 2.80$$

The number of degrees of freedom for the numerator is $v_1 = 10 - 1 = 9$, and for the denominator is $v_2 = 8 - 1 = 7$. In the F table (see Appendix B), we find, for a 5% significance level, that the F statistic is equal to 3.68 for the appropriate numbers of degrees of freedom. Since the calculated value is less than the table value, we accept H_0 and conclude there is no significant difference in variances, and proceed to the test for means.

SQL/Query

Now, let us see how we can accomplish this in SQL. First we create a table called Balls that contains two fields. The first field is a Type field identifying the player as either a basketball player or a football player. The second field is the weight of the player. Query 4_8, Query 4_9, and Query 4_10 establish the Balls table. We have made the assumption that the weights of the two teams are maintained in separate database tables called Basketball and Football.

Query 4_8:

```
CREATE TABLE Balls(Type Text, ID Long, Weight Long);
```

Query 4_9:

```
SELECT "B" as Type, Weight
FROM Basketball
UNION ALL
SELECT "F" as Type, Weight
FROM Football;
```

Query 4_10:

```
INSERT INTO Balls ( Type, Weight )
SELECT [Query 4_9].[Type], [Query 4_9].[Weight]
FROM [Query 4_9];
```

In order to calculate the *F* and *t* statistics, we need the means and variances for the two teams. Query 4_11 produces this information. Notice the usage of the Var function for calculating the variance. (If you do not have a variance function, you can simply square the standard deviation.)

Query 4_11:

```
SELECT Balls.Type, Count(Balls.Weight) AS N,
Round(Avg([Weight]), 1) AS Mean,
Round(Var([Weight]), 1) AS Var
FROM Balls
GROUP BY Balls.Type;
```

Query 4_12 calculates the *F* ratio. Notice how we used the Max function to find the larger variance for the numerator, and the Min function to find the smaller variance for the denominator. We are able to do this because we put all the means and variances into one table.

Query 4_12:

```
SELECT Round(Max([Var]) / Min([Var]), 2) AS F
FROM [Query 4_11];
```

In this case, we compute the *t* statistic shown in the left box of Figure 4-2.

$$t = \frac{|142.9 - 220.9|}{\sqrt{\frac{(749.9)(9)+(268.1)(7)}{10+8-2} \cdot \frac{10+8}{(10)(8)}}} = 7.08$$

The number of degrees of freedom for this statistic (v) is equal to the sum of the two sample sizes less two, or 16 in this instance. A *t* table is used to look up the comparable value for 16 degrees of freedom and 5% significance. The result is 2.12. Since the calculated *t* value exceeds the table value, we reject the null hypothesis $H_0 : \bar{x}_1 = \bar{x}_2$ and accept $H_1 : \bar{x}_1 \neq \bar{x}_2$, thereby concluding that the means of the two samples of athletes' weights are significantly different.

SQL/Query

You can use the following SQL queries to calculate the *t* value. Queries 4_13 and 4_14 obtain the number of data values, the means, and variances. The complicated-looking formula for *t* is in Query 4_15.

Query 4_13:

```
SELECT Count([Weight]) AS N1,
Round(Avg([Weight]),1) AS Mean1,
Round(Var([Weight]),1) AS Var1
FROM Basketball;
```

Query 4_14:

```
SELECT Count([Weight]) AS N2,
Round(Avg([Weight]),1) AS Mean2,
Round(Var([Weight]),1) AS Var2
FROM Football;
```

Query 4_15:

```
SELECT Abs([Mean1] - [Mean2]) /
Sqr((([var1]*([n1] - 1) + [var2]*([n2] - 1)) * ([n1] + [n2])) /
(([n1] + [n2] - 2) * [n1] * [n2])) AS t
FROM [Query 4_13], [Query 4_14];
```

Now let's look at an example where the variances differ significantly. The manager of a parcel delivery service is planning methods to help drivers who appear habitually late in returning after their daily deliveries are complete. She feels that there may be problems with the routes, with the customers, with the sizes of the parcels, with the physical limitations of the delivery people, and so forth. However, she wants to be sure that the performance of an individual is significantly different from the norm to warrant special assistance. To this end, she has chosen a driver she feels is representative of the typical delivery person. This is a person who has worked for the firm for several years, has shown reasonable ability, and has generally been able to complete his route on time. His daily delivery times (from the time he leaves the facility until he returns with an empty truck) are recorded for a period of many weeks. To keep the scale of the example small, the following times (in hours) are taken from a three-week period:

6.8 6.7 6.8 5.9 6.6 6.8 6.4 6.5 6.4 6.7 6.9 6.2 6.5 6.6 6.9

A (perceived) slower driver's times are clocked for the same period, with the following results:

7.3 7.6 7.0 7.3 6.7 6.9 7.5 8.8 7.7 7.1 6.3 8.2 7.8 7.6 8.3

In this example, we can again calculate the sample means and variances.

$$\bar{x}_1 = 6.58\,\text{hr.} \quad s_1^2 = 0.076\,\text{hr.}^2 \quad \bar{x}_2 = 7.47\,\text{hr.} \quad s_2^2 = 0.421\,\text{hr.}^2$$

We first compare the variances by the F ratio test, as in the previous example, remembering that the larger variance estimate (in this case s_2^2) goes in the numerator.

$$s_2^2/s_1^2 = 0.421/0.076 = 5.54$$

The table F statistic for a 5% significance level and $v_1 = v_2 = 14$ degrees of freedom for both the numerator and denominator is 2.48. Since the calculated value exceeds the table value, we conclude there is a significant difference between the variance estimates. In other words, we reject $H_0 : s_2^2 = s_1^2$ and accept $H_1 : s_2^2 > s_1^2$. As a result, we have to modify the t test for comparing means by calculating (from the right box in Figure 4-2)

$$t' = \frac{|7.47 - 6.58|}{\sqrt{\dfrac{0.076}{15} + \dfrac{0.421}{15}}} = 4.89$$

Determining the number of degrees of freedom in this application is a bit more involved, as seen in Figure 4-2. The result is

$$v = \frac{\left(\dfrac{0.076}{15} + \dfrac{0.421}{15}\right)^2}{\dfrac{\left(\dfrac{0.076}{15}\right)^2}{15-1} + \dfrac{\left(\dfrac{0.421}{15}\right)^2}{15-1}} = 19$$

As before, we now look up the t statistic in its table for a 5% significance level and 19 degrees of freedom. The value is 2.093. In this case, the calculated t' value exceeds the table value, so the means are judged to be significantly different. We therefore reject $H_0 : \bar{x}_1 = \bar{x}_2$ and accept $H_1 : \bar{x}_1 \neq \bar{x}_2$. The conclusion is that the second driver does indeed appear to have problems completing his route in a timely fashion, and therefore needs some help or additional training.

SQL/Query

We can use SQL queries to determine the values of t' and v. First, Query 4_16 and Query 4_17 are executed to obtain the means and variances. Query 4_18 calculates the value for t', and Query 4_19 calculates the value for v.

Query 4_16:

```
SELECT Count([Typical Driver].[Hours]) AS N1,
Round(Avg([Hours]), 2) AS Mean1, Round(Var([Hours]), 3) AS Var1
FROM [Typical Driver];
```

Query 4_17:

```
SELECT Count([Slower Driver].[Hours]) AS N2,
Round(Avg([Hours]), 2) AS Mean2, Round(Var([Hours]), 3) AS Var2
FROM [Slower Driver];
```

Query 4_18:

```
SELECT Abs([Mean1] - [Mean2]) / Sqr([var1] / [n1] + [var2] / [n2]) AS t
FROM [Query 4_17], [Query 4_16];
```

Query 4_19:

```
SELECT (([var1]/[n1] + [var2]/[n2])^2) /
(((([var1] / [n1])^2) / ([n1]-1) + (([var2] / [n2])^2) / ([n2]-1))) AS V
FROM [Query 4_16], [Query 4_17];
```

John's Jewels
Testing the Hypothesis That the Corn Dries the Same

My wife and I own an 1873 water-powered mill and museum called Falls Mill, located near Belvidere, Tennessee. We buy and mill local corn and wheat, so we have to insure that the moisture content of the grain is low enough to grind properly on traditional millstones. Generally, the corn is allowed to dry in the field before it is harvested. Occasionally, however, due to weather changes, it has to be cut early and placed in large drying silos, where gas heat is used to lower the moisture content. One of the farmers we deal with was interested to know the relative efficiencies of two identical drying bins. He placed shelled white corn from the same field in the same quantities in the two bins. He dried them an equal amount of time, and then we pulled 20 samples from each bin. We checked each sample on a moisture meter, then I ran a test of variances and means. The average moisture content of the samples from the first bin was 13.2%, and from the second bin was 15.7%. Our test showed no significant difference in variances between the two sets of samples, but did show a significant difference in mean moisture levels. After emptying the bin with the higher moisture content, it was found that the grates in the bottom of the bin were partially clogged with dust and needed cleaning. This had apparently restricted the degree to which the drying air was able to blast up through the corn. Once the grates were pulled out, cleaned, and replaced, the bin was again loaded and tested. This second trial showed comparable results between the two bins.

Comparisons of More Than Two Samples

Several methods are available for comparing more than two samples for significant differences. The one we are going to illustrate makes use of our old pal the chi-square (χ^2) statistic. It goes under the name of a *contingency test*, and actually tests for independence among the samples. The conclusion is that either there is or is not significant difference from one sample to another. As before, one example speaks louder than a thousand words, so here we go.

Four different makes of sport-utility vehicles (called A, B, C, and D), having similar weights and body styles, were tested for fuel efficiency. The following results were obtained (in miles per gallon of gasoline) for a three-day test period:

Table 4-3. Vehicle Mileage Data

	DAY 1	DAY 2	DAY 3
A	16.8	16.4	16.7
B	18.5	19.0	18.8
C	23.2	22.8	22.9
D	17.6	17.5	17.5

Does the mileage from day to day vary significantly (at the 5% level) by vehicle type? We can answer this question by computing expected frequencies for each cell of the table under the assumption of independence (i.e., no significant variation). To do this, we sum the rows and columns of the table first, obtaining the results shown in Table 4-4.

Table 4-4. Vehicle Mileage Data Showing Row and Column Totals

	DAY 1	DAY 2	DAY 3	ROW TOTAL
A	16.8	16.4	16.7	49.9
B	18.5	19.0	18.8	56.3
C	23.2	22.8	22.9	68.9
D	17.6	17.5	17.5	52.6
Column Total	76.1	75.7	75.9	227.7

We can now use the row and column totals (and grand total of 227.7) to calculate the expected frequencies for each of the 12 cells in the table. For Vehicle A on Day 1, this is accomplished by dividing the row total for Vehicle A by the grand total, then multiplying the result by the column total for Day 1. This results in

$$(49.9/227.7)(76.1)=16.7$$

The expected frequencies for all other cells are calculated in the same manner. We thus obtain the results shown in Table 4-5, where the expected frequencies are shown in parentheses.

Table 4-5. Vehicle Mileage Data Showing Expected Frequencies

	DAY 1	DAY 2	DAY 3
A	16.8 (16.7)	16.4 (16.6)	16.7 (16.6)
B	18.5 (18.8)	19.0 (18.7)	18.8 (18.8)
C	23.2 (23.0)	22.8 (22.9)	22.9 (23.0)
D	17.6 (17.6)	17.5 (17.5)	17.5 (17.5)

SQL/Query

The values shown in Table 4-5 can be generated by executing a sequence of SQL queries. Query 4_20 and Query 4_21 determine the row and column totals, and Query 4_22 calculates the overall total. Using the results of these queries we can execute Query 4_23 to obtain the expected frequencies that are shown in Table 4-5.

Query 4_20:

```
SELECT Mileage.Vehicles, Round(Sum([MPG]), 1) AS [Row Total]
FROM Mileage
GROUP BY Mileage.Vehicles;
```

Query 4_21:

```
SELECT Mileage.Day, Round(Sum([MPG]), 1) AS [Column Total]
FROM Mileage
GROUP BY Mileage.Day;
```

Query 4_22:

```
SELECT Round(Sum([MPG]), 1) AS Total
FROM Mileage;
```

Query 4_23:

```
SELECT Mileage.Vehicles, Mileage.Day, Mileage.MPG,
Round((([Row Total] / [Total]) * [Column Total]), 1) AS [Exp Freq]
FROM [Query 4_22], Mileage, [Query 4_20], [Query 4_21]
WHERE Mileage.Vehicles = [Query 4_20].Vehicles
AND Mileage.Day = [Query 4_21].Day;
```

Now the χ^2 statistic may be computed in a similar way to the goodness of fit test in Chapter 3. For each cell in the table, we take the difference between the observed and expected frequencies, square the result, and then divide it by the expected frequency for the cell. Once all 12 terms have been calculated, they are summed. For the first cell, we have

$$\frac{(16.8-16.7)^2}{16.7} = 0.00060$$

Performing this calculation successively over all cells, the sum of the result is 0.016 and is obtained by calculating

$$\chi^2 = \frac{(16.8-16.7)^2}{16.7} + \frac{(16.4-16.6)^2}{16.6} + L + \frac{(17.5-17.5)^2}{17.5} = 0.016$$

This is so small, we really don't even have to look up the χ^2 table value, but we better anyway. One reason for this is that we haven't told you how the degrees of freedom are determined in this application. They're always equal to the product of the number of rows less one, and the number of columns less one. In this example the result is 6. The χ^2 table value for 5% significance and 6 degrees of freedom is 12.592. The calculated value is much less, so we conclude there is no significant difference in the day-to-day variation of vehicle mileage relative to the make of vehicle.

SQL/Query

We can also use SQL to calculate the χ^2 statistic. Query 4_24 performs the calculation.

Query 4_24:

```
SELECT Sum(([MPG] - [Exp Freq])^2 / [Exp Freq]) AS ChiSq
FROM [Query 4_23];
```

Conclusion

These are but a few of the tests of hypothesis available to you. Most of the others are applicable to more specific or obscure situations and are not presented here. They are accessible in most statistics texts, however. We have chosen for our examples a 5% significance level. If you want to alter the chance of an erroneous conclusion in your test, you might go to a 10% level, or even a 1% level. In practice, 5% works pretty well most of the time.

T-SQL Source Code

Below are four T-SQL source code listings for performing the hypothesis tests described in this chapter. The first listing is called Calculate_T_Statistic. Calculate_T_Statistic calculates the Student's *t* statistic for a given set of sample data when the population mean is known and conducts the appropriate test. The second listing is called Calculate_Z_Statistic. The Calculate_Z_Statistic determines the *z* statistic (normal variate) for a given sample and accomplishes the normal test. The next listing is the Compare_Means_2_Samples procedure. This procedure compares the variances and means of two samples at a significance level of α using the F and *t* tests. The last listing is the procedure called Contingency_Test. The contingency test is used to compare more than two samples. The test uses the chi-square statistic to test for independence among the samples. Following the procedure listings is a section illustrating the call statements to each procedure with parameter specifications.

Calculate_T_Statistic

```
SET QUOTED_IDENTIFIER OFF
GO
SET ANSI_NULLS OFF
GO


ALTER PROCEDURE Calculate_T_Statistic
@SrcTblName     Varchar(50) = 'Internet Usage',
@SrcColName Varchar(50) = 'Hours',
@PopMean Float = 26.0,
@t Float OUTPUT
AS


/**********************************************************/
/*                                                    */
/*              CALCULATE_t_STATISTIC                 */
/*                                                    */
/*  This procedure calculates the t statistic.        */
/*                                                    */
/* INPUTS:                                            */
/*   SrcTblName - table containing sample data        */
/*   SrcColName - column containing sample data values */
/*   PopMean - population mean                         */
/* OUTPUTS:                                           */
/*   t - the t statistic                              */
/*                                                    */
/**********************************************************/
```

```
/* Local Variables */
DECLARE @N Int                  /* Sample Size */
DECLARE @Mean Float             /* Sample mean */
DECLARE @SD Float               /* Sample standard deviation */
DECLARE @tCalc Float            /* t calculated */
DECLARE @Q varchar(200)         /* query string */

/* Build query to obtain the sample size, mean, and SD */
Set @Q = 'SELECT Count(' + @SrcColName + ') AS N, ' +
     'Round(Avg(' + @SrcColName + '),1) AS Mean, ' +
     'Round(StDev(' + @SrcColName + '),1) AS SD ' +
     'INTO ##tmpTable ' +
     'FROM [' + @SrcTblName + '] ' +
     'WHERE (([' + @SrcTblName + '].[ID])<=20) '

/* Execute the query */
EXEC(@q)

/* Get sample size, mean, and SD */
SELECT @N = N,
     @Mean = Mean,
     @SD = SD
     FROM ##tmpTable

/* Calculate the t statistic */
SELECT @tCalc =(@Mean-@PopMean)/(@SD/Sqrt(@N))

/* Return the t statistic */
SET @t = @tCalc

GO
SET QUOTED_IDENTIFIER OFF
GO
SET ANSI_NULLS ON
GO
```

Calculate_Z_Statistic

```
SET QUOTED_IDENTIFIER OFF
GO
SET ANSI_NULLS OFF
GO

ALTER PROCEDURE Calculate_Z_Statistic
@SrcTblName VarChar(50) = 'Internet Usage',
@SrcColName Varchar(50) = 'Hours',
@PopMean Float = 26.0,
@Z Float OUTPUT
AS

/**********************************************************/
/*                                                      */
/*                CALCULATE_Z_STATISTIC                 */
/*                                                      */
/*  This procedure calculates the Z statistic.          */
/*                                                      */
/* INPUTS:                                              */
/*    SrcTblName - table containing sample data          */
/*    SrcColName - column name contains sample values    */
/*    PopMean - population mean                          */
/* OUTPUTS:                                             */
/*    Z - the Z statistic                               */
/*                                                      */
/**********************************************************/

/* Local Variables */
DECLARE @N Int              /* Sample Size */
DECLARE @Mean Float         /* Sample mean */
DECLARE @SD Float           /* Sample standard deviation */
DECLARE @Zcalc Float        /* Z calculated */

/* Get sample size, mean and SD */
DECLARE @Q varchar(200) /* query string */

/* Build query to obtain sample size, mean, and SD */
Set @Q = 'SELECT Count(' + @SrcColName + ') AS N, ' +
    'Round(Avg(' + @SrcColName + '),1) AS Mean, ' +
    'Round(StDev(' + @SrcColName + '),1) AS SD ' +
    'INTO ##tmpZTable ' +
    'FROM [' + @SrcTblName + '] '
```

```
/* Execute the query */
EXEC(@Q)

/* Get sample size, mean, and SD */
SELECT @N = N,
    @Mean = Mean,
    @SD = SD
    FROM ##tmpZTable

/* Calculate the Z statistic */
SELECT @Zcalc =(@Mean-@PopMean)/(@SD/Sqrt(@N))

/* Return the Z statistic */
SET @z = @Zcalc

GO
SET QUOTED_IDENTIFIER OFF
GO
SET ANSI_NULLS ON
GO
```

Compare_Means_2_Samples

```
SET QUOTED_IDENTIFIER OFF
GO
SET ANSI_NULLS ON
GO

ALTER PROCEDURE [Compare_Means_2_Samples]
@Alpha Float = 0.05,
@Src1TblName Varchar(50) = 'Basketball',
@Src1ColName Varchar(50) = 'Weight',
@Src2TblName Varchar(50) = 'Football',
@Src2ColName Varchar(50) = 'Weight'
AS
```

```
/**************************************************************/
/*                                                          */
/*                 COMPARE_MEANS_2_SAMPLES                  */
/*                                                          */
/*  This procedure compares the means of two samples at a   */
/*  significance level of Alpha using the t test.           */
/*                                                          */
/* INPUTS:                                                  */
/*    Alpha - significance level for comparison             */
/*    Src1TblName - table containing sample one data         */
/*    Src1ColName - column name contains sample one values   */
/*    Src2TblName - table containing sample two data         */
/*    Src2ColName - column name contains sample two values   */
/* OUTPUTS:                                                 */
/*    Z - the Z statistic                                   */
/*                                                          */
/* TABLES:                                                  */
/*    F_table - table of F statistical values               */
/*       Alpha - significance level                         */
/*       v - degrees of freedom                             */
/*       F - the F statistic for given Alpha and v          */
/*    t_table - table of t statistical values               */
/*       Alpha - significance level                         */
/*       v - degrees of freedom                             */
/*       t - the t statistic for given Alpha and v          */
/*                                                          */
/**************************************************************/

/* Temporary table and query variables */
DECLARE @Q varchar(200) /* query string */

/* Temp variables for swapping sample values */
DECLARE @SN Int          /* Sample Size */
DECLARE @SxBar Float      /* Sample mean */
DECLARE @SVar Float       /* Sample variance */

/* Basic measures for first (large variance) sample */
DECLARE @N1 Int          /* Sample Size */
DECLARE @xBar1 Float      /* Sample mean */
DECLARE @Var1 Float       /* Sample variance */
DECLARE @v1 Int           /* Degrees of freedom */
```

```
/* Basic measures for second (smaller variance) sample */
DECLARE @N2 Int          /* Sample Size */
DECLARE @xBar2 Float     /* Sample mena */
DECLARE @Var2 Float      /* Sample variance */
DECLARE @v2 Int          /* Degrees of freedom */

/* Local variables for calculated statistics */
DECLARE @v Float         /* combine degrees of freedom */
DECLARE @FCalc Float     /* F calculated */
DECLARE @Ftbl Float      /* F table value */
DECLARE @tCalc Float     /* t calculated */
DECLARE @tTbl Float      /* t table value */

/* Factors for intermediate calculations */
DECLARE @Nfac1 Float     /* numerator factor */
DECLARE @Nfac2 Float     /* numerator factor */
DECLARE @Nfac3 Float     /* numerator factor */
DECLARE @Dfac1 Float     /* denominator factor */
DECLARE @Dfac2 Float     /* denominator factor */

/* For the first sample get */
/* sample size, mean, and SD */
SET @Q = 'SELECT Count(' + @Src1ColName + ') AS N, ' +
    'Avg(CONVERT(Float, ' + @Src1ColName + ')) AS xBar, ' +
    'Var(CONVERT(Float, ' + @Src1ColName + ')) AS S2 ' +
    'INTO ##temp1Sample FROM [' + @Src1TblName + '] '
EXEC(@Q)
SELECT @N1 = N,
    @xBar1 = xBar,
    @Var1 = S2
    FROM ##temp1Sample

/* For the second sample get */
/* sample size, mean, and SD */
SET @Q = 'SELECT Count(' + @Src2ColName + ') AS N, ' +
    'Avg(CONVERT(Float, ' + @Src2ColName + ')) AS xBar, ' +
    'Var(CONVERT(Float, ' + @Src2ColName + ')) AS S2 ' +
    'INTO ##temp2Sample FROM [' + @Src2TblName + '] '
EXEC(@Q)
SELECT @N2 = N,
    @xBar2 = xBar,
    @Var2 = S2
    FROM ##temp2Sample
```

```
/* Check to see if the larger variance is the */
/* second sample. If so, swap the two samples */
IF @var1 < @var2
Begin
      /* Swap the two samples so that the sample      */
      /* with the larger variance is the first sample */
      Set @SN = @N1
      Set @SxBar = @XBar1
      Set @SVar = @Var1

      Set @N1 = @N2
      Set @xBar1 = @XBar2
      Set @Var1 = @Var2

      Set @N2 = @SN
      Set @xBar2 = @SXBar
      Set @Var2 = @SVar
End

/* Determine degrees of freedom */
SELECT @v1 = @N1 - 1
SELECT @v2 = @N2 - 1

/* Calculate F statistic */
SELECT @Fcalc = @Var1/@var2

/* Get table F */
SELECT @Ftbl = (SELECT F
      FROM F_Table
      WHERE Alpha = convert(varchar(10), @Alpha)
      And v1 = convert(varchar(10), @v1)
      And V2 = convert(varchar(10), @v2))
```

```
/* Compare calculated F to table F */
IF @Fcalc < @Ftbl
Begin
    /* No significant difference between variances */

    /* Calculate t statistic */
    Set @Nfac1 = @Var1*convert(float,(@N1-1))+@Var2*convert(float,(@N2-1))
    Set @DFac1 = @Nfac1/convert(float,(@N1+@N2-2))
    Set @DFac2 = convert(float,(@N1+@N2))/convert(float,(@N1*@N2))
    Set @Tcalc = abs(@xBar1 - @xBar2)/sqrt(@Dfac1*@Dfac2)

    /* Calculate degrees of freedom */
    Set @v = @N1+@N2-2
End

ELSE
Begin
    /* Significant difference between variances */

    /* Calculate t statistic */
    Set @Dfac1 = @Var1/convert(float, @N1)
    Set @Dfac2 = @Var2/convert(float, @N2)
    Set @Tcalc = abs(@xBar1 - @xBar2)/sqrt(@Dfac1+@Dfac2)

    /* Calculate degrees of freedom */
    Set @Nfac1 = (@Dfac1+@Dfac2)*(@Dfac1+@Dfac2)
    Set @Dfac1 = (@Dfac1*@Dfac1)/convert(float,@N1-1)
    Set @Dfac2 = (@Dfac2*@Dfac2)/convert(float, @N2-1)
    Set @v = round(@Nfac1/(@Dfac1+@Dfac2),0)

End

/* Get t table value */
SELECT @tTbl = (SELECT t
    FROM t_Table
    WHERE V = convert(varchar(10), @v)
    AND Alpha = convert(varchar(10), @Alpha))
```

```
/* Compare calculated t and table t */
IF @tCalc >= @tTbl
Begin
      print 'Since the calculated t value (' + convert(varchar(10), @tCalc) + ')'
      print 'exceeds the table t value (' + convert(varchar(10), @tTbl) + '),'
      print 'the means of the two samples '
      print 'are significantly different'
      print 'at the ' + convert(varchar(10), @Alpha*100) + '% significance level.'
      print ' '
End

ELSE
begin
      print 'Since the calculated t value (' + convert(varchar(10), @tCalc) + ')'
      print 'is less than the table t value (' + convert(varchar(10), @tTbl) + '),'
      print 'the means of the two samples '
      print 'are NOT significantly different'
      print 'at the ' + convert(varchar(10), @Alpha*100) + '% significance level.'
      print ' '
End

GO
SET QUOTED_IDENTIFIER OFF
GO
SET ANSI_NULLS ON
GO
```

Contingency_Test

```
SET QUOTED_IDENTIFIER OFF
GO
SET ANSI_NULLS ON
GO

ALTER     PROCEDURE [Contingency_Test]
@Alpha float = 0.05,
@Samp1TblName varchar(50) = 'Mileage',
@Samp1Col1Name varchar(50) = 'Vehicles',
@Samp1Col2Name varchar(50) = 'Day',
@Samp1Col3Name varchar(50) = 'MPG'

AS
```

```
/*********************************************************/
/*                                                       */
/*                 CONTINGENCY_TEST                      */
/*                                                       */
/*   The contingency test is used to perform comparisons of  */
/*   more than two samples. The test uses the chi square     */
/*   statistic to test for independence among the samples.   */
/*                                                       */
/* INPUTS:                                               */
/*    Alpha - significance level for chi square          */
/*    Samp1TblName - table containing sample data        */
/*    Samp1Col1Name - column identifying each sample type  */
/*                 (e.g., Vehicle in Table 4_3 of book)  */
/*    Samp1Col2Name - column identifying each sample taken  */
/*                 (e.g., Day in Table 4_3 of book)      */
/*    Samp1Col3Name - column containing sample value     */
/*                                                       */
/* TABLES:                                               */
/*   Chi_Sq_table - table of chi square values           */
/*     Alpha - significance level                        */
/*     v - degrees of freedom                            */
/*     ChiSq - the chi square for given Alpha and v      */
/*                                                       */
/*********************************************************/


/* Local Variables */
DECLARE @Q varchar(300)          /* Query string */
DECLARE @v Float                 /* degrees of freedom */
DECLARE @ChiSqCalc Float         /* calculated Chi Sq */
DECLARE @ChiSqTbl Float          /* table Chi Sq */

/* Tally by sample type */
SET @Q = 'SELECT [' + @Samp1Col1Name + '] AS SampType, ' +
    'Round(Sum([' + @Samp1Col3Name + ']),1) AS SampTypeTotal ' +
    'INTO ##temp1Sample ' +
    'FROM ['+ @Samp1TblName + '] ' +
    'GROUP BY [' + @Samp1Col1Name + '] '
EXEC(@Q)
```

```
/* Tally by sample taken */
SET @Q = 'SELECT [' + @Samp1Col2Name + '] AS SampTaken, ' +
    'Round(Sum([' + @Samp1Col3Name + ']),1) AS SampTakenTotal ' +
    'INTO ##temp2Sample ' +
    'FROM [' + @Samp1TblName + '] ' +
    'GROUP BY [' + @Samp1Col2Name + '] '
EXEC(@Q)


/* Tally all values */
SET @Q = 'SELECT Round(Sum([' + @Samp1Col3Name + ']),1) AS TotalAll ' +
    'INTO ##temp3Sample ' +
    'FROM [' + @Samp1TblName + '] '
EXEC(@Q)


/* Determine expected frequency */

SET @Q = 'SELECT  [' + @Samp1Col1Name + '], ' +
    '[' + @Samp1Col2Name + '], ' +
    '[' + @Samp1Col3Name + '], ' +
    'Round((([SampTypeTotal]/[TotalAll])*[SampTakenTotal]),1) AS [Exp Freq] ' +
    'INTO ##temp4Sample ' +
    'FROM ##temp1Sample, ##temp2Sample, ' +
        '##temp3Sample, [' + @Samp1TblName + '] ' +
    'WHERE ((([' + @Samp1TblName + '].[' + @Samp1Col1Name + ']) = ' +
        '(##temp1Sample.[SampType]))' +
    'And (([' + @Samp1TblName + '].[' + @Samp1Col2Name + ']) = ' +
        '(##temp2Sample.[SampTaken])))'
EXEC(@Q)


/* Determine calculated chi square */
SET @Q = 'SELECT Sum(Power((([' + @Samp1Col3Name +
    ']-[Exp Freq]),2)/[Exp Freq]) ' +
    'AS ChiSq_Calc ' +
    'INTO ##temp5Sample FROM ##temp4Sample'
EXEC(@Q)
SELECT @ChiSqCalc = (SELECT ChiSq_Calc FROM ##temp5Sample)


/* Calculate the degrees of freedom */
SET @Q = 'SELECT COUNT(DISTINCT [' + @Samp1Col1Name + ']) AS N1, ' +
    'COUNT(DISTINCT [' + @Samp1Col2Name + ']) AS N2 ' +
    'INTO ##temp6 FROM [' + @Samp1TblName + '] '
EXEC(@Q)
Select @V = (Select (N1-1) * (N2-1) from ##temp6)
```

```
/* Get chi square table value */
SELECT @ChiSqTbl = (SELECT ChiSq
    FROM Chi_Sq_Table
    WHERE V = convert(varchar(10), @v)
    AND Alpha = convert(varchar(10), @Alpha))

/* Compare calculated chi square and table chi square */
IF @ChiSqCalc >= @ChiSqTbl
Begin
    print 'Since the calculated chi square value (' +
        convert(varchar(10), @ChiSqCalc) + ')'
    print      'exceeds the table chi square value (' +
        convert(varchar(10), @ChiSqTbl) + '),'
    print 'the samples are significantly different'
    print 'at the ' + convert(varchar(10), @Alpha*100) + '% significance level.'
    print ' '
End

ELSE
Begin
    print 'Since the calculated chi square value (' +
        convert(varchar(10), @ChiSqCalc) + ')'
    print      'is less than the table chi square value (' +
        convert(varchar(10), @ChiSqTbl) + '),'
    print 'the samples are NOT significantly different'
    print      'at the ' + convert(varchar(10), @Alpha*100) + '% significance
level.'
    print ' '
End

GO
SET QUOTED_IDENTIFIER OFF
GO
SET ANSI_NULLS ON
GO
```

Procedure Calls

Below are the call statements for calling the procedures Calculate_t_Statistic, Calculate_Z_Statistic, Compare_Means_2_Samples, and Contingency_Test.

```
DECLARE @RC int
DECLARE @SrcTblName varchar(50)
DECLARE @SrcColName varchar(50)
DECLARE @PopMean float
DECLARE @t float
EXEC @RC = [CH4].[dbo].Calculate_t_Statistic @SrcTblName='Internet Usage',
    @SrcColName='Hours', @PopMean=26.0, @t=@t OUTPUT


DECLARE @RC int
DECLARE @SrcTblName varchar(50)
DECLARE @SrcColName varchar(50)
DECLARE @PopMean float
DECLARE @z float
EXEC @RC = [CH4].[dbo].Calculate_Z_Statistic @SrcTblName='Internet Usage',
    @SrcColName='Hours', @PopMean=26.0, @z=@z OUTPUT


DECLARE @RC int
DECLARE @Alpha float
DECLARE @Src1TblName varchar(50)
DECLARE @Src1ColName varchar(50)
DECLARE @Src2TblName varchar(50)
DECLARE @Src2ColName varchar(50)
EXEC @RC = [CH4].[dbo].Compare_Means_2_Samples @Alpha=0.05,
    @Src1TblName='Basketball', @Src1ColName='Weight',
    @Src2TblName='Football', @Src2ColName='Weight'


DECLARE @RC int
DECLARE @Alpha float
DECLARE @Samp1TblName varchar(50)
DECLARE @Samp1Col1Name varchar(50)
DECLARE @Samp1Col2Name varchar(50)
DECLARE @Samp1Col3Name varchar(50)
EXEC @RC = [CH4].[dbo].[Contingency_Test] @Alpha=0.05,
    @Samp1TblName='Mileage', @Samp1Col1Name='Vehicles',
    @Samp1Col2Name='Day', @Samp1Col3Name='MPG'
```

CHAPTER 5
Curve Fitting

Curve Fit

SO FAR WE'VE LOOKED AT methods that use statistical measures or distributions to characterize databases. These have included the central tendency and dispersion of the data, the shape of the distribution of observations, and the comparison of the data to known parameters or other databases. These databases have displayed a common thread; that is, they all involved one variable (usually denoted by *x*) that represented the observations. There are other situations that occur, however, when data collections include results from the observation of two or more variables. In the two-variable case, for example, the value of one of the variables typically varies, and its effect on the other variable is studied. It is desirable to know if some relationship exists between the factors.

In the diagnostic tree in Figure 1-3, the last branch represents this situation. Sometimes a data gathering effort is specifically designed for a certain analysis technique. In the present chapter, however, we are concerned with the relationship among variables, or how one variable may be influenced by others.

Let's consider the case where a truck's gas mileage is influenced by the amount of weight the vehicle is carrying. If we want to develop a relationship between mileage degradation and payload, we might approach the problem as a simple experiment. All we need to do is drive the truck over a preset distance under varying load conditions, and clock the resulting fuel consumption under each payload. Maybe we end up with the sample data shown in Table 5-1.

Table 5-1. Truck Mileage Degradation with Increasing Payload

PAYLOAD IN POUNDS	MILEAGE IN MILES PER GALLON
0	19.3
500	18.8
1,000	18.6
1,500	17.9
2,000	16.8

In this instance, two variables are involved. We'll let *x* represent the payload values and *y* represent the mileage values. We might ask, "Which variable can we control?" The obvious answer is the payload values. We can set them at any value we want. If we're using concrete blocks for the payload, we can put in the number of blocks we desire to reach a particular weight. Therefore, the variable *x* (the payload) is called the *independent variable*. What we find for the resultant mileage depends on the payload (we're assuming). So you guessed it, the variable *y* (the mileage) is called the *dependent variable*.

Now we could complicate the problem even further if we wanted to, by throwing in a *third* variable. Maybe the mileage also degrades with changes in speed as well as payload. Here again, we can control the speed of the vehicle in order to conduct the test, so speed becomes a second *independent variable*. The table of results would then have three columns of numbers. We might even want to know which of the two independent variables (payload or speed) has the greater effect on mileage. We'll investigate the three variable case later. For now, we'll stick with two variables.

If you are faced with a similar set of circumstances, your first step should be to examine the relationship between the dependent and independent variable(s). Does the dependent variable seem to vary linearly with the independent one, or is their relationship nonlinear? And how strong is this relationship? These questions may be answered by using procedures from the areas of statistics known as *regression* and *correlation*.

Progress, Digress, and Regress

The term "regression" as it applies to statistical data analysis found its origin in the work of Sir Francis Galton (1822–1911), an English scientist, psychologist, and explorer. Galton was a cousin of Charles Darwin, and became interested in the study of heredity and intelligence. Galton believed that talent and physique were transmitted in families, and undertook numerous studies to establish this theory. For example, he cited the Bach family of musicians and their attainment of great eminence. As regarded physical characteristics, Galton set about in the 1880s to measure the heights of many sets of parents and their adult offspring. After he plotted the results, he found that, on the average, the children of tall parents were not so tall as their parents, while the children of short parents were not so short as their parents. In other words, human height appeared to "regress" back to a common type or base. The term *regression* came to mean simply an average relationship between (or among) variables. Galton also developed the theory of correlation between variables.

Linear Regression in Two Variables

The term regression simply refers to fitting a curve (which may be a straight line) to the graph of the data. The curve is defined in terms of a functional expression, such as

$y = 2x + 5$ (linear) or $y = 4x^2 - 3x + 2$ (nonlinear)

Once the equation of the fitted curve is determined (regression), the degree to which it "fits" the graph may be calculated (correlation). Then you can determine if this is a reasonably good fit or a poor one. If the correlation is good, the equation may be used to predict intermediate values in the data, or can project the data beyond the largest or smallest value of the independent variable. For example, in the truck mileage case, maybe we'd like to estimate the mileage for a 750-pound payload. This was not one of the values of the independent variable for which we tested the mileage (see Table 5-1). But if we have a good mathematical relationship between payload and mileage, we can substitute $x = 750$ pounds into our regression equation and derive the mileage (y) that we would expect. The trick is in developing this equation.

To get us moving in that direction, let's refer back to the sample data in Table 5-1 and graph it on a simple x-y coordinate system. It is customary to show the values of the independent variable x on the horizontal axis, and the values of the dependent variable y on the vertical axis. The result is shown in Figure 5-1.

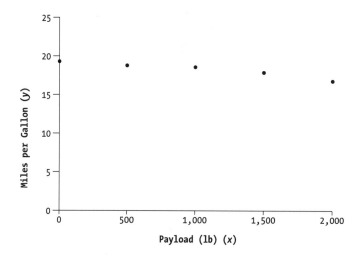

Figure 5-1. Graph of truck mileage degradation with increasing payload

Notice from the graph that the points representing the mileage values for the corresponding payloads appear to very closely follow a straight line. This is indeed fortunate, for the regression of a line in two variables is by far the easiest to handle. Our goal is to draw this straight line so that it passes as closely as possible to all the points. In other words, we want to minimize the sum of the distances that the points lie from the fitted line. Or do we? If we draw this line by eye, we might end up with the result shown in Figure 5-2.

If we consider a point above the line as having a positive distance from the line, we would have to agree that a point below the line has a negative distance from it. Otherwise, we have no way of specifying which side of the line the point falls on. Now if we try to minimize the sum of these distances in order to fit the line, we might come up with nothing usable, because the positive and negative values would tend to just cancel each other out. What a dilemma!

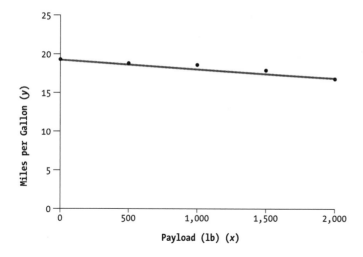

Figure 5-2. Graph of truck mileage showing "best guess" line

What we really need to do is ensure that we retain the orientation of the points above or below the line, but work only with positive distances. Long years ago, someone of superior intelligence discovered that the best fit is achieved when you minimize the sum of the *squares* of the distances. By squaring each distance, we avoid the positive versus negative trap. So regression is based on the *principle of least squares*.

The general equation for a line that fits data of this sort is

$$y = a + bx$$

where (again) x is the independent variable, y is the dependent variable, and a and b are the constants that must be determined to place the line in its proper position on the graph. Until the widespread availability of curve fitting software, the determination of a and b was somewhat painful, and required the use of certain principles from (shudder) calculus (namely, partial derivatives). The derivation of the equations that determine a and b is contained in almost any statistics text, so we omit the derivation here. We are only interested in the results, shown in the following equations.

$$(n)(a) + (\Sigma x)(b) = \Sigma y \quad \text{and}$$
$$(\Sigma x)(a) + (\Sigma x^2)(b) = \Sigma xy$$

These are called the *normal equations* for the linear regression model. Solving these equations in terms of a and b yields

$$a = \frac{(\Sigma y)(\Sigma x^2) - (\Sigma x)(\Sigma xy)}{(n)(\Sigma x^2) - (\Sigma x)^2}$$

$$b = \frac{(n)(\Sigma xy) - (\Sigma x)(\Sigma y)}{(n)(\Sigma x^2) - (\Sigma x)^2}$$

What does all this mean, and how do we use it? Well, any factor shown in the preceding equations that has the summation symbol (Σ) in front of it, simply means we add up all the values represented by the variable name(s) accompanying the summation sign. For example, Σx or Σy means add up all the x or y values in the data table; Σx^2 or Σy^2 means add up all the squares of the x or y values in the table; Σxy means multiply each pair of x and y values together (in their proper order in the data table), and add the results. The symbol n represents the number of pairs (in this case) of data values we have. When the a and b values are calculated, they are inserted into the equation relating y to x to provide the straight line of best fit for the sample data. One point needs to be clarified, however. Although this is the straight line of best fit, it doesn't necessarily mean that a straight line is the most appropriate representation of the data. Rather, a curved line of some sort might more closely weave its way among the points. Only a test of correlation (which is yet another test of hypothesis) can establish the equation of "best" fit. We'll get to that eventually.

For right now, let's use the data in Table 5-1 to "regress" a straight line through the points on the graph in Figure 5-1. The calculations we need are shown in Table 5-2. The column sums are given in the last row.

Table 5-2. Calculations for the Straight Line Fit in the Gas Mileage Example

	PAYLOAD IN POUNDS (X)	MILEAGE IN MPG (Y)	X^2	Y^2	XY
	0	19.3	0	372.49	0
	500	18.8	250,000	353.44	9,400
	1,000	18.6	1,000,000	345.96	18,600
	1,500	17.9	2,250,000	320.41	26,850
	2,000	16.8	4,000,000	282.24	33,600
Totals	5,000	91.4	7,500,000	1,674.54	88,450

The column sums in the table are what we use to find a and b, by the formulas presented previously. These results are shown next.

$$a = \frac{(91.4)(7,500,000) - (5,000)(88,450)}{(5)(7,500,000) - (5,000)^2} = 19.46000$$

$$b = \frac{(5)(88,450) - (5,000)(91.4)}{(5)(7,500,000) - (5,000)^2} = -0.00118$$

SQL/Query

We can obtain the values of *a* and *b* by executing Query 5_1 and Query 5_2. Query 5_1 calculates the totals shown in the last row of Table 5-2. Query 5_2 uses the totals from Query 5_1 to compute the values of *a* and *b*.

Query 5_1:

```
SELECT Count(Payload) AS n,
Sum(Payload) AS Sx,
Sum(Mileage) AS Sy,
Sum([Payload]^2) AS Sx2,
Sum([Mileage]^2) AS Sy2,
Sum([Payload]*[Mileage]) AS Sxy
FROM [Table 5_1];
```

Query 5_2:

```
SELECT ([sy]*[sx2] - [sx]*[sxy]) / ([n]*[sx2] - [sx]^2) AS a,
([n]*[sxy] - [sx]*[sy]) / ([n]*[sx2] - [sx]^2) AS b
FROM [Query 5_1];
```

The values of *a* and *b* are then substituted into the general equation for a straight line shown previously to yield the specific straight line that fits the sample data. This equation is

$$y = 19.46000 - 0.00118x$$

The value of *a* is the point on the vertical (*y*) axis on the graph where the line intersects that axis. The *b* value is the slope of the line. It is small because the line has only a very gradual slope downward, and it is negative because the slope is declining. By substituting any nonzero value of *x* into the equation, we can locate a second point that will determine the line as shown in Figure 5-3.

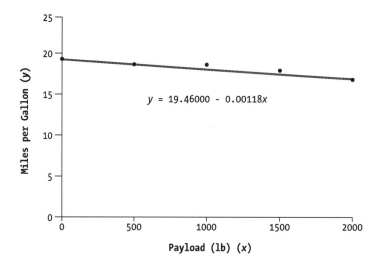

Figure 5-3. Graph of truck mileage showing regression line

SQL/Query

Sometimes the *x* values may be too far away from the *y*-axis and plotting the $(0, a)$ point to draw the line might render a visually undesirable graph. So, an alternative is to use the minimum and the maximum *x* values for the two points of the line. Query 5_3 through Query 5_6 find the minimum and maximum *x* values, calculate the respective *y* values using the line equation, and place the coordinates of the points in a database table called LinePoints.

Query 5_3:

```
SELECT Min(Payload) AS MinX, Max(Payload) AS MaxX
FROM [Table 5_1];
```

Queries 5_4a and 5_4b are used to manage the table LinePoints. If the table LinePoints does not exist, create it by executing Query 5_4a. Once the table is created, use Query 5_4b to clear the table of old data.

Query 5_4a:

```
CREATE TABLE LinePoints (X Single, Y Single);
```

Query 5_4b:

```
DELETE *
FROM LinePoints;
```

Query 5_5:

```
INSERT INTO LinePoints ( X, Y )
SELECT MinX AS X, (a + b*MinX) AS Y
FROM [Query 5_2], [Query 5_3];
```

Query 5_6:

```
INSERT INTO LinePoints ( X, Y )
SELECT MaxX AS X, (a + b*MaxX) AS Y
FROM [Query 5_2], [Query 5_3];
```

> **NOTE** *For convenience, you might wish to incorporate Query 5_3 into Query 5_1.*

This line equation may be used to estimate mileage values between those known from the experiment, or to project fuel consumption beyond the payloads tested. However, in this situation, there is probably a practical limit for payload. In other words, the truck may not be able to physically carry more than 2,000 pounds. To answer the earlier question about the projected mileage at 750 pounds payload, we simply substitute 750 into the regression equation for x and calculate y, as follows:

$$y = 19.46 - (0.00118)(750) = 18.575 \text{ miles per gallon}$$

Linear Correlation in Two Variables

Before we proceed to the more complicated regression models, let's discuss how we determine whether or not we have a good fit with our little equation. The standard method involves calculating what is called the *correlation coefficient* for the curve fit. For a two-variable linear regression, this coefficient, commonly denoted by r, is given by

$$r = \frac{(n)(\Sigma xy) - (\Sigma x)(\Sigma y)}{\sqrt{\left[(n)(\Sigma x^2) - (\Sigma x)^2\right]\left[(n)(\Sigma y^2) - (\Sigma y)^2\right]}}$$

The number of degrees of freedom for this statistic is n – 2, since this is a two-variable situation. The value of r may be either positive or negative, but its absolute value will always be between 0 and 1. If it isn't, you've made a calculation error somewhere. If r is 0, there is no correlation between x and y. Hence, y is not dependent on x, and x cannot be used as a predictor for y. If r is 1, there is perfect correlation between x and y, and y is completely dependent on x, which would be virtually unheard of in a practical problem. The closer the actual (absolute) value of r is to 1, the better the correlation, or the better the fit. This is what we generally desire.

Let's calculate the correlation coefficient for our example. (Now you see why we totaled the y^2 values in Table 5-2; we didn't need this sum for the regression equation, but we do for the calculation of r.) The result is as follows:

$$r = \frac{(5)(88,450)-(5,000)(91.4)}{\sqrt{\left[(5)(7,500,000)-(5,000)^2\right]\left[(5)(1,674.54)-(91.4)^2\right]}} = -0.964$$

SQL/Query

We can also use the SQL Query 5_7 to calculate r.

Query 5_7:

```
SELECT ([n]*[sxy] - [sx]*[sy]) / Sqr(([n]*[sx2] - [sx]^2) * ([n]*[sy2] - [sy]^2)) AS r
FROM [Query 5_1];
```

The reason the correlation coefficient is negative is because the line decreases with increasing x values (in other words, fuel efficiency declines as payload goes up). Now it's obvious that the absolute value of r is very close to 1, but is it "close enough," in the statistical sense? We can only answer this question by establishing a significance level for a test of hypothesis, and comparing r to a table value at this level. The number of degrees of freedom in this example is $5 - 2 = 3$. If we use 5% significance, we find in the table of values for the correlation coefficient (see Appendix B) that $r = 0.878$. Since the absolute value of our calculated value exceeds this table value, we conclude that we have good correlation. As a result, our regression equation seems to be a good fit for the mileage data.

There is an alternative way to calculate the correlation coefficient that doesn't involve such a complicated formula. It also has the advantage of being applicable to the nonlinear (sometimes called curvilinear) regression models discussed later. The basic formula is

$$|r| = \sqrt{1 - \frac{s_{y-\hat{y}}^2}{s_y^2}}$$

where s_y^2 is the variance of the sample y values (assumed to be the dependent variable), and $s_{y-\hat{y}}^2$ is the variance of the regression errors $y - \hat{y}$ over all sample points. y is the observed value of the dependent variable, and $\hat{y}$ is the value predicted by the regression equation. Therefore, $y - \hat{y}$ tells us how close the fitted line or curve is to the actual value observed. It is called a "residual."

To apply this alternative method to our example, we have to substitute each value of x into the regression equation $\hat{y} = 19.46000 - 0.00118x$ and find the predicted $\hat{y}$, then subtract that from the observed y. Table 5-3 summarizes the calculations.

Table 5-3. Comparison of Observed to Predicted Mileage Values

PAYLOAD IN POUNDS (X)	MILEAGE IN MPG (Y)	$\hat{y} = 19.46000 - 0.00118x$	$y - \hat{y}$
0	19.3	19.46	-0.16
500	18.8	18.87	-0.07
1000	18.6	18.28	0.32
1500	17.9	17.69	0.21
2000	16.8	17.10	-0.30

SQL/Query

By executing Query 5_8 we obtain the values shown in Table 5-3.

Query 5_8:

```
SELECT [Table 5_1].ID, [Table 5_1].Payload, [Table 5_1].Mileage,
[a]+[b]*[payload] AS Pred_y, [Mileage] - ([a]+[b]*[payload]) AS Diff

FROM [Query 5_2], [Table 5_1];
```

The variance of the five y and the $y - \hat{y}$ values are calculated from the standard formula given in Chapter 2. They are

$$s_y^2 = 0.93700 \text{ and } s_{y-\hat{y}}^2 = 0.06675$$

We can now find the correlation coefficient value:

$$|r| = \sqrt{1 - \frac{s_{y-\hat{y}}^2}{s_y^2}} = \sqrt{1 - \frac{0.06675}{0.93700}} = 0.964$$

This is the same result as before, except this method only returns the absolute value of r.

SQL/Query

Query 5_9 shows us how to calculate |r| through Query 5_8.

Query 5_9:

```
SELECT Sqr(1 - (Var([diff]) / Var([mileage]))) AS r
FROM [Query 5_8];
```

Robert's Rubies
The Creeping Menace of Round-Off Errors

In regression problems, as well as other applications involving several multiplications and divisions, round-off errors occur in the intermediate and final calculations. The degree to which this happens depends on the number of significant digits the computer or hand calculator is carrying in intermediate calculations. We have tried in our examples throughout this book to rectify these situations for the sake of consistency. There is a whole field of error analysis (as you can imagine), but for most practical statistical applications, it is not a significant problem. However, there are some cases where a conclusion of significance is reversed simply by adding a few more digits to a standard deviation value!

Polynomial Regression in Two Variables

Often the graph of the data will not be linear, so a linear regression may be inaccurate. If the curve moves downward, turns, and then moves upward (or vice versa), it may exhibit what is known as a *parabolic* shape. A parabola has the following general equation that characterizes it:

$$y = a + bx + cx^2$$

Notice that x is still considered the independent variable, y the dependent variable, and now we have three constants (a, b, and c) to determine from the data. The procedure for developing the so-called normal equations for the determination of these constants is the same as for the linear model, but a bit more involved. The three normal equations that result are given below.

$$(n)(a) + (\Sigma x)(b) + (\Sigma x^2)(c) = \Sigma y$$
$$(\Sigma x)(a) + (\Sigma x^2)(b) + (\Sigma x^3)(c) = \Sigma xy$$
$$(\Sigma x^2)(a) + (\Sigma x^3)(b) + (\Sigma x^4)(c) = \Sigma x^2 y$$

Notice this time that we have to calculate the sums of the third and fourth powers of x, as well as the sum of the products of x^2 and y. We could again solve these equations for a, b, and c, but it begins to become more complicated. The alternative is to substitute the appropriate values for the sums and then solve the three equations simultaneously. To illustrate this model, consider the following example:

A new Web site is keeping track of the number of "hits" per hour over a 9-hour period (workday). Observations for a month yielded the averages shown in Table 5-4. The hours are coded from 0 through 8.

Table 5-4. Average Number of "Hits" per Hour on a Web Site

HOURS	HITS
0	120
1	105
2	100
3	80
4	70
5	80
6	75
7	85
8	90

The first step is to graph the data and look at the shape. The graph is shown in Figure 5-4. Note that the hour is denoted by x, and the number of "hits" by y.

It is obvious from the graph that the numbers decline through the morning, reaching a low around the lunch hour, then gradually increase again into the afternoon. The shape is not linear, but more closely resembles a parabola. Therefore, a second degree polynomial model may be the best fit for the data. Let's try this.

We need several sums to substitute into the normal equations. These are shown in Table 5-5.

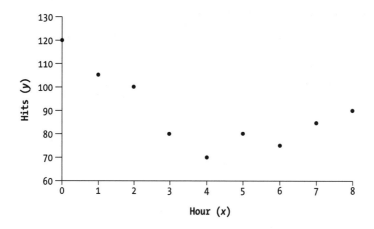

Figure 5-4. Graph of the average number of "hits" per hour

SQL/Query

We can use an SQL query like Query 5_10 to obtain the sums shown in Table 5-5.

Query 5_10:

```
SELECT Count([Hours]) AS n,
Sum([Hours]) AS Sx,
Sum([Hits]) AS Sy,
Sum([Hours]^2) AS Sx2,
Sum([Hours]^3) AS Sx3,
Sum([Hours]^4) AS Sx4,
Sum([Hours]*[Hits]) AS Sxy,
Sum([Hours]^2*[Hits]) AS Sx2y
FROM [Table 5_3];
```

Table 5-5. Sums Needed for the Regression Equation

HOUR (X)	HITS (Y)	X^2	X^3	X^4	XY	X^2Y
0	120	0	0	0	0	0
1	105	1	1	1	105	105
2	100	4	8	16	200	400
3	80	9	27	81	240	720
4	70	16	64	256	280	1,120
5	80	25	125	625	400	2,000
6	75	36	216	1,296	450	2,700
7	85	49	343	2,401	595	4,165
8	90	64	512	4,096	720	5,760
Total 36	805	204	1,296	8,772	2,990	16,970

We are now in a position to substitute these sums from the last row of Table 5-5 into the normal equations for the second degree polynomial regression. It is then necessary to solve these three equations simultaneously to find a, b, and c. Several standard methods, including *Gaussian elimination*, are available to accomplish this (see Appendix D for a Visual Basic procedure for Gaussian elimination). However, we'll do it manually as an illustration.

$$9a + 36b + 204c = 805$$
$$36a + 204b + 1,296c = 2,990$$
$$204a + 1,296b + 8,772c = 16,970$$

We can solve these by taking pairs of equations and eliminating one variable at a time. For example, multiply the first equation through by 4 and subtract the result term by term from the second equation:

$$
\begin{array}{rl}
36a &+204b+1,296c = 2,990 \\
-36a &-144b- 816c =-3,220 \\
\hline
& 60b + 480c =-230
\end{array}
$$

Then multiply the first equation through by 22.6667 and subtract the third equation from it:

$$
\begin{array}{rl}
204a &+ 816b + 4,624c = 18,246.6667 \\
-204a &-1,296b - 8,772c =-16,970.000 \\
\hline
&-480b - 4,148c = 1,276.6667
\end{array}
$$

Now we have two equations in *b* and *c* only. Multiply the first through by 8 and add it to the second:

$$480b + 3,840c = -1,840$$
$$\underline{-480b - 4,148c = 1,276.6667}$$
$$-308c = -563.3333$$

This gives *c* = 1.829. Now we can find *b* from

$$60b + 480(1.829) = -230$$

As a result, *b* = -18.465. Then *b* and *c* may be substituted into the first equation to find *a*:

$$9a + 36(-18.465) + 204(1.829) = 805$$

This yields *a* = 121.848. Therefore, the regression equation we are looking for is

$$y = 121.848 - 18.465x + 1.829x^2$$

When this equation is plotted on the graph of data points in Figure 5-4, the result appears in Figure 5-5.

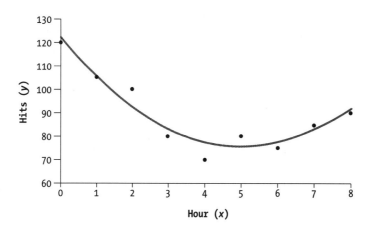

Figure 5-5. Regression curve plotted on the graph of data points

SQL/Query

We can use an SQL query along with a Visual Basic procedure to determine the coefficients of the regression equation. Query 5_11 below creates a matrix-like layout of the normal equations. Notice how Query 5_11 uses the results of Query 5_10. Using the results of this query and the Visual Basic procedure called Database_Gauss (see Appendix D), we calculate the values for *a*, *b*, and *c*. These values are saved in a database table called "Polynomial Coefficients" by the Database_Gauss procedure.

Query 5_11:

```
SELECT n, Sx, Sx2, Sy
FROM [Query 5_10]
UNION ALL
SELECT Sx, Sx2, sx3, Sxy
FROM [Query 5_10]
UNION ALL
SELECT Sx2, Sx3, sx4, Sx2Y
FROM [Query 5_10];
```

The correlation coefficient for nonlinear regression models may be calculated as it was for the linear cases. We might point out, however, that extensive software is available to not only fit various models to input data, but also calculate the appropriate correlation coefficients (e.g., Total VB Statistics by FMS, Inc.). As a result, it is only necessary to provide the data to the software package. Then the regression equation that has the best correlation coefficient may be chosen as the model for the data.

For now, though, let's go ahead and find the correlation coefficient for the second-degree polynomial curve $\hat{y} = 121.848 - 18.465x + 1.829x^2$. Again, we set up a table of observed and predicted y values and find their differences (see Table 5-6), or residuals. As before, we find s_y^2 and $s_{y-\hat{y}}^2$. These are $s_y^2 = 259.0278$ and $s_{y-\hat{y}}^2 = 20.0271$. Then

$$|r| = \sqrt{1 - \frac{20.0271}{259.0278}} = 0.9606$$

SQL/Query

We can also determine $|r|$ by executing Query 5_12 to obtain the values shown in Table 5-6 and then execute Query 5_13 to determine $|r|$.

Query 5_12:

```
SELECT [Table 5_4].ID, [Table 5_4].Hours, [Table 5_4].Hits,
[coef1] + [coef2] * [hours] + [coef3] * [hours] ^ 2 AS Pred_y,
[hits] - ([coef1] + [coef2] * [hours] + [coef3] * [hours] ^ 2) AS Dif
FROM [Table 5_4], [Polynomial Coefficients];
```

Query 5_13:

```
SELECT Sqr(1 - Var([Dif]) / Var([hits])) AS r
FROM [Query 5_12];
```

The number of degrees of freedom in this case is 9 – 2 = 7. At the 5% significance level, the table r value is 0.666. Since the calculated r is considerably greater, we conclude we have good correlation and accept the polynomial fit.

Table 5-6. Comparison of Observed to Predicted Hits

HOURS(X)	HITS (Y)	$\hat{y} = 121.848 - 18.465x + 1.829x^2$	$y - \hat{y}$
0	120	121.848	-1.848
1	105	105.212	-0.212
2	100	92.234	7.766
3	80	82.914	-2.914
4	70	77.252	-7.252
5	80	75.248	4.752
6	75	76.902	-1.902
7	85	82.214	2.786
8	90	91.184	-1.184

The development of higher-order polynomial regression fits follows the same basic steps as the second degree case shown above. A third degree model has four normal equations, a fourth degree has five, and so forth.

Other Nonlinear Regression Models

Numerous other nonlinear models are available for curve fitting, if they seem appropriate to the situation. In order to develop normal equations for determining the parameters (or coefficients) that provide the specific curve fit, it is generally necessary to linearize the model by some trick or transformation of variables. For example, the exponential model is often encountered. It may be of the form

$$y = ab^x$$

and have a graph that looks something like the one in Figure 5-6.

Developing the normal equations would be a painful (if not impossible) experience until we realize that we can transform this model into a linear form. By taking the logarithms of both sides of the equation (either natural or common logs, but natural are usually used), we obtain

$$\ln(y) = \ln(ab^x)$$

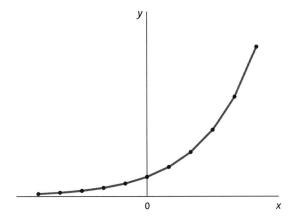

Figure 5-6. A typical exponential curve, this one having a negative portion of its range

which then reduces to

$$\ln(y) = \ln(a) + \left[\ln(b)\right]x$$

Voila! Now we have a linear form. If we let $y' = \ln(y)$, $a' = \ln(a)$, and $b' = \ln(b)$, the equation becomes

$$y' = a' + b'x$$

This should have a familiar look to it. It is exactly like the two-variable linear model we discussed earlier, and its normal equations may be developed in terms of the substituted variables y', a', and b', then converted back to the original variables. The normal equations may again be solved in terms of a and b in the original exponential model. The results are given below. The entire expression in brackets is the exponent of "e."

$$a = e^{\left[\dfrac{\left(\Sigma \ln y\right)\left(\Sigma x^2\right) - \left(\Sigma x\right)\left(\Sigma x \ln y\right)}{n\Sigma x^2 - \left(\Sigma x\right)^2}\right]}$$

$$b = e^{\left[\dfrac{\left(n\right)\left(\Sigma x \ln y\right) - \left(\Sigma x\right)\left(\Sigma \ln y\right)}{n\Sigma x^2 - \left(\Sigma x\right)^2}\right]}$$

Notice that we would need to sum the logarithms of the *y* data values in order to solve these equations for *a* and *b*. Let's consider the following example:

The number of registered voters in a certain precinct has increased over the last five years, with the results shown in Table 5-7. The graph of the data is also presented in Figure 5-7.

Table 5-7. Increase in Number of Registered Voters Over Five Years

YEAR	NUMBER OF VOTERS
1	148,324
2	149,887
3	150,641
4	154,208
5	154,815

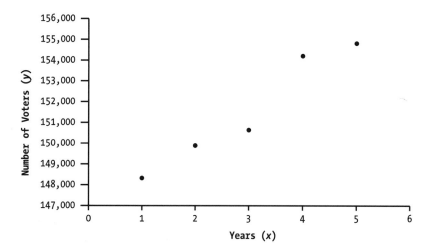

Figure 5-7. Graph of the voter data in Table 5-5

The sums necessary for the normal equations are calculated in Table 5-8. Notice that the natural logarithm is determined for each y value (representing the number of voters), then these are totaled.

Table 5-8. Sums for the Regression Equation

YEAR (X)	NO. OF VOTERS (Y)	X^2	LN(Y)	(X)[LN(Y)]
1	148,324	1	11.907	11.907
2	149,887	4	11.918	23.835
3	150,641	9	11.923	35.768
4	154,208	16	11.946	47.784
5	154,815	25	11.950	59.750
Totals 15		55	59.644	179.044

Now we are ready to substitute the sums into the normal equations to find a and b for our regression model. The results follow.

$$a = e^{\left[\frac{(59.644)(55) - (15)(179.044)}{(5)(55) - (15)^2}\right]} = e^{11.8952} = 146,561.44$$

$$b = e^{\left[\frac{(5)(179.044) - (15)(59.644)}{(5)(55) - (15)^2}\right]} = e^{0.0112} = 1.01$$

$$y = 146,561.44(1.01)^x$$

SQL/Query

We can find a and b by running two SQL queries. We first run Query 5_14 to obtain the sums given in the last row of Table 5-8. Using the results of Query 5_14 we can execute Query 5_15 to determine the values of a and b. Notice the use of the built-in mathematical functions `log` and `exp` that are available in Microsoft Access and SQL Server.

Query 5_14:

```
SELECT Count([Table 5_7].ID) AS N,
Sum([Table 5_7].Year) AS Sx,
Sum([Table 5_7].Voters) AS Sy,
Sum([Year]^2) AS Sx2,
Sum(Log([Voters])) AS Slny,
Sum([Year]*Log([Voters])) AS Sxlny
FROM [Table 5_7];
```

Query 5_15:

```
SELECT Exp((([slny]*[sx2] - [sx]*[sxlny]) / ([n]*[sx2] - [sx]^2)) AS a,
Exp((([n]*[sxlny] - [sx]*[slny]) / ([n]*[sx2] - [sx]^2)) AS b
FROM [Query 5_14];
```

When the exponential curve is drawn on the graph of data points, the result appears as in Figure 5-8. Notice that this exponential curve is almost a straight line in this example.

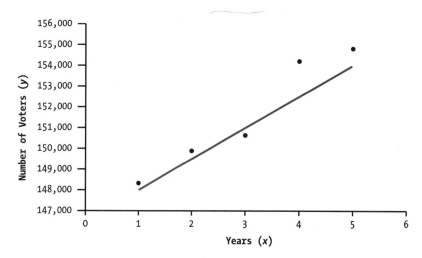

Figure 5-8. Exponential curve of voters by year

The correlation coefficient may again be calculated as before. Table 5-9 contains the necessary values. In this case, $s_y^2 = 7,930,247.5$ and $s_{y-\hat{y}}^2 = 571,871.20$, and

$$|r| = \sqrt{1 - \frac{571,871.20}{7,930,247.5}} = 0.9633$$

This exceeds the table $r = 0.878$ for 5% significance and 3 degrees of freedom, so the exponential model seems to be a good fit. We can use a pair SQL queries similar to Query 5_8 and Query 5_9 (or like Query 5_12 and Query 5_13) to calculate the value of $|r|$.

Table 5-9. Comparison of Observed and Predicted Numbers of Voters

YEAR (X)	NO. OF VOTERS (Y)	$y = 146,561.44(1.01)^x$	$y - \hat{y}$
1	148,324	148,027.05	296.95
2	149,887	149,507.32	379.68
3	150,641	151,002.40	-361.40
4	154,208	152,512.42	1,695.58
5	154,815	154,037.55	777.45

Linear Regression in More Than Two Variables

Now we may approach the case where we have more than two variables involved in the database. For the three-variable situation, two of the variables are independent. You may encounter two different formats for data presentation. To illustrate, let's suppose we again visit the mileage versus payload example given earlier in this chapter. We stated that a second independent variable for the experiment might be the speed at which the vehicle was driven to obtain the mileage results. We had five payloads for which the experiment was conducted. Let's suppose that, for each payload, we also had the option of driving the vehicle at 40, 50, or 60 miles per hour. Now the experiment has the potential of yielding 15 distinct mileage results. The data may be presented in a tabular or columnar format. Table 5-10 illustrates the tabular format and Table 5-11 illustrates the columnar format.

Table 5-10. Tabular Format for the Three-Variable
Linear Regression Model

PAYLOAD (LB) (X)	SPEED (MILES/HOUR) (Y)		
	40	50	60
0	19.4	19.3	19.1
500	18.4	18.8	18.9
1,000	18.2	18.6	18.6
1,500	16.0	17.9	17.8
2,000	15.7	16.8	16.6

Table 5-11. Columnar Format for the Three-Variable Linear Regression Model

PAYLOAD (LB) (X)	SPEED (MILES/HOUR) (Y)	MILEAGE (MILES/GALLON) (Z)
0	40	19.4
500	40	18.4
1,000	40	18.2
1,500	40	16.0
2,000	40	15.7
0	50	19.3
500	50	18.8
1,000	50	18.6
1,500	50	17.9
2,000	50	16.8
0	60	19.1
500	60	18.9
1,000	60	18.6
1,500	60	17.8
2,000	60	16.6

Notice in this situation that the columnar format has no particular advantage, because so many of the payload and speed values have to be repeated. However, there are experiments where, for each value of the dependent variable obtained, there is both a unique first and second independent variable. If that was the case here, we could have 15 unique payloads as well as 15 unique speeds. Then the tabular format would be infeasible.

The regression model for the three-variable linear case is the equation of a plane, so we have to have three coefficients and three variables. The form is shown below.

$$z = a + bx + cy$$

As before, z is the dependent variable (mileage in our example), and x and y are the independent variables. The normal equations are developed in a manner similar to the two-variable case, and are given below.

$$(n)(a) + (\Sigma x)b + (\Sigma y)c = \Sigma z$$

$$(\Sigma x)(a) + (\Sigma x^2)b + (\Sigma xy)c = \Sigma xz$$

$$(\Sigma y)(a) + (\Sigma xy)b + (\Sigma y^2)c = \Sigma yz$$

Now we're ready to fit a regression plane to the mileage data in Table 5-10 (or Table 5-11). As in previous examples, we have to add the x and the y values, add their squared values, and so forth. However, we have to be careful this time. If we are using the data in the tabular format, we have to account for the fact that we have one unique payload and speed pair for each mileage value in the table. Therefore, when we sum the payloads, we have to multiply the result by 3, since there are three speeds associated with each of the five payloads. Likewise, we have to multiply the sum of the speeds by 5. Now in a situation like this, it becomes apparent that the columnar presentation of the data may be safer to work with. We merely sum the columns and don't worry about multiplying by any factors. This is how we show the results in Table 5-12. The last row contains the column sums.

Table 5-12. Sums of the Terms and Factors Needed for the Regression Equation

PAYLOAD (X)	SPEED (Y)	MILEAGE (Z)	X^2	Y^2	Z^2	XY	XZ	YZ
0	40	19.4	0	1,600	376.36	0	0	776
500	40	18.4	250,000	1,600	338.56	20,000	9,200	736
1,000	40	18.2	1,000,000	1,600	331.24	40,000	18,200	728
1,500	40	16.0	2,250,000	1,600	256.00	60,000	24,000	640
2,000	40	15.7	4,000,000	1,600	246.49	80,000	31,400	628
0	50	19.3	0	2,500	372.49	0	0	965
500	50	18.8	250,000	2,500	353.44	25,000	9,400	940
1,000	50	18.6	1,000,000	2,500	345.96	50,000	18,600	930
1,500	50	17.9	2,250,000	2,500	320.41	75,000	26,850	895
2,000	50	16.8	4,000,000	2,500	282.24	100,000	33,600	840
0	60	19.1	0	3,600	364.81	0	0	1,146
500	60	18.9	250,000	3,600	357.21	30,000	9,450	1,134
1,000	60	18.6	1,000,000	3,600	345.96	60,000	18,600	1,116
1,500	60	17.8	2,250,000	3,600	316.84	90,000	26,700	1,068
2,000	60	16.6	4,000,000	3,600	275.56	120,000	33,200	996
15,000	750	270.1	22,500,000	38,500	4,883.57	750,000	259,200	13,538

Remembering that there are 15 data values ($n = 15$), we can substitute these sums into the normal equations and solve simultaneously to obtain the coefficients for the regression equation. The normal equations are

$$(15)(a)+(15,000)(b)+(750)(c)=270.1$$

$$(15,000)(a)+(22,500,000)(b)+(750,000)(c)=259,200$$

$$(750)(a)+(750,000)(b)+(38,500)(c)=13,538$$

Simultaneous solution of these equations yields $a = 17.810000$, $b = -0.001453$, and $c = 0.033000$. Therefore, the regression equation would be

$$z = 17.810000 - 0.001453x + 0.033000y$$

To see if this is a good model for the data, we can calculate the multiple correlation coefficient as we did in the previous examples.

We again need the residuals, or differences between the observed dependent variable value (z in this case) and the predicted value. Table 5-13 provides this information.

Table 5-13. Observed and Predicted Mileage Values, with Residuals

PAYLOAD (X)	SPEED (Y)	MILEAGE (Z)	PREDICTED MILEAGE ($\hat{z}$)	RESIDUAL ($z - \hat{z}$)
0	40	19.4	19.13	0.27
500	40	18.4	18.4035	-0.0035
1,000	40	18.2	17.677	0.523
1,500	40	16.0	16.9505	-0.9505
2,000	40	15.7	16.224	-0.524
0	50	19.3	19.46	-0.16
500	50	18.8	18.7335	0.0665
1,000	50	18.6	18.007	0.593
1,500	50	17.9	17.2805	0.6195
2,000	50	16.8	16.554	0.246
0	60	19.1	19.79	-0.69
500	60	18.9	19.0635	-0.1635
1,000	60	18.6	18.337	0.263
1,500	60	17.8	17.6105	0.1895
2,000	60	16.6	16.884	-0.284

The correlation coefficient again requires the variance of the z values and the variance of the residuals. These are

$$s_z^2 = 1.42638 \text{ and } s_{z-\hat{z}}^2 = 0.21707$$

The correlation coefficient is

$$|r| = \sqrt{1 - \frac{0.21707}{1.42638}} = 0.9208$$

This exceeds the table value of 0.627 for 3 variables, 12 degrees of freedom (15-3), and a 5% confidence level, so we conclude that the regression fit is good.

Although we have a good fit, we really don't know at this point which of the two independent variables (payload and speed) is more influential on mileage. We could conduct a t test on each of the regression coefficients of payload and speed and compare their values to the table t for 5% significance and 12 degrees of freedom (15 mileage values less 3 variables). This table t value is 2.179 (see Appendix B). The formulas for determining the t statistic for the coefficients b and c require calculation of the mean $\bar{x}$ of the payloads and the mean $\bar{y}$ of the speeds. These are

$$\bar{x} = \frac{15,000}{15} = 1,000 \text{ and } \bar{y} = \frac{750}{15} = 50$$

The formulas look awful, but are as follows:

$$t_b = \frac{|b|}{\sqrt{s_{z-\hat{z}}^2 \left(\frac{n-1}{n-3}\right) \left\{ \frac{\sum y^2 - n(\bar{y})^2}{\left[\sum x^2 - n(\bar{x})^2\right]\left[\sum y^2 - n(\bar{y})^2\right] - \left[\sum xy - n(\bar{x})(\bar{y})\right]^2} \right\}}}$$

$$t_c = \frac{|c|}{\sqrt{s_{z-\hat{z}}^2 \left(\frac{n-1}{n-3}\right) \left\{ \frac{\sum x^2 - n(\bar{x})^2}{\left[\sum y^2 - n(\bar{y})^2\right]\left[\sum x^2 - n(\bar{x})^2\right] - \left[\sum xy - n(\bar{x})(\bar{y})\right]^2} \right\}}}$$

We'll now substitute all the values from Table 5-12 and calculate the two results. We'll also need $s_{z-\hat{z}}^2 = 0.21707$ that we calculated earlier. And, of course, n = 15.

$$t_b = \frac{\left|-0.00145333\right|}{\sqrt{0.21707\left(\dfrac{15-1}{15-3}\right)\left\{\dfrac{38,500-(15)(50)^2}{\left[22,500,000-(15)(1,000)^2\right]\left[38,500-(15)(50)^2\right]-\left[750,000-(15)(1,000)(50)\right]^2}\right\}}}$$

$$= 7.90899$$

$$t_c = \frac{\left|0.033\right|}{\sqrt{0.21707\left(\dfrac{15-1}{15-3}\right)\left\{\dfrac{22,500,000-(15)(1,000)^2}{\left[38,500-(15)(50)^2\right]\left[22,500,000-(15)(1,000)^2\right]-\left[750,000-(15)(1,000)(50)\right]^2}\right\}}}$$

$$= 2.07367$$

When compared to the table t value of 2.179, we see that payload significantly influences mileage, but speed doesn't. This is very important knowledge if we plan to report the results or continue the experiment.

In problems involving more than three variables, the procedures are similar, except the tests for significance of the coefficients of regression variables become more complex. Now we must not only test the contribution, if any, of each individual coefficient, but that of combinations of the coefficients as well. In addition, when the model contains more than five variables, the table of correlation coefficient values for test comparisons in Appendix B can no longer be used. It will only accommodate up to five variables. In these situations, we can still test the significance of the regression of the dependent variable on all the independent variables by use of the correlation coefficient, but we must use a different test statistic. Specifically, we calculate a statistic that is F distributed (remember the F tests from Chapter 4), and compare it to the table F statistic for the appropriate degrees of freedom. The statistic is

$$F = \frac{r^2(n-k-1)}{k(1-r^2)}$$

Where r is again the correlation coefficient, n is the number of data values, and k is the number of *independent* variables in the problem. The number of degrees of freedom for the table F statistic is $v_1 = k$ and $v_2 = n - k - 1$. In fact, this relationship will work for any number of variables in the multiple linear regression case. For example, if we apply it to the auto mileage example, we recall that $|r| = 0.9208$, so $r^2 = 0.8479$. There were

15 data values, so $n = 15$, and two independent variables x (payload) and y (speed), so $k = 2$. Therefore,

$$F = \frac{0.8479(15-2-1)}{2(1-0.8479)} = 33.448$$

The table F value for $v_1 = k = 2$ and $v_2 = n - k - 1 = 12$ is 3.89 (see Appendix B), so obviously the calculated value exceeds the table value. Therefore, the correlation between mileage (z) and at least one of the independent variables of payload and speed is significant, as we concluded earlier.

Conclusion

The methods illustrated may be extended to the case of four or more variables. The most commonly encountered situation, however, is the two-variable case, for which either a linear or nonlinear model might be appropriate. As we stated earlier, much software is now available to fit and check these models. It is only necessary for the data miner to provide the data, and the software will do the rest. Ah, for the good old days, when we did it all by hand.

T-SQL Source Code

What follows are four T-SQL source code listings for performing the regressions described in this chapter along with two additional support procedures. The first listing is called Linear_Regression_2_Variables. The Linear_Regression_2_Variables procedure takes two numerical columns of a table and determines the a and b coefficients of a linear equation that best fits these two columns of data. The second procedure is a support procedure called Gaussian_Elimination and is used as a general-purpose procedure to solve a system of linear equations. The third procedure is called Array_2D and is a support procedure primarily used by the Gaussian_Elimination procedure. The Array_2D procedure takes a table of data alone with a row number and column number and returns the table value at the intersection of the row and column. The fourth procedure is called Polynomial_Regression and is used to fit a polynomial of degree n to a set of data. The fifth listing is a procedure called Exponential_Model and is used to fit an exponential curve through a set of points that can be any two numerical columns in a table. The last listing is called Multiple_Linear_Regression and is used to fit a linear equation through an n-tuple of data points. Each regression routine also calculates the correlation coefficient and provides a conclusion.

Linear_Regression_2_Variables

```
SET QUOTED_IDENTIFIER OFF
GO
SET ANSI_NULLS ON
GO

CREATE PROCEDURE [Linear_Regression_2_Variables]
@SrcTable Varchar(50)='Table 5_1',
@SrcColX Varchar(50)='Payload',
@SrcColY Varchar(50)='Mileage',
@a Float = 0 OUTPUT,
@b Float = 0 OUTPUT,
@r Float = 0 OUTPUT
AS


/************************************************************/
/*                                                          */
/*              Linear_Regression_2_Variables               */
/*                                                          */
/*   This procedure performs a linear regression in two     */
/*   variables. 'x' denotes one variable, and 'y' denotes the */
/*   other variable.  The regression equation of fit is of  */
/*   the form y=a+bx where a and b are coefficients. The    */
/*   degree to which the equation fits is called the        */
/*   correlation and is denoted by 'r'.                     */
/*                                                          */
/* INPUTS:                                                  */
/*    SrcTable - name of table containing sample data       */
/*    SrcColX - column containing sample data values for x  */
/*    SrcColY - column containing sample data values for y  */
/* OUTPUTS:                                                 */
/*    a -- coefficient a                                    */
/*    b -- coefficient b                                    */
/*    r -- correlation coefficient                          */
/*                                                          */
/************************************************************/
```

```
/* Local variables */
DECLARE @Q varchar(500)      /* query string */
DECLARE @n Int               /* number of sample values */
DECLARE @Sx Float            /* Sum of x */
DECLARE @Sx2 Float           /* Sum of x squared */
DECLARE @Sy Float            /* Sum of y */
DECLARE @Sy2 Float           /* Sum y squared */
DECLARE @Sxy Float           /* Sum of x times y */


/* Remove temporary tables if they exist */
If exists (SELECT * FROM tempdb..sysobjects
    where Type = 'U' and Name like "temp1LR2V")
    drop table ##temp1LR2V


If exists (SELECT * FROM tempdb..sysobjects
    where Type = 'U' and Name like "temp2LR2V")
    drop table ##temp2LR2V


If exists (SELECT * FROM tempdb..sysobjects
    where Type = 'U' and Name like "temp2LR2V")
    drop table ##temp3LR2V



/* EQUATION COEFFICIENTS */

/* Build query */
Set @Q = 'SELECT Count([' + @srcColX + ']) AS N, ' +
    'Sum([' + @srcColX + ']) AS Sx, ' +
    'Sum([' + @srcColX + '] * ['+ @srcColX + ']) AS Sx2, ' +
    'Sum([' + @srcColY + ']) AS Sy, ' +
    'Sum([' + @srcColY + '] * ['+ @srcColY + ']) AS Sy2, ' +
    'Sum([' + @srcColX + '] * ['+ @srcColY + ']) AS Sxy ' +
    'INTO ##temp1LR2V FROM [' + @SrcTable + ']'
print @Q

/* Execute the query */
EXEC(@Q)

/* Get query result */
SELECT @n = N,
    @SX = Sx,
    @Sx2 = Sx2,
    @Sy = Sy,
    @Sy2 = Sy2,
    @Sxy = Sxy
    FROM ##TEMP1lr2V
```

```
/* Calculate the two coefficients "a" and "b" */
Set @a = (@Sy * @Sx2 - @Sx * @Sxy) / (@n * @Sx2 - @Sx * @Sx)
Set @b = (@n * @Sxy - @Sx * @Sy) / (@n * @Sx2 - @Sx * @Sx)


/* CORRELATION COEFFICIENT */
/* Build query */
Set @Q = 'SELECT [' + @SrcColY + '] AS Yval, ' +
    '[' + @SrcColY + '] - (' +
    convert(varchar(20), @a) + ' + ' +
    convert(varchar(20), @b) + ' * ' +
    '[' + @SrcColX + ']) AS Diff ' +
    'INTO ##temp2LR2V FROM [' + @SrcTable + ']'
print @Q

/* Execute query */
EXEC(@Q)

/* Figure Correlation coefficient */
Set @Q = 'SELECT Sqrt(1 - (Var(Diff) / Var(Yval))) AS CorrelCoef ' +
    'INTO ##temp3LR2V FROM ##temp2LR2V'
EXEC(@Q)
Select @r = CorrelCoef FROM ##temp3LR2V


GO
SET QUOTED_IDENTIFIER OFF
GO
SET ANSI_NULLS ON
GO
```


Gaussian_Elimination

```
SET QUOTED_IDENTIFIER OFF
GO
SET ANSI_NULLS OFF
GO

CREATE PROCEDURE Gaussian_Elimination
@Num_Variables Int = 3,
@SrcMatrix varchar(50)='Matrix',
@RstlVector varchar(50) = 'Solution'
AS
```

```
/***********************************************************/
/*                                                       */
/*              Gaussian_Elimination                     */
/*                                                       */
/*    This procedure implements the Gaussian elimination  */
/* method for solving a set of n linear equations with n  */
/* unknowns. Basically, the procedure uses a matrix of size */
/* n rows by n+2 columns. Each row represents a linear    */
/* equation. The first n columns are the x sub i values   */
/* (coefficients of the unknowns) for i =1,2,...,n, and   */
/* the n+1 column is y. The last column is called 'Rid' and */
/* serves as a row identifier that simply numbers the     */
/* equations from 1 to n. In this implementation, a copy of */
/* the input matrix, @ScrMatrix, is placed into a temporary */
/* table called ##Matrix.                                 */
/*                                                       */
/* INPUTS:                                                */
/*    Num_Variables - the value of n                      */
/*    SrcMatrix - the name of the table composed of n rows */
/*        by n+2 columns matrix                           */
/*    RstlVector - the name for a two column table to be   */
/*        filled with the Rid (i) and the solution (x sub i) */
/*                                                       */
/***********************************************************/


/* Local variables */
DECLARE @Q varchar(5000)        /* query string */
DECLARE @i Int                  /* loop index */
DECLARE @j Int                  /* loop index */
DECLARE @k Int                  /* loop index */
DECLARE @yi Int                 /* index pointer to y */
DECLARE @xi varchar(50)         /* an x sub i column name */
DECLARE @Axi varchar(50)        /* x in above equation */
DECLARE @Bxi varchar(50)        /* x in below equation */
DECLARE @av Float               /* a coef in above equation */
DECLARE @ay Float               /* right side of equation */
DECLARE @a varchar(50)          /* a coef in above equation */
DECLARE @b varchar(50)          /* a coef in below equation */
DECLARE @bv Float               /* a coef in below equation */
DECLARE @by Float               /* right side of equation */
DECLARE @m varchar(50)          /* multipliers as text */
DECLARE @mv Float               /* multipliers as number */
DECLARE @S Float                /* row sum */
DECLARE @Rid Int                /* row index */
```

```
DECLARE @r Int                          /* row index for row swap */
DECLARE @BigRid Int                     /* row index for row swap */
DECLARE @MaxVal Float                   /* max value for row swap */

/* Make sure we have at least 2 variables */
IF @Num_Variables < 2
Begin
    print 'Number of variables must be two or more.'
    print 'Gaussian_Elimination terminated.'
    Return (222)
End

/* Establish temporary work copy of the matrix */
SET @Q = 'CREATE TABLE ##Matrix ('
SET @i = 1
WHILE @i <= @Num_Variables
Begin
    SET @Q = @Q + 'x' + convert(varchar(5), @i) + ' FLOAT, '
    SET @i = @i + 1
End
SET @Q = @Q + 'y FLOAT, ' + 'Rid Int)'
EXEC(@Q)

/* Populate matrix with data */
SET @Q = 'INSERT INTO ##Matrix SELECT * FROM ' + @SrcMatrix
EXEC(@Q)

/* Initialize */
SET @yi = @Num_Variables + 1

/* Go through the matrix row by row and */
/* triangulate the lower left to all zeros */

SET @k = 1
WHILE @k < @Num_Variables
Begin

    /* Do row swap */
    SET @Axi = 'x' + convert(varchar(5),@k)
```

```
/* Get the Rid of the row with the largest value */
SET @Q = 'SELECT Min(Rid) AS MaxRid ' +
     'INTO ##Temp1 ' +
     'FROM ##Matrix ' +
     'WHERE ABS([' +  @Axi + ']) = ' +
          '(SELECT Max(Abs([' + @Axi + '])) ' +
          'FROM ##Matrix ' +
          ' WHERE Rid >= ' + convert(varchar(10), @k) + ')'

EXEC(@Q)
SELECT @r = (SELECT MaxRid FROM ##Temp1)

/* Remove temporary table */
DROP TABLE ##Temp1

/* Check to see if we need to swap rows */
IF @r <> @k
Begin
     /* Swap the two rows by */
     /* swapping their Rids. */
     SET @BigRid = @Num_Variables + 2
     UPDATE ##MAtrix
     SET Rid = @BigRid
     WHERE Rid = @k

     UPDATE ##MAtrix
     SET Rid = @k
     WHERE Rid = @r

     UPDATE ##MAtrix
     SET Rid = @r
     WHERE Rid = @BigRid
End


/* DO ELIMINATION */

SET @i = @k + 1
WHILE @i <= @Num_Variables
Begin
```

```
/* Handle first row by dividing */
/* the first row by x1 */
IF @k = 1
Begin
     SET @j = 2
     SET @Q = 'UPDATE ##Matrix SET x1 = 1 '
     WHILE @j <= @Num_Variables
     Begin
          SET @Bxi = 'x' + convert(varchar(5),@j)
          SET @Q = @Q + ', [' + @Bxi + ']=[' + @Bxi + ']/[x1]'
          SET @j = @j + 1
     End
     SET @Q = @Q + ', [y]=[y]/[x1]'
     SET @Q = @Q + 'WHERE Rid = 1'
     EXEC(@Q)
End


/* Get above row value */
EXEC Array_2D '##Matrix', 'Rid', @k, @k, @av OUTPUT
SET @a = convert(varchar(20), @av)

/* Get below row value */
EXEC Array_2D '##Matrix', 'Rid', @i, @k, @bv OUTPUT
SET @b = convert(varchar(20), @bv)

/* Obtain multiplier */
SET @mv = @bv / @av
SET @m = convert(varchar(20),@mv)

/* Begin building update query */
SET @Bxi = 'x' + convert(varchar(5),@k)
SET @Q = 'UPDATE ##Matrix '
SET @Q = @Q + 'SET [' + @Bxi + '] = 0'

/* Append additional columns to be updated */
SET @j = @k + 1
WHILE @j <= @Num_Variables
Begin
     EXEC Array_2D '##Matrix', 'Rid', @k, @j, @av OUTPUT
     SET @a = convert(varchar(20), @av)
     SET @Bxi = 'x' + convert(varchar(5),@j)
```

```
            SET @Q = @Q + ', '
            SET @Q = @Q + '[' + @Bxi + ']=[' + @Bxi + ']-(' + @a + '*' + @m + ')'
            SET @j = @j + 1
       End      /* loop for @j */


       /* Append last column */
       EXEC Array_2D '##Matrix', 'Rid', @k, @yi, @ay OUTPUT
       SET @Q = @Q + ', '
       SET @Q = @Q + '[y] = [y] - (' + convert(varchar(20), @ay) + '*' + @m + ') '

       /* Identify which row (below row) to apply update to */
       SET @Q = @Q + 'WHERE [Rid] = ' + convert(varchar(10), @i)

       /* Do the update to the below equation (row) */
       EXEC(@Q)


       SET @i = @i + 1

End      /* loop for @i */


/* Build an update query to divide */
/* the k-th row by x sub k. Thus making */
/* x sub k diagonal element one. */
SET @j = @k+1
SET @Axi = 'x' + convert(varchar(5),@k)
SET @Q = 'UPDATE ##Matrix SET [' + @axi + '] = 1 '
WHILE @j <= @Num_Variables
Begin
       SET @Bxi = 'x' + convert(varchar(5),@j)
       SET @Q = @Q + ', [' + @Bxi + ']=[' + @Bxi + ']/[' + @Axi + ']'
       SET @j = @j + 1
End
SET @Q = @Q + ', [y]=[y]/[' + @Axi + ']'
SET @Q = @Q + 'WHERE Rid = ' + convert(varchar(10), @k)
EXEC(@Q)


SET @k = @k + 1

End      /* Loop for @k */
```

```
/* Divide the last row  by the diagonal term */
EXEC Array_2D '##Matrix', 'Rid', @Num_Variables, @Num_Variables, @bv OUTPUT
EXEC Array_2D '##Matrix', 'Rid', @Num_Variables, @yi, @by OUTPUT

SET @av = @by / @bv
SET @a = convert(varchar(20), @av)

SET @Q = 'UPDATE ##Matrix ' +
    'SET [' + @Bxi + '] = 1, ' +
    '[y] = ' + @a + ' ' +
    'WHERE [Rid] = ' + convert(varchar(10), @Num_Variables)
EXEC(@Q)


/*  BACK SUBSTITUTION  */

/* Initialize loop index so that we can */
/* work our way back up the matrix. */
SET @i = @Num_Variables - 1
WHILE @i >= 1
Begin

    /* Sum the terms after the main diagonal term */
    SET @S = 0
    SET @j = @i + 1
    WHILE @j <= @Num_Variables
    Begin
        EXEC Array_2D '##Matrix', 'Rid', @i, @j, @bv OUTPUT
        EXEC Array_2D '##Matrix', 'Rid', @j, @yi, @av OUTPUT

        SET @S = @S + @bv * @av
        SET @j = @j + 1
    End

    EXEC Array_2D '##Matrix', 'Rid', @i, @i, @bv OUTPUT
    EXEC Array_2D '##Matrix', 'Rid', @i, @yi, @by OUTPUT

    /* Calculate the solution value for row i */
    SET @av = (@By - @S) / @bv
    SET @a = convert(varchar(20), @av)
```

```
            /* Save the solution value under y */
        SET @Q = 'UPDATE ##Matrix ' +
            'SET [y] = ' + @a + ' ' +
            'WHERE [Rid] = ' + convert(varchar(10), @i)
        EXEC(@Q)

        SET @i = @i - 1

End     /* Loop for @k */


/* SAVE FINAL SOLUTION */

/* If result table exists, then drop it */
IF exists (SELECT id FROM tempdb..sysobjects
            WHERE name = @RstlVector) or
        exists (SELECT id FROM ..sysobjects
            WHERE name = @RstlVector)
Begin
        SET @Q = 'DROP TABLE ' + @RstlVector
        EXEC(@Q)
End

/* Establish the solution table */
SET @Q = 'CREATE TABLE ' + @RstlVector + '(i Int, xi Float)'
EXEC(@Q)

/* Store the solution */
        SET @Q = 'INSERT ' + @RstlVector + '(i, xi) ' +
            'SELECT Rid AS i, y AS xi FROM ##Matrix ' +
            'ORDER BY Rid'
        EXEC(@Q)

GO
SET QUOTED_IDENTIFIER OFF
GO
SET ANSI_NULLS ON
GO
```

Array_2D

```
SET QUOTED_IDENTIFIER OFF
GO
SET ANSI_NULLS OFF
GO

CREATE PROCEDURE Array_2D
@SrcTable varchar(50) = 'Matrix',
@SrcColRid Varchar(50) = 'Rid',
@Rid Int = 1,
@Cid Int = 1,
@Val Float = 0 OUTPUT
AS


/**********************************************************/
/*                                                        */
/*                      Array_2D                          */
/*                                                        */
/*  In this procedure ScrTable is viewed as being a two   */
/*  dimensional (2D) array. An element in the array is    */
/*  addressed by a row index and a column index.  The row */
/*  index is a column in the table that simply numbers    */
/*  the rows (1,2,3,..).  The column names of the table may */
/*  be any valid column name.  The column indexing is     */
/*  through the 'Colid' in 'syscolumns' table.  Using     */
/*  Rid and Cid as row and column index numbers of the    */
/*  array, the procedure returns the value at the         */
/*  intersection of the row denoted by Rid and the column */
/*  that has colid equal to Cid.  To keep element (1,1) as */
/*  the first column and first row, the column containing */
/*  Rid values should be the last column in the ScrTable. */
/*                                                        */
/* INPUTS:                                                */
/*   SrcTable - name of table containing 2-D array        */
/*   SrcColRid - name of column containing the Rid values */
/*   Rid - row index number                               */
/*   Cid - column index number                            */
/* OUTPUTS:                                               */
/*   V -- value in array at (Rid, Cid)                    */
/*                                                        */
/**********************************************************/
```

```
/* Local variables */
DECLARE @Q Varchar(500)          /* Query array */
DECLARE @Tid Int                 /* Table ID number */
DECLARE @ColName Varchar(50)     /* column name*/


/* Determine if we are working with a */
/* base table or a temporary table */
IF substring(@SrcTable,1,1)  = '#'

Begin
    /* Get Table ID number */
    SELECT @Tid = (SELECT id FROM tempdb..sysobjects
        WHERE name = @SrcTable)


    /* Get Column Name */
    Select @ColName = (Select name FROM tempdb..syscolumns
        WHERE id = @Tid And Colid = @Cid)
End


Else
Begin
    /* Get Table ID number */
    SELECT @Tid = (SELECT id FROM ..sysobjects
        WHERE name = @SrcTable)


    /* Get Column Name */
    Select @ColName = (Select name FROM ..syscolumns
        WHERE id = @Tid And Colid = @Cid)
End


/* Does temporary table exist? */
IF not exists (SELECT * FROM tempdb..sysobjects
    WHERE Name like "##TempT")


Begin
    /* Create temporary table */
    CREATE TABLE ##TempT (Array_Value Float)
End


ELSE
Begin
/* Clear temporary table */
    DELETE FROM ##TempT
End
```

```
/* Get array element value */
INSERT INTO ##TempT (Array_Value)
    EXEC('SELECT ' + @ColName + ' ' +
    'FROM '+ @SrcTable + ' ' +
    'WHERE ' + @SrcColRid + ' = ' +  @Rid + ' ')
SELECT @Val = Array_Value FROM ##TempT

GO
SET QUOTED_IDENTIFIER OFF
GO
SET ANSI_NULLS ON
GO
```

Polynomial_Regression

```
SET QUOTED_IDENTIFIER OFF
GO
SET ANSI_NULLS OFF
GO


CREATE    PROCEDURE [Polynomial_Regression]
@Degree Int = 2,
@SrcTable varchar(50) = 'Table 5_4',
@xCol varchar(50) = 'Hours',
@yCol varchar(50) = 'Hits',
@Alpha Float = 0.05
AS


/***********************************************************/
/*                                                       */
/*                Polynomial_Regression                  */
/*                                                       */
/*   This procedure performs the polynomial regression on */
/* the set of data given in @SrcTable.  In general, the  */
/* polynomial equation is of the form:                   */
/*                                                       */
/*      y = a0 + a1x + a2x^2 + a3x^3 + ... + anx^n        */
/*                                                       */
/* The independent variable x in the equation corresponds */
/* to the @xCol in the @SrcTable, and the dependent      */
/* variable y in the equation corresponds to the @yCol in */
/* the @ScrTable.  The degree of the polynomial is given  */
/* as @Degree.                                           */
```

```
/*   Basically, the procedure generates a set of linear    */
/* equations and then uses the Gaussian elimination to     */
/* find the coefficients a0, a1, a2, ... an of the         */
/* polynomial. Afterwards, the correlation coefficient     */
/* test is performed to test the goodness of fit.          */
/*   The procedure begins by creating two temporary tables */
/* called ##Temp_Xs and ##Temp_Ys.  The table ##Temp_Xs    */
/* contains all the sums needed for the left side of the   */
/* linear equations, and the table ##Temp_Ys contains the  */
/* right side of the linear equations.  A matrix table     */
/* called ##MatrixTbl is created, and each row of the table */
/* corresponds to a linear equation.  The ##MatrixTbl is   */
/* passed to the procedure Gaussian_Elimination which      */
/* solves the linear equations and returns the solution.   */
/*                                                         */
/* INPUTS:                                                 */
/*   Degree - degree of polynomial                         */
/*   SrcTable - the name of the table containing the data  */
/*   xCol - the name for the column denoting x             */
/*   yCol - the name for the column denoting y             */
/*   Alpha - significance level                            */
/*                                                         */
/**********************************************************/


/* Local Variables */
DECLARE @i Int                     /* loop index */
DECLARE @j Int                     /* loop index */
DECLARE @iStart Int                /* Lower loop limit */
DECLARE @iEnd Int                  /* Upper loop limit */
DECLARE @col_i varchar(50)         /* @i as text */

DECLARE @Q varchar(5000)           /* query string */
DECLARE @Q2 varchar(5000)          /* query string */
DECLARE @EQ varchar(5000)          /* the polynomial */

DECLARE @N Int                     /* number of samples */
DECLARE @x Float                   /* a coefficient */
DECLARE @y Float                   /* Right side of Equation */
DECLARE @Col Varchar(50)           /* Column name */
DECLARE @xTid Int                  /* x Table ID number */
DECLARE @yTid Int                  /* y Table ID number */
```

```
DECLARE @v Int                      /* degrees of freedom */
DECLARE @CorrCalc Float             /* Calculated corr. coef. */
DECLARE @CorrTbl  Float             /* table corr. coef. */

DECLARE @Num_Variables Int          /* number of x variables */
DECLARE @SrcMatrix Varchar(50)      /* Gaussian source table */
DECLARE @RstlVector Varchar(50)     /* Gaussian results */

/* CALCULATE TERMS ON LEFT SIDE OF THE EQUATIONS */
/* AND SAVE THE TERMS IN TABLE ##Temp_Xs */

/* Get the number of samples */
SET @Q = 'SELECT Count(' + @Xcol + ') AS NS ' +
    'INTO ##Temp1 ' +
    'FROM [' + @SrcTable + ']'
EXEC(@Q)
Select @N = NS From ##Temp1
DROP TABLE ##Temp1

/* Begin building a query to */
/* get the sum of the x's */
SET @Q = 'SELECT Sum([' + @xCol + '])'
SET @Q = @Q + ' AS Sx1'

/* Append to the query's SELECT clause */
/* term for the sum of the power of x */
SET @i = 2
WHILE @i <= 2 * @Degree
Begin
    SET @col_i = convert(varchar(10), @i)
    SET @Q = @Q + ', '
    SET @Q = @Q + 'Sum(Power([' + @xCol +'], ' + @col_i + '))'
    SET @Q = @Q + ' AS Sx' + @col_i
    SET @i = @i + 1
End

/* Append the INTO and FROM clauses to the query */
SET @Q = @Q + ' INTO ##Temp_Xs '
SET @Q = @Q + 'FROM [' + @SrcTable + ']'

/* Execute the query to obtain */
/* the sums of the powers of x */
EXEC(@Q)
```

```
/* CALCULATE TERMS ON RIGHT SIDE OF EQUATIONS */


/* Begin building a query to */
/* get the sum of the y's */
SET @Q = 'SELECT Sum([' + @yCol + '])'
SET @Q = @Q + ' AS Sy1'


/* Append to the query's SELECT clause */
/* term for the sum of the power of x */
SET @i = 1
WHILE @i <= @Degree
Begin
     SET @col_i = convert(varchar(10), @i)
     SET @Q = @Q + ', '
     SET @Q = @Q + 'Sum(Power([' + @xCol +'], ' + @col_i + ') * [' + @yCol + '])'
     SET @Q = @Q + ' AS Sx' + @col_i + 'y'
     SET @i = @i + 1
End


/* Append the INTO and FROM clauses to query */
SET @Q = @Q + ' INTO ##Temp_Ys '
SET @Q = @Q + 'FROM [' + @SrcTable + ']'


/* Execute the query to obtain the sums */
/* and to popluate the ##Temp_Ys table */
EXEC(@Q)



/* GENERATE THE MATRIX FOR GAUSSIAN ELIMINATION */
/* Establish a work table */
CREATE TABLE ##TempVal (vi Float)


/* Establish the matrix table for */
/* Gaussian elimination procedure */
SET @Q = 'CREATE TABLE ##MatrixTbl ('
SET @i = 1
WHILE @i <= @Degree + 1
Begin
     SET @Col = 'x' + convert(varchar(10), @i)
     SET @Q = @Q + @Col + ' Float, '
     SET @i = @i + 1
End
SET @Q = @Q + 'y Float, Rid Int)'
EXEC(@Q)
```

```
/* Get the table ID number */
SELECT @xTid = (SELECT id FROM tempdb..sysobjects WHERE name = "#TempVal")

/* HANDLE FIRST EQUATION */

/* Set up to build INSERT query for first equation */
SET @Q = 'INSERT INTO ##MatrixTbl VALUES(' + convert(varchar(20), @N)

/* Loop for each term of equation */
SET @i = 1
WHILE @i <= @Degree
Begin

    /* Obtain sum of xi value */
    SET @col_i = 'Sx' + convert(Varchar(10), @i)
    SET @Q2 = ' INSERT INTO ##TempVal (vi) SELECT ' +
        @Col_iI + ' AS vi FROM ##Temp_Xs'
    EXEC(@Q2)
    SELECT @x = vi FROM ##TempVal

    /*Append the sum of xi value */
    SET @Q = @Q + ', ' + convert(varchar(20), @x)

    /* Prepare for next pass through loop */
    DELETE FROM ##TempVal
    SET @i = @i +1

End     /* End of @i loop */

/* Append right side */
SET @Q2 = 'INSERT INTO ##TempVal (vi) SELECT Sy1 AS vi FROM ##Temp_Ys'
EXEC(@Q2)
SELECT @y = vi FROM ##TempVal
SET @Q = @Q + ', ' + convert(varchar(20), @y)
DELETE FROM ##TempVal

/* Append last value and complete the Insert query */
SET @Q = @Q + ',1)'
EXEC(@Q)
```

```
/* HANDLE REMAINING EQUATIONS */

/* Let @j denote the equation and */
/* let @i denote the terms of the equation */
SET @iStart = 1
SET @iEnd = @Degree + 1
SET @j = 1

/* Equation loop */
WHILE @j <= @Degree
Begin

    /* Set up to build INSERT query */
    SET @i = @iStart
    SET @Q = 'INSERT INTO ##MatrixTbl VALUES('

    /* Terms of equation loop */
    WHILE @i <= @iEnd
    Begin

        /*Append the sum of xi value */
        SET @col_i = 'Sx' + convert(Varchar(10), @i)
        SET @Q2 = ' INSERT INTO ##TempVal (vi) SELECT ' +
            @Col_i + ' AS vi FROM ##Temp_Xs'
        EXEC(@Q2)
        SELECT @x = vi FROM ##TempVal
        SET @Q = @Q + convert(varchar(20), @x)

        /* Prepare for next pass through loop */
        IF @i < @iEnd SET @Q = @Q + ', '
        DELETE FROM  ##TempVal
        SET @i = @i + 1

    End     /* End of @i loop */

    /* Append right side */
    SET @col_i = 'Sx' + convert(varchar(10), @j) + 'y'
    SET @Q2 = ' INSERT INTO ##TempVal (vi) SELECT ' +
        @Col_i + ' AS vi FROM ##Temp_Ys'
    EXEC(@Q2)
    SELECT @y = vi FROM ##TempVal
    SET @Q = @Q + ', ' + convert(varchar(20), @y)
    DELETE FROM ##TempVal
```

```
        /* Append equation id number */
        SET @Q = @Q + ', ' + convert(varchar(10), @j+1) + ')'
        EXEC(@Q)

        /* Prepare for next equation */
        SET @iStart = @iStart + 1
        SET @iEnd = @iEnd + 1
        SET @j = @j + 1

End     /* End of @j loop */


/* Do Gaussian Elimination */
SET @Num_Variables = @Degree + 1
SET @SrcMatrix = '##MAtrixTbl'
SET @RstlVector = 'Solution'
EXEC [CH5]..[Gaussian_Elimination] @Num_Variables, @SrcMatrix, @RstlVector


/* BUILD THE COMPARISON TABLE */

/* Establish the comparison table */
CREATE TABLE ##Compare (x Float, y Float, y_eq Float, Diff Float)

/* Fill comparison table with x and y values */
SET @Q = 'INSERT INTO ##Compare SELECT ' +
    @xCol + ', ' + @yCol + ', 0, 0 FROM [' + @SrcTable + ']'
EXEC(@Q)

/* Build the polynomial equation term by term */
SET @i = 1
WHILE @i <= @Degree + 1
Begin

        /* Get equation term coefficient from solution table */
        SET @Q = 'INSERT INTO ##TempVal SELECT xi ' +
            'FROM ' + @RstlVector + ' ' +
            'WHERE i = ' + convert(varchar(10), @i)
        EXEC(@Q)
        SELECT @x = vi FROM ##TempVal

        /* Add term to the equation */
        /* @Q2 is the equation for the query */
        /* @EQ is the equation for printing */
```

```
        IF @i = 1
        Begin
            SET @Q2 = '(' + convert(varchar(20), @x) + ') '
            SET @EQ = convert(varchar(20), @x) + ' '
        End
        ELSE
        Begin
            IF @i = 2
            Begin
                SET @Q2 = @Q2 + '+ (' + convert(varchar(20), @x) + '* [x]) '
                IF @x >= 0 SET @EQ = @EQ + '+ ' + convert(varchar(20), @x) + 'x '
                IF @x < 0 SET @EQ = @EQ + '- ' + convert(varchar(20), abs(@x)) + 'x '
            End
            ELSE
            Begin
                SET @Q2 = @Q2 + '+ (' + convert(varchar(20), @x) + '* ' +
                    'Power([x], ' + convert(varchar(10), @i-1) + ')) '
                IF @x >= 0
                Begin
                SET @EQ = @EQ + '+ ' + convert(varchar(20), @x) + '*x^' +
                    convert(varchar(10), @i-1) + ' '
                End
                ELSE
                Begin
                SET @EQ = @EQ + '- ' + convert(varchar(20), abs(@x)) + '*x^' +
                    convert(varchar(10), @i-1) + ' '
                End
            End
        End


    /* Prepare for next pass through loop */
    DELETE FROM ##TempVal
    SET @i = @i + 1

End     /* End while loop */

/* Update the compare with the equation values */
SET @Q = 'UPDATE ##Compare SET [y_eq] = ' + @Q2
EXEC(@Q)

/* Obtain the differences */
UPDATE ##Compare SET [Diff] = [y] - [y_eq]
```

```
/* Calculate the correlation coefficient */
SELECT @CorrCalc = (SELECT Sqrt(1 - (Var([Diff]) / (Var([y])))))
FROM ##Compare)

/* Calculate degrees of freedom */
SET @v = @N-2

/* Get table correlation value */
SELECT @CorrTbl = (SELECT CorrCoef FROM [Corr_Coef_Table]
    WHERE v = convert(varchar(10), @v)
    AND Alpha = convert(varchar(10), @Alpha)
    AND NumVars = 2)

/* Compare the correlation coefficient */
IF @CorrCalc > @CorrTbl
Begin
    Print 'Since the calculated correlation coefficient value (' +
        convert(varchar(10), @CorrCalc) + ') '
    Print 'exceeds the table correlation coefficient value (' +
        convert(varchar(10), @CorrTbl) + '), '
    Print 'the polynomial is considered to be a good fit '
    Print     'at the ' + convert(varchar(10), @Alpha*100) +
        '% significance level.'
End

ELSE
Begin
    Print 'Since the calculated correlation coefficient value (' +
        convert(varchar(10), @CorrCalc) + ') '
    Print 'is less than the table correlation coefficient value (' +
        convert(varchar(10), @CorrTbl) + '), '
    Print 'the polynomial is NOT a good fit '
    Print 'at the ' + convert(varchar(10), @Alpha*100) +
        '% significance level.'
End

/* Display the polynomial equation */
Print 'THE POLYNOMIAL IS:'
Print '     ' + @EQ

GO
SET QUOTED_IDENTIFIER OFF
GO
SET ANSI_NULLS ON
GO
```

Exponential_Model

```
SET QUOTED_IDENTIFIER OFF
GO
SET ANSI_NULLS OFF
GO

CREATE  PROCEDURE [Exponential_Model]
@SrcTable Varchar(50) = 'Table 5_7',
@xCol Varchar(50) = 'Year',
@yCol Varchar(50) = 'Voters',
@Alpha Float = 0.05
AS


/*************************************************************/
/*                                                         */
/*                  Exponential_Model                      */
/*                                                         */
/*   This procedure performs the exponential regression on */
/* the set of data given in @SrcTable.  In general, the    */
/* exponential equation is of the form:                    */
/*                                                         */
/*       y = ab^x                                          */
/*                                                         */
/* The independent variable x in the equation corresponds  */
/* to the @xCol in the @SrcTable, and the dependent         */
/* variable y in the equation corresponds to the @yCol in  */
/* the @ScrTable.                                           */
/*   Basically, the procedure calculates the values of 'a' */
/* and 'b' of the exponential.  Afterwards, the correlation */
/* coefficient test is performed to test the goodness of fit. */
/*                                                         */
/* INPUTS:                                                 */
/*   SrcTable - the name of the table containing the data  */
/*   xCol - the name for the column denoting x             */
/*   yCol - the name for the column denoting y             */
/*   Alpha - significance level                            */
/*                                                         */
/*************************************************************/
```

```
/* Local Variables */
DECLARE @Q Varchar(5000)          /* query string */
DECLARE @EQ Varchar(5000)         /* exponential model */
DECLARE @n Int                    /* number of samples */
DECLARE @a Float                  /* a in equation */
DECLARE @b Float                  /* b in equation */

DECLARE @CorrCalc Float           /* calculated Corr. Coef. */
DECLARE @CorrTbl Float            /* table Corr. Coef. */
DECLARE @v Int                    /* degrees of freedom */

/* Build query to obtain sums */
SET @Q = 'SELECT Count([' + @xCol + ']) AS N, ' +
    'Sum([' + @xCol + ']) AS Sx, ' +
    'Sum([' + @yCol + ']) AS Sy, ' +
    'Sum([' + @xCol + ']*[' + @xCol + ']) AS Sx2, ' +
    'Sum(Log([' + @yCol + '])) AS Slny, ' +
    'Sum([' + @xCol + ']* Log([' + @yCol + '])) AS Sxlny ' +
    'INTO ##ExpTerms ' +
    'FROM [' + @SrcTable + ']'
EXEC(@Q)


/* Get number of samples */
SELECT @n = (SELECT N FROM ##ExpTerms)

/* Find the value of a and b */
SET @Q = 'SELECT Exp(([Slny]*[Sx2]-[Sx]*[Sxlny]) / ' +
    '([N]*[Sx2]-[Sx]*[Sx])) AS a, ' +
    'Exp(([N]*[Sxlny]-[Sx]*[Slny]) / ' +
    '([N]*[Sx2]-[Sx]*[Sx])) AS b ' +
    'INTO ##a_and_b ' +
    'FROM ##ExpTerms'
EXEC(@Q)


/* BUILD COMPARISON TABLE */

/*Establish the comparison table */
CREATE TABLE ##Compare (x Float, y Float, y_eq Float, Diff Float)

/* Fill comparison table with x and y values */
SET @Q = 'INSERT INTO ##Compare SELECT [' +
    @xCol + '], [' + @yCol + '], 0, 0 FROM [' + @SrcTable + ']'
EXEC(@Q)
```

```
/* Build the exponential equation */
SET @a = (SELECT a FROM ##a_and_b)
SET @b = (SELECT b FROM ##a_and_b)
SET @Q = 'UPDATE ##Compare SET [y_eq] = (' +
    convert(varchar(20), @a) + ') * ' +
    'Power(' + convert(varchar(20), @b) + ', [x])'
EXEC(@Q)


/* Obtain the difference */
UPDATE ##Compare SET [Diff] = [y]-[y_eq]


/* Calculate the correlation coefficient */
SELECT @CorrCalc = (SELECT Sqrt(1.0 - (Var([Diff]) / Var([y])))
    FROM ##Compare)
print 'Calculated Corr = ' + convert(varchar(20), @CorrCalc)


/* Calculate degrees of freedom */
SET @v = @n - 2


/* Get the table correlation coefficient
SELECT @CorrTbl = (SELECT CorrCoef FROM Corr_Coef_Table
    WHERE v = convert(varchar(10), @v)
    AND NumVars = 2
    AND Alpha = convert(varchar(10), @Alpha))
print 'Table Corr = ' + convert(varchar(20), @CorrTbl)


/* Compare the correlation coefficients */
IF @CorrCalc > @CorrTbl
Begin
    Print 'Since the calculated correlation coefficient value (' +
        convert(varchar(10), @CorrCalc) + ') '
    Print 'exceeds the table correlation coefficient value (' +
        convert(varchar(10), @CorrTbl) + '), '
    Print 'the exponential model is considered to be a good fit '
    Print 'at the ' + convert(varchar(10), @Alpha*100) +
        '% significance level.'
End
```

```
ELSE
Begin
    Print 'Since the calculated correlation coefficient value (' +
        convert(varchar(10), @CorrCalc) + ') '
    Print 'is less than the table correlation coefficient value (' +
        convert(varchar(10), @CorrTbl) + '), '
    Print 'the exponential model is NOT a good fit '
    Print 'at the ' + convert(varchar(10), @Alpha*100) +
        '% significance level.'
End

/* Display the exponential model */
Print 'THE EXPONENTIAL MODEL IS:'
SET @EQ = 'y = ' + convert(varchar(20), @a) + ' * ' +
    convert(varchar(20), @b) + '^x'
Print '      ' + @EQ

GO
SET QUOTED_IDENTIFIER OFF
GO
SET ANSI_NULLS ON
GO
```

Multiple_Linear_Regression

```
SET QUOTED_IDENTIFIER OFF
GO
SET ANSI_NULLS OFF
GO

CREATE  PROCEDURE [Multiple_Linear_Regression]
@Num_Variables Int = 2,
@SrcTable Varchar(50) = 'Table 5_11',
@Alpha Float = 0.05
AS
```

```
/**********************************************************/
/*                                                        */
/*            Multiple_Linear_Regression                  */
/*                                                        */
/*    This procedure performs the multiple linear regression */
/*  on the set of data given in @SrcTable.   In general, the */
/* regression equation is of the form:                    */
/*                                                        */
/*       y = a0 + a1x1 + a2x2 + a3x3 + ... + anxn         */
/*                                                        */
/* The independent variables x1, x2, x3, ..., xn in the   */
/* equation correspond to the first n columns of          */
/* @SrcTable, respectively. The n+1 (last) column of      */
/* @SrcTable correspond to y. The number of independent   */
/* variables is denoted by @Num_Variables.                */
/*    Basically, the procedure generates a temporary table */
/* called ##A. Each row in ##A corresponds to a summation */
/* linear equation for Gaussian elimination. After        */
/* creating the ##A matrix, the ##A matrix is passed to the */
/* Gaussian_Elimination procedure which solves the linear */
/* equations and returns the solution.                    */
/*    To test for significance of the regression of the   */
/* dependent variable on all the independent variables by */
/* use of the correlation coefficient, the F test is      */
/* applied.                                                */
/*                                                        */
/* INPUTS:                                                 */
/*   Num_Variables - Number of independent variables      */
/*   SrcTable - the name of the table containing the data  */
/*   Alpha - significance level                            */
/*                                                        */
/**********************************************************/


/* Local Variables */
DECLARE @i Int                          /* loop index */
DECLARE @j Int                          /* loop index */
DECLARE @x float                        /* x sub i value */
DECLARE @xi varchar(50)                 /* x sub i as text */
DECLARE @xj varchar(50)                 /* x sub j as text */
DECLARE @as varchar(50)                 /* Select 'AS' name */

DECLARE @Q varchar(5000)                /* query string */
DECLARE @Q2 varchar(5000)               /* query string */
DECLARE @EQ varchar(5000)               /* the regression equation */
```

```
DECLARE @N Int                              /* number of samples */

DECLARE @Num_Gauss_Vars Int                 /* number variables */
DECLARE @SrcMatrix Varchar(50)              /* Gaussian source table */
DECLARE @RstlVector Varchar(50)             /* Gaussian results */

DECLARE @v1 Int                             /* degrees of freedom */
DECLARE @v2 Int                             /* degrees of freedom */
DECLARE @CorrCalc Float                     /* Calculated corr. coef. */
DECLARE @Fcalc  Float                       /* Calculated F statistic */
DECLARE @Ftbl  Float                        /* Table F statistic */

/* Get the number of samples */
SET @Q = 'SELECT Count(x1) AS NS ' +
    'INTO ##Temp1 ' +
    'FROM [' + @SrcTable + ']'
EXEC(@Q)
SELECT @N = NS FROM ##Temp1
DROP TABLE ##Temp1

/* BUILD THE ##A MATRIX */

/* Establish the ##A matrix */
SET @Q = 'CREATE TABLE ##A ('
SET @i = 1
WHILE @i <= @Num_Variables+1
Begin
    SET @xi = 'x' + convert(varchar(10), @i)
    SET @Q = @Q + @xi + ' Float, '
    SET @i = @i + 1
End
SET @Q = @Q + 'y Float, Rid Int)'
EXEC(@Q)

/* Handle first row of ##A matrix */
SET @Q = 'INSERT INTO ##A SELECT Count([x1]) AS N, '

/* Append remaining x sub j columns */
SET @j = 1
WHILE @j <= @Num_Variables
```

```
Begin
     SET @xj = 'x' + convert(varchar(10), @j)
     SET @as = 'Sx1_' + convert(varchar(10), @j)
     SET @Q= @Q + 'Sum(' + @xj + ') AS ' + @as + ', '
     SET @j = @j + 1
End
SET @Q = @Q + 'Sum(y) AS Sy, 1 AS Rid FROM [' + @SrcTable + ']'
EXEC(@Q)


/* Append remaining rows to ##A */
/* @i indicate the row and @j the column */
SET @i = 1
WHILE @i <= @Num_Variables

Begin
     /* Begin a new row */
     SET @xi = 'x' + convert(varchar(10), @i)
     SET @Q = 'INSERT INTO ##A SELECT Sum(' + @xi + '), '

     /* Append columns values */
     SET @j = 1
     WHILE @j <= @Num_Variables
     Begin
          SET @xj = 'x' + convert(varchar(10), @j)
          SET @as = 'Sx' + convert(varchar(10), @i) + '_' +
               convert(varchar(10), @j)
          SET @Q = @Q + 'Sum([' + @xi +'] * [' + @xj +']) AS ' + @as + ', '
          SET @j = @j + 1
     End      /* end column loop */

     /* Append right side of equation and row id number */
     SET @as = 'Sx' + convert(varchar(10), @i) + '_y'
     SET @Q = @Q + 'Sum([' + @xi + '] * [y]) AS ' + @as +
          ', ' + convert(varchar(10), @i+1) + ' AS Rid ' +
          'FROM [' + @SrcTable + ']'
     EXEC(@Q)
     SET @i = @i + 1
End      /* end row loop */


/* DO GAUSSIAN ELIMINATION */

SET @Num_Gauss_Vars = @Num_Variables + 1
SET @SrcMatrix = '##A'
SET @RstlVector = 'Solution'
EXEC [CH5]..[Gaussian_Elimination] @Num_Gauss_Vars, @SrcMatrix, @RstlVector
```

```
/* BUILD THE COMPARISON TABLE */


/*Establish the comparison table */
SET @Q = 'CREATE TABLE ##Compare ('
SET @i= 1
WHILE @i <= @Num_Variables
Begin
    SET @xi = 'x' + convert(varchar(10), @i)
    SET @Q = @Q + @xi + ' Float, '
    SET @i = @i + 1
End
SET @Q = @Q + 'y Float, y_eq Float, Diff Float)'
EXEC(@Q)


/* Fill comparison table with x and y values */
SET @Q = 'INSERT INTO ##Compare ' +
    'SELECT [' + @SrcTable + '].*, 0, 0 ' +
    'FROM [' + @SrcTable + ']'
EXEC(@Q)


/* Establish a work table */
CREATE TABLE ##TempValue (vi Float)


/* Build the regression equation term by term */
SET @i = 1
WHILE @i <= @Num_Variables + 1
Begin

    /* Get equation coefficients from solution table */
    SET @Q = 'INSERT INTO ##TempValue SELECT xi ' +
        'FROM [' + @RstlVector + '] ' +
        'WHERE i = ' + convert(varchar(10), @i)
    EXEC(@Q)
    SELECT @x = vi FROM ##TempValue

    /* Append term onto the equation */
    /* @Q2 is the equation for the query */
    /* @EQ is the equation for printing */
    IF @i = 1
    Begin
        SET @Q2 = '(' + convert(varchar(20), @x) + ') '
        SET @EQ = convert(varchar(20), @x)
    End
```

```
        ELSE
        Begin
            SET @Q2 = @Q2 + '+(' + convert(varchar(20), @x) +
                '* [x' + convert(varchar(10), @i-1) + ']) '
            IF @x >= 0
            Begin
                /* coefficient is positive, so print '+' term */
                SET @EQ = @EQ + ' + ' + convert(varchar(20), @x) +
                    '*x' + convert(varchar(10), @i-1) + ' '
            End
            ELSE
            Begin
                /* coefficient is negative, so print '-' term */
                SET @EQ = @EQ + ' - ' + convert(varchar(20), ABS(@x)) +
                    '*x' + convert(varchar(10), @i-1) + ' '
            End
        End

    /* Prepare for next pass through loop */
    DELETE FROM ##TempValue
    SET @i = @i + 1

End     /* end while @i loop */

/* Update the compare table with equation values */
SET @Q = 'UPDATE ##Compare SET [y_eq] = ' + @Q2
EXEC(@Q)

/* Calculate the difference between the y's */
UPDATE ##Compare SET [Diff] = [y] - [y_eq]

/* Calculate the correlation coefficient */
SELECT @CorrCalc = (SELECT Sqrt(1.0 - (Var([Diff]) /
    (Var([y]))))) FROM ##Compare)

/* Calculate degrees of freedom for F test */
SET @v1 = @Num_Variables
SET @v2 = @N - @Num_Variables - 1

/* Calculate F value */
SET @Fcalc = (@CorrCalc * @CorrCalc * @v2) /
    (@v1 * (1.0 - @CorrCalc * @CorrCalc))
```

177

```
/* Get table F value */
/* Notice the 'Round' function in the WHERE */
/* clause which ensures an equality match. */
SELECT @Ftbl = (SELECT F FROM F_Table
    WHERE V1 = @v1
    AND V2 = @v2
    AND Round(Alpha,2) = Round(@Alpha,2))


/* Compare the calculated F and the table F values */
IF @Fcalc > @Ftbl
Begin
    PRINT 'Since the calculated F statistic value (' +
        convert(varchar(10), @Fcalc) + ')'
    PRINT 'exceeds the table F statistic value (' +
        convert(varchar(10), @Ftbl) + ')'
    PRINT 'the correlation coefficient value (' +
        convert(varchar(10), @CorrCalc) + ') between '
    PRINT 'the dependent variable (y) and at least one of the '
    PRINT 'independent variables (x1, x2, ..., xn) is considered '
    PRINT 'significant at the ' +
        convert(varchar(10), @Alpha * 100) +
        '% level of significance.'
End


ELSE
Begin
    PRINT 'Since the calculated F statistic value (' +
        convert(varchar(10), @Fcalc) + ')'
    PRINT 'is less than the table F statistic value (' +
        convert(varchar(10), @Ftbl) + ')'
    PRINT 'the correlation coefficient value (' +
        convert(varchar(10), @CorrCalc) + ') between '
    PRINT 'the dependent variable (y) and NONE of the '
    PRINT 'independent variables (x1, x2, ..., xn) is considered '
    PRINT 'significant at the ' +
        convert(varchar(10), @Alpha * 100) +
        '% level of significance.'
End
```

```
/* Print regression equation */
SET @EQ = 'y = ' + @EQ
PRINT ' '
PRINT 'THE MULTIPLE LINEAR REGRESSION EQUATION IS:'
PRINT '      ' + @EQ
PRINT ' '

GO
SET QUOTED_IDENTIFIER OFF
GO
SET ANSI_NULLS ON
GO
```

Procedure Calls

Below are the call statements for calling the procedures
Linear_Regression_2_Variables, Gaussian_Elimination, Polynomial_Regression,
Exponential_Model, and Multiple_Linear_Regression.

```
DECLARE @SrcTable varchar(50)
DECLARE @SrcColX varchar(50)
DECLARE @SrcColY varchar(50)
DECLARE @a float
DECLARE @b float
DECLARE @r float

SET @SrcTable = 'Table 5_1'
SET @SrcColX = 'Payload'
SET @SrcColY = 'Mileage'
SET @a = 0
SET @b = 0
SET @r = 0

EXEC CH5..Linear_Regression_2_Variables
@SrcTable, @SrcColX, @SrcColY, @a OUTPUT , @b OUTPUT , @r OUTPUT


DECLARE @Num_Variables int
DECLARE @SrcMatrix varchar(50)
DECLARE @RstlVector varchar(50)
```

```
SET @Num_Variables = 3
SET @SrcMatrix = 'Matrix'
SET @RstlVector = 'Solution'

EXEC CH5..Gaussian_Elimination
@Num_Variables, @SrcMatrix, @RstlVector


DECLARE @Degree int
DECLARE @SrcTable varchar(50)
DECLARE @xCol varchar(50)
DECLARE @yCol varchar(50)
DECLARE @Alpha float

SET @Degree = 2
SET @SrcTable = 'Table 5_4'
SET @xCol = 'Hours'
SET @yCol = 'Hits'
SET @Alpha = 0.05

EXEC CH5..Polynomial_Regression
@Degree, @SrcTable, @xCol, @yCol, @Alpha


DECLARE @SrcTable varchar(50)
DECLARE @xCol varchar(50)
DECLARE @yCol varchar(50)
DECLARE @Alpha float

SET @SrcTable = 'Table 5_7'
SET @xCol = 'Year'
SET @yCol = 'Voters'
SET @Alpha = 0.05

EXEC CH5..Exponential_Model @SrcTable, @xCol, @yCol, @Alpha


DECLARE @Num_Variables int
DECLARE @SrcTable varchar(50)
DECLARE @Alpha float

SET @Num_Variables = 2
SET @SrcTable = 'Table 5_11'
SET @Alpha = 0.05

EXEC CH5..Multiple_Linear_Regression @Num_Variables, @SrcTable, @Alpha
```

Control Charting

Out of Control

WE LIVE IN A WORLD OF CHANGE. Each day we are faced with decisions, and our choices are generally made by "seat of the pants" judgments. This mindset carries over into business as well. Some managers pride themselves on making snap decisions, whether the outcomes are good or bad. In most cases, there exist underlying phenomena at work that cause the variations or changes that require decisions. These phenomena are usually measurable (as attributes or variables). Statistics is the language of variation, and it may be used to analyze these measurable phenomena so that knowledgeable judgments may be made.

Dr. W. Edwards Deming, the famous statistician who educated the Japanese in statistical process control methods following World War II, believed that the failure to understand variation and act accordingly was the greatest fault of management in the United States. He spent his entire career, up until his death at age 93, trying to educate managers in this simple principle. One of his most important tools in analyzing process variation was the *control chart*. Although Dr. Deming did not conceive the control chart, he presented it to a wider audience than probably any other practicing statistician, over a career spanning more than 65 years. Although the control chart is fairly elementary to set up and use, it is one of the most powerful tools for guiding decision-makers. In this chapter, we examine four of the most common types of control charts.

John's Jewels
The Man Who Advocated Honest Coat Hangers

I first met Dr. W. Edwards Deming in 1989, when he was 89 years old. He invited me to attend one of his 4-day seminars in Minneapolis, which I did. Although age had taken its toll, he put all his effort into his subject, which was trying to save American management from itself. Dr Deming preached some radical ideas, including the abolishment of all types of performance rating systems (even grading in classrooms). He also believed that if certain "monopolies" acted for the good of the public, they should be left alone. One of his favorite lampoons was the telephone system in the United States. In his deep bass voice, he would lament in a sad monotone "Our phone system was once the envy of the world. Now it's a disgrace!" He also possessed a particular disdain for the lack of what he termed "honest coat hangers" in hotel rooms. He abhorred the two-piece, pilfer-proof kind. He even convinced some hotel managers to change. Dr. Deming left an indelible mark on the world of management, statistics, and quality improvement. The great man died in 1993.

Common and Special Causes of Variation

Every evening on the news we hear the stock market report. We witness the reactions of all those folks who try to buy when the market is down and sell when it's up. They seem to react to the small day-to-day variations as if they were wide swings. If we looked at these daily ups and downs over a period of several months, we could plot the changes as points on a graph, connecting lines for visual clarity. This is called a "run" chart. We might observe a chart similar to the one shown in Figure 6-1, assuming there were no major outside influences during the period (such as a war, an economic collapse, or whatever).

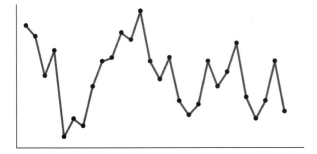

Figure 6-1. Run chart of daily stock index values

We could associate numbers with each of the points on the graph in Figure 6-1. These might be the Standard & Poor's (S & P's) Index values at the close of each day. We could calculate their mean and standard deviation if we wanted to. The mean, we know, is a single value, and is a measure of central tendency of the index values. Therefore, it could be indicated as a line on the graph, as shown in Figure 6-2. Obviously, since this line represents the mean of the values, we'd expect about half to fall above the line and about half below the line.

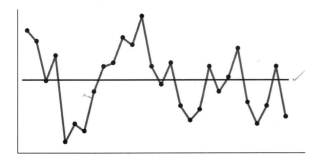

Figure 6-2. Run chart of stock index values showing mean or central line

If the daily closing S & P Index values for the year (say about 250) are used to set up a histogram, the shape of that chart depends on how stable the market was for the period. However, let's suppose instead of looking at the daily values, we calculate a weekly *average* closing value. In general, five values would contribute to each of these means (although some weeks the market would be open only four days). If we developed a histogram for these 52 weekly means, its shape would almost always be normal, or bell-shaped, whether or not the 250 or so daily numbers are normally distributed. This phenomenon is called the *Central Limit Theorem*. In addition, the standard deviation of the 52 means is less than that of all the 250-plus individual values. This is because the means of the weekly index numbers tend to smooth out the wilder ups and downs that happen daily, so there is consequently less variation among the means than among the individual values. In fact, it has been shown that the relationship between the standard deviation of the individual values (say s), and the standard deviation of the means (say $s_{\bar{x}}$), is given by the following expression:

$$s_{\bar{x}} \cong \frac{s}{\sqrt{n}}$$

where n is the size of the sample data that led to the calculation of each weekly mean value (in this case 5, for 5 days per week).

If the weekly means of the S & P's data follow a normal distribution, we would expect 99.74% of the mean values to fall within 3 standard deviation units of the overall mean (which is the mean of the 52 means). This is evident from the normal distribution table (see Appendix B). If you look up a z variate value of +3.0 in the table, the corresponding area under the normal curve up to this value is 0.9987, leaving 0.0013 in the right tail of the distribution. Similarly, at –3.0 the area is also 0.0013 in the left tail. The sum of these two areas is equal to 0.0026, leaving 0.9974 or 99.74% of the area under the normal curve between ±3 standard deviations (commonly denoted as ±3σ, as shown in Figure 6-3).

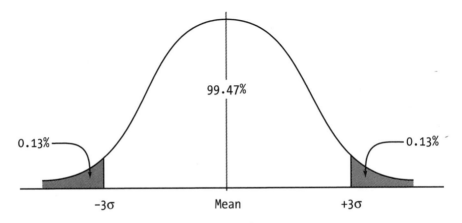

Figure 6-3. Percent of areas under the normal curve between and beyond 3 standard deviation limits

If the process generating the S & P's means is stable, we could expect a chance of only 0.26% that any mean value observed falls outside the band of 3 standard deviations. But what does "stable" mean? This means that the occurrence of mean values is completely random. In other words, we cannot predict what the next mean will be, knowing what the current one is. An example is the roll of a fair die. We know the result will be a number between one and six, but we can't predict the next roll knowing what the current roll produced. In the diagnostic tree of Figure 1-3, the third branch is concerned with changes in data over time. If the stability of the process is of interest to the data miner, this chapter provides a methodology for assessing stability. Since the control chart is the tool, we need to study its construction and application.

Now let's take the normal distribution of S & P's weekly means and turn it sideways, as shown in Figure 6-4. The points we'll be plotting are different than those on the run chart in Figure 6-1, because now we're concerned with weekly averages rather than daily values. We'll draw the line representing the overall mean (the mean of the weekly means) through the high point of the normal curve, as shown. Let's call this the *central line (CL)*, as is customary. Next we draw the two lines through the tails of the normal curve representing the 3 standard deviation values. We'll show these as dashed lines on the figure. They are called the *control limits*. The top one is the *upper control limit (UCL)* and the bottom one is the *lower control limit (LCL)*.

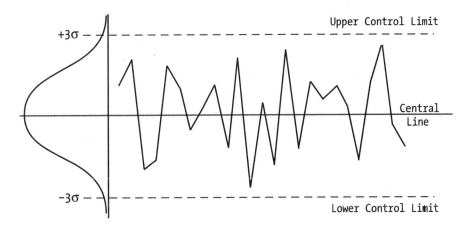

Figure 6-4. Control chart for weekly stock index averages

The chart in Figure 6-4 is called a *control chart*, because it is an indicator of whether or not the process that generates the data points plotted on the chart is "in control" (operating randomly) or not. If the process is stable or in control, as is the case in Figure 6-4, it is subject to only common causes of variation. It is performing at its best capability, and only an outside force (in this case, war, economic collapse, a run on Wall Street, or a pessimistic remark by the President) can disrupt it. There is in this situation no way of knowing whether to buy or sell based on the random ups and downs of the market. But most people ignore this. When they see a jump below the central line, they think the market is on a downturn, so they panic and sell. When the point jumps upward, they become optimistic that it will continue to climb, so they buy. So in effect they themselves can even become disruptive forces to the market. By exhibiting this behavior, we say that the investors are continually making Type I errors (see Chapter 4) without knowing it. These are the most common errors in judgment committed day to day. They occur when we act as if a stable process is actually unstable.

Off to the Milky Way

Dr. W. Edwards Deming (1900–1993) was a pioneer in the use of statistical techniques to help management do a better job. The control chart was one of his favorite tools, and he was always quick to give due credit to his friend Dr. Walter Shewhart of the Bell Telephone Laboratories as the originator of this simple yet powerful device (in 1924). Dr. Deming achieved fame by presenting a series of lectures to Japanese managers and engineers following World War II, in which he outlined quality improvement through statistical methodology and common sense. Out of these lectures came his famous "14 Points for Management." In the 1980s, Dr. Deming's philosophy became widely studied, and he launched a series of seminars throughout the United States to attempt to save management from itself. A contemporary of Dr. Deming is Dr. Joseph Juran (born in 1904), who has also published extensively in the field of quality assurance. Dr. Juran had a somewhat different view and delivery than Dr. Deming. Although friends, they occasionally exchanged barbs when they appeared together on the lecture circuit. Dr. Deming enjoyed presenting, in his characteristic deep bass voice, a demonstration of the "funnel experiment." His point was to show that tampering with a stable system can push it into instability, with results going "off to the Milky Way," as he put it. Dr. Juran, it is said, once followed this admonition to management as he stepped to the lectern and pronounced, "Now it's time for us to get back down to Earth!"

On the other hand, unstable processes are also widespread. Let's now look at the S & P's weekly closing index averages for the entire year of 2000 (Table 6-1). This was an unusual year for the stock market. The data was obtained from www.Barchart.com on the Internet.

SQL/Query

Suppose the S & P's daily closing index values are in a database table called S_and_P, which contains two columns: Date and Closing. Query 6_1 and Query 6_2 show how to obtain the weekly closing index average from the daily closing index values. Query 6_1 assigns a week ID number to each of the 251 days. Those days that are in the same week are assigned the same weekly ID number. We use the Format function with the date format code of "ww" to get the week number for the date. Query 6_2 uses the results of Query 6_1 to calculate the weekly averages and ranges by grouping the daily closing index values with respect to the weekly ID number. Notice in Query 6_2 how we used the Max function to find the last date of the week. We had to qualify Date with respect to Query 6_1 so that it would not be ambiguous (or form a circular reference) with respect to the AS clause which also uses the name Date. The results of Query 6_2 are the values shown in Table 6-1.

Query 6_1:

```
SELECT Date, CInt(Format([Date],"ww")) AS WeekID, Close
FROM S_and_P;
```

Query 6_2:

```
SELECT Max([Query 6_1].Date) AS [Date],
Avg(Close) AS [Weekly Average],
Max(Close) - Min(Close) AS Range
FROM [Query 6_1]
GROUP BY WeekID
ORDER BY WeekID;
```

> **NOTE** *The database associated with this chapter on the Apress Web site contains extended query names. For example, Query 6_1 is called "Query 6_1 Adds weekID column" in the Apress Web database file.*

Table 6-1. Standard & Poor's Weekly Closing Index Averages for Year 2000

DATE	WEEKLY AVERAGE	RANGE
01/07/00	1411.613	42.05
01/14/00	1448.648	32.90
01/21/00	1449.493	14.54
01/28/00	1394.950	49.87
02/04/00	1412.440	30.51
02/11/00	1416.308	54.63
02/18/00	1382.800	55.96
02/25/00	1349.913	27.33
03/03/00	1376.918	61.12
03/10/00	1382.072	46.07
03/17/00	1411.570	105.32
03/24/00	1501.190	70.83
03/31/00	1505.322	35.94
04/07/00	1501.152	28.98
04/14/00	1454.008	147.15
04/20/00	1426.265	40.17
04/28/00	1457.068	47.28
05/05/00	1434.330	58.87
05/12/00	1409.626	41.12
05/19/00	1442.072	59.09
05/26/00	1386.634	26.86
06/02/00	1442.280	56.66
06/09/00	1463.090	14.41
06/16/00	1465.808	32.60
06/23/00	1466.948	44.52
06/30/00	1451.522	12.86
07/07/00	1462.780	32.67

Table 6-1. Standard & Poor's Weekly Closing Index Averages for Year 2000 (Continued)

DATE	WEEKLY AVERAGE	RANGE
07/14/00	1491.048	34.36
07/21/00	1492.370	30.30
07/28/00	1452.138	54.58
08/04/00	1444.624	32.10
08/11/00	1473.418	22.56
08/18/00	1488.726	16.22
08/25/00	1503.670	10.18
09/01/00	1512.994	18.18
09/08/00	1499.085	14.83
09/15/00	1480.286	22.04
09/22/00	1450.704	15.39
09/29/00	1437.522	31.72
10/06/00	1428.456	27.29
10/13/00	1371.302	72.25
10/20/00	1370.482	54.80
10/27/00	1380.566	33.69
11/03/00	1420.858	30.74
11/10/00	1407.892	66.21
11/17/00	1372.858	38.78
11/24/00	1338.525	24.99
12/01/00	1331.434	34.02
12/08/00	1353.282	51.57
12/15/00	1352.890	68.05
12/22/00	1294.778	58.00
12/29/00	1324.653	19.03

There are 52 values in all. Since some of the weekly mean values are based on 4 days in the week rather than 5 (when the markets were closed for a holiday), it's not exactly correct to calculate the overall mean by adding the 52 weekly means and dividing by 52. Rather, we should, to be more accurate, find the overall mean by adding all 251 daily values and dividing by 251. The resulting mean is equal to 1,427.10. The standard deviation of the 251 daily values is 56.03. This standard deviation may be used to estimate the standard deviation of the 52 weekly means ($s_{\bar{x}}$), as follows:

$$s_{\bar{x}} \cong \frac{56.03}{\sqrt{5}} = 25.06$$

SQL/Query

We can use Query 6_3 to calculate the mean, standard deviation, and the estimated standard deviation of the 52 weekly means.

Query 6_3:

```
SELECT Avg(Close) AS Mean,
StDev(Close) AS StdDev,
StDev(Close) / Sqr(5) AS [Weekly StdDev]
FROM S_and_P;
```

Perhaps it would be a bit more accurate to find an average sample size instead of using 5, because 9 of the 52 weeks were only 4 days long. This average sample size (actually a weighted mean) would be

$$\left[(9)(4)+(43)(5)\right]/52 \text{ or just } 251/52 = 4.83 \text{ days/week}$$

SQL/Query

We can obtain the average sample size by executing Queries 6_4, 6_5, and 6_6. Query 6_4 determines how many days are in each week by going back to Query 6_1 and using the weekly ID number. Query 6_5 counts the number of weeks with the same number of days. Query 6_6 calculates the average sample size.

Query 6_4:

```
SELECT WeekID, Count(WeekID) AS [Days Per Week]
FROM [Query 6_1]
GROUP BY WeekID;
```

Query 6_5:

```
SELECT [Days Per Week], Count([Days Per Week]) AS [Num of Weeks]
FROM [Query 6_4]
GROUP BY [Days Per Week];
```

Query 6_6:

```
SELECT Sum([Days Per Week] * [Num of Weeks]) / 52 AS [Avg Sample Size]
FROM [Query 6_5];
```

Using this average sample size, we have

$$s_{\bar{x}} \cong \frac{56.03}{\sqrt{4.83}} = 25.49$$

SQL/Query

Query 6_7 calculates this by using the results of Queries 6_3 and 6_6. Notice that Query 6_7 has no join condition since the result of both Query 6_3 and Query 6_6 is a single row. Also, notice that Query 6_7 calculates the value for three standard deviations.

Query 6_7:

```
SELECT [StdDev] / Sqr([Avg Sample Size]) AS [More Accurate SD],
3 * [StdDev] / Sqr([Avg Sample Size]) AS [3SD]
FROM [Query 6_3], [Query 6_6];
```

Now we'll multiply $s_{\bar{x}}$ by 3 to establish control limits for a control chart for the weekly means. The result is $3 \times 25.49 = 76.47$. If we add this to the overall mean (1,427.10), we obtain the upper control limit, as follows:

UCL = 1,427.10 + 76.47 = 1,503.57

If we subtract the result from the mean, we have the lower control limit:

LCL = 1,427.10 − 76.47 = 1,350.63

SQL/Query

We can obtain the central line and control limits by executing Query 6_8.

Query 6_8:

```
SELECT [Mean] + [3SD] AS UCL,
Mean as CL,
[Mean] - [3SD] AS LCL
FROM [Query 6_7], [Query 6_3];
```

We can plot the mean as the central line on a control chart, and show the control limits above and below this line. Then, using the scale on the left-hand side of the chart, we can plot the weekly averages from Table 6-1. The result is shown in Figure 6-5.

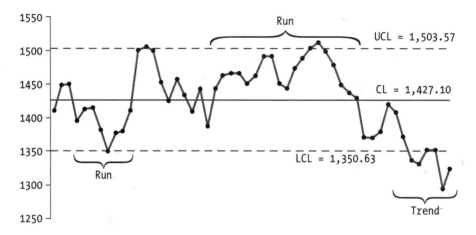

Figure 6-5. Control chart for weekly S & P's closing index averages

SQL/Query

The data for the graph in Figure 6-5 can be obtained by executing Query 6_9. No join condition is needed since Query 6_8 produces a single row.

Query 6_9:

```
SELECT [Weekly Average], UCL, CL, LCL
FROM [Query 6_2], [Query 6_8];
```

Now we have cause for concern. There are eight points outside our control limits. There are also three sequences (runs) of several points either above or below the central line. In addition, the third run also seems to contain a downward trend of points toward the end of the year. There are so many points in a row below the original central line in this period that we would suspect an entire shift in the process average. All this indicates that the process is quite unstable, or nonrandom in nature. Now we can be more confident in predicting what the next weekly average will be if we know the previous seven or so were declining. This is an example of the influence of *special causes* (or "assignable" causes) of variation. Some outside source (or sources) has caused the market to decline. Maybe it was the uncertainty over the outcome of the 2000 Presidential election. In addition, special causes have contributed to the earlier points outside the control limits, and the runs of points. We could probably search the newspapers during those periods to have an idea of what was happening to cause the instability in the market.

In this case, we do have reason to react. We might want to sell stock if we think the market will continue to decline, as it seems to toward year's end. Or, we might want to buy stock if we think the prices are lower. If we do nothing, we make a Type II error (see Chapter 4), because we act as if the process is stable when it really isn't.

John's Jewels
Yesterday's Data

During my teaching and consulting years, I have been privileged to visit a wide array of manufacturing and service facilities. One of my earliest training sessions was in a carpet mill in Dalton, Georgia, where I was asked for help in control charting and process improvement techniques. The mill, as was typical in those days, had data so voluminous as to be worthless to management. I remember sitting across the desk from the man who had hired me, talking about the training materials and outline. I looked on the corner of his desk and noticed a 3-inch thick printout from the big old mainframe computer in the plant. I said stupidly, "Maybe we can use that. It looks like last month's data." He answered rather soberly, "That was *yesterday's* data!" Then I knew his dilemma. Fortunately the control chart provided a means of condensing a great deal of that massive amount of data into a simple little visual tool that could actually tell him something useful about his process.

Dissecting the Control Chart

The statistical control chart is a graphical tool for assessing process variation and indicating whether or not special causes are present. Most control charts plot sample statistics (like means and ranges) rather than individual measurements. The stock market example plots weekly means of the Standard & Poor's closing index values. If the underlying data are measured quantities, like the stock values, there are variables control charts available to analyze the process generating the data. Likewise, attributes control charts are also available for data representing "yes/no" type decisions.

Every standard control chart has a central line, representing the mean or average of all sample measurements. The upper control limit is commonly set a distance of 3 standard deviations above the central line, while the lower control limit is 3 standard deviations below the central line. These control limits are, on almost all control charts, straight lines, but there are exceptions. By convention, the central line is indicated by a solid line, and the control limits are indicated by dashed lines. The points are usually plotted left to right in order of observation (by time), although this is not always the case (they may indicate occurrences in a spatial context rather than a time context). These points typically represent sample averages, sample ranges, sample "fraction nonconforming," or sample "number of nonconformities." These types of charts are illustrated later. The vertical line on the left of the chart is a scale corresponding to the sample statistics plotted.

There are three common indicators on the chart of "lack of statistical control," or special causes present in the process. We saw all three in the stock market data control chart example. These are

1. Points above the upper and/or below the lower control limits.

2. At least seven points in a row either above or below the central line.

3. Trends, or gradual movements of points upward or downward.

In addition, there are other "weird" things that sometimes show up on a control chart and reveal a lack of control. Cycles, or periodic swings up, then down, then up again, may indicate special causes of variation. Also, a similar but wilder seesaw effect with very few points near the central line is usually an indication of instability.

Special causes of variation show themselves in many different ways. They may be changes in political conditions, environmental conditions, or economic conditions. They may be changes in materials, workers, work methods, machine settings, maintenance practices, wear and tear, or training. It is important to use the control chart as a means of determining whether these conditions are present in a process or system, then attempting to identify and correct any problems the special causes might reveal. However, we need to keep in mind that the presence of special causes may not always be detrimental to the operation of the process. There are "good" as well as "bad" special causes of variation. Good causes are those that shift the process average to a more desirable level, through enhanced profits, working conditions, productivity, and so forth. They need to be incorporated permanently into the process or system. The bad ones should be eliminated wherever possible, and this is usually a responsibility of the management of the organization.

Control Charts for Sample Range and Mean Values

The most common control charts in use for analyzing processes that generate variables data are the range and mean charts. In quality control lingo, these are called the R chart (sample range chart) and the $\bar{X}$ chart (sample mean chart). When we investigate processes for which measured data can be obtained, it is customary to set up the R chart first. This is because the control limits for the $\bar{X}$ chart depend on the stability of the sample ranges. Therefore, if the ranges are out of control, this tends to inhibit correct analysis of the chart for means. It's the practice, then, to try to bring the process under control according to its sample ranges first, then set up the $\bar{X}$ chart. The R and $\bar{X}$ charts are so common that a special table has been developed for aid in calculating the control limits for both charts, and is given in Appendix B. This table contains factors that may be applied to the average range and mean of the sample values to find the upper and lower control limits. It is advantageous because it alleviates the necessity of calculating the standard deviation values, as was the case in the stock market example.

The following question often arises in using these charts: What's an appropriate sample size and frequency of sampling? In many cases, this is answered by the circumstances surrounding the data generation. In the stock market data, it was natural to use a sample size of 5 because the typical week was 5 days in duration. Also, in this situation the data were readily available to us every week. However, in other cases, there may be so much data generated by the process that we have to settle for random sampling with a frequency restricted by the personnel and economics of the situation. Several complicated statistical rules for sample sizes have been proposed, but a good rule of thumb is the following: Use a sample size of 4 or 5 observations whenever possible, and sample as often as is practically possible, using a random selection technique. Generally speaking, try to obtain at least 25 samples before the first control chart is set up. Once the initial set of sample data becomes available for calculating a central line and control limits, these may be extended into the future sampling period when the process is brought under control. Then points are plotted as soon as the sample data are available. Now the chart becomes a real-time tool for assessing process variation, so frequent samples prove more likely to detect process shifts or instability, and quicker action is possible to avoid out-of-control situations. The following example is designed to make all this clearer.

Example: Training Course

In this example, the data miner is working for a company that provides a user training course for a new software package it manufactures and markets. Normally, there is one person in the buyer's firm who will be using the package, so the training is one-on-one and tailored to the firm's requirements and the user's experience. Sometimes the user requires only a few hours to learn the product, but ordinarily the training time seems to be closer to 20 hours. The manufacturer would like to develop a more generic training program, with appropriate instruction materials, so that its trainers can standardize their preparation and procedures. The first step is to pull some records from the files and randomly select some customers who bought the product. From this database, the times for instruction may be extracted and analyzed. Over the past 15 weeks, 5 customers per week are selected randomly and their instruction times (in hours) recorded. For each week's sample data, the range (difference in longest and shortest times) and mean are calculated. The sample data and these results appear in Table 6-2.

Table 6-2. Software Training Times (in Hours) for Randomly Selected Customers Over a 15-Week Period

SAMPLE NUMBER	ITEM NUMBER 1	2	3	4	5	SAMPLE RANGE	SAMPLE MEAN
1	21	29	26	29	29	8	26.8
2	29	24	23	20	23	9	23.8
3	20	29	20	23	22	9	22.8
4	21	21	20	22	21	2	21.0
5	17	20	19	20	16	4	18.4
6	13	22	20	20	22	9	19.4
7	17	19	19	20	18	3	18.6
8	20	18	17	17	18	3	18.0
9	22	20	19	17	19	5	19.4
10	20	18	19	18	18	2	18.6
11	17	15	25	20	25	10	20.4
12	4	22	20	16	19	18	16.2
13	14	17	20	20	12	8	16.6
14	17	18	19	19	22	5	19.0
15	19	19	19	18	19	1	18.8

SQL/Query

Suppose the initial sampling data are collected and saved in the table called Initial Samples with three columns: Sample Number, Item Number, Sample Value. It is important that the Sample Number and Item Number are consecutive integers beginning with 1 because we shall make use of this in subsequent queries. We use Query 6_10 to find the range and mean of each sample as shown in Table 6-2.

Query 6_10:

```
SELECT [Sample Number], Max([Sample Value]) - Min([Sample Value]) AS Range,
Avg([Sample Value]) AS [Sample Mean]
FROM [Initial Samples]
GROUP BY [Sample Number]
ORDER BY [Sample Number];
```

A reduction in variability is desirable to bring the process generating the training times under control, so better planning is possible. Toward this end, a control chart plotting sample ranges is set up first. The table in Appendix B, Table B-7, *Factors for Determining Control Limits for R and $\bar{X}$ Charts*, will be used to help establish the control limits for both the R and the $\bar{X}$ charts. For convenience, a portion of this table is reproduced in Table 6-3.

Table 6-3. Portion of Table from Appendix B for Use in Calculating Control Limits for $\bar{X}$ and R charts

NUMBER OF OBSERVATIONS IN SAMPLE N	LOWER CONTROL LIMIT FACTOR FOR R CHART D_3	UPPER CONTROL LIMIT FACTOR FOR R CHART D_4	FACTOR FOR BOTH CONTROL LIMITS FOR $\bar{X}$ CHART A_2
2	0	3.27	1.88
3	0	2.57	1.02
4	0	2.28	0.73
5	0	2.11	0.58

To use this table, we first need to calculate the average range (denoted $\bar{R}$) for the 15 samples. The sum of the ranges is 96, so we divide 96/15 and obtain $\bar{R} = 6.4$ hours. This will be our central line for the R chart. The upper and lower control limits (denoted UCL and LCL) are calculated by looking up the appropriate factors in the table (see Table 6-3) for a sample size of 5. (Cautionary note: Don't confuse the sample size with the total number of samples when looking up these factors.) The calculations are

$$\text{LCL} = (D_3)(\bar{R}) = (0)(6.4) = 0 \text{ hours}$$
$$\text{UCL} = (D_4)(\bar{R}) = (2.11)(6.4) = 13.5 \text{ hours}$$

SQL/Query

To accomplish the calculations for $\bar{R}$ (CL), LCL, and UCL in SQL, we need to execute a sequence of queries. First we need to find the sample size by executing Query 6_11. Next we determine $\bar{R}$ by running Query 6_12. Performing Query 6_13 give us the values of LCL, CL, and LCL. The table Control Limits used in Query 6_13 is Table B-7, *Factors for Determining Control Limits for R and $\bar{X}$ Charts* in Appendix B.

Query 6_11:

```
SELECT Max([Item Number]) AS [Sample Size]
FROM [Initial Samples];
```

Query 6_12:

```
SELECT Avg(Range) AS Rbar
FROM [Query 6_10];
```

Query 6_13:

```
SELECT [Rbar] * [LCL Factor for R Chart] AS LCL,
[Rbar] AS CL,
[Rbar] * [UCL Factor for R Chart] AS UCL
FROM [Query 6_11], [Query 6_12], [Control Limits]
WHERE [Sample Size] = [Number of Observations];
```

Now we have all the information we need to draw the *R* chart. The scale on the left of the chart represents hours, ranging from 0 to (at least) the largest range value we need to accommodate (in this case 18). Then the central line (CL) and control limits are indicated, and the sample ranges plotted on the chart in the order of sampling. The chart is shown in Figure 6-6.

SQL/Query

Query 6_14 provides the data for making the Chart shown in Figure 6-6. Again no join condition is needed since Query 6_13 produces a single row.

Query 6_14:

```
SELECT [Sample Number], Range, LCL, CL, UCL
FROM [Query 6_10], [Query 6_13];
```

Note from the chart that the circled point corresponding to Sample 12 is above the upper control limit, indicating that the range value for that sample is much higher than would be expected in random variation. Sample 12 should therefore be investigated for

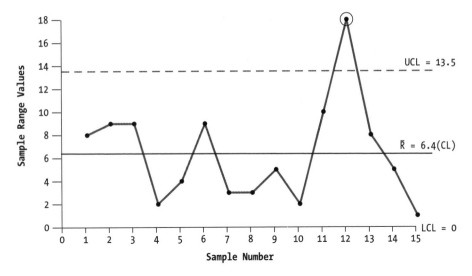

Figure 6-6. The R chart for the training data

special causes. If we return to the original data in Table 6-2 and look at Sample (or week) 12, we see that there is a very low training time of 4 hours for that sample. This low value caused the range to be higher than normal. On investigating the files, the data miner finds that the customer/trainee in that situation had already learned the software at another company, so her training time was minimal. Since this was an extraordinary occurrence and not expected to recur, it was decided to eliminate Sample 12 from the database and calculate new central line and control limits for a revised *R* chart. (It should be pointed out that this may only be done if the special cause can be identified and eliminated, or if advantageous to the process, incorporated.) The new central line is found by taking the old range sum of 96 and subtracting Sample 12's range of 18, then dividing the result by 14, as follows:

New CL $= (96-18)/14 = 78/14 = 5.57$ hours

The new lower control limit will remain at 0, but the upper control limit changes to

UCL $= (2.11)(5.57) = 11.75$ hours

SQL/Query

We can eliminate sample 12 by executing Query 6_15. We use the NOT IN condition in the WHERE clause of the query to eliminate sample 12. We did this so that if we happen to have a second sample to eliminate we could simply include it in the list by placing a comma after the 12 and by following the comma with the second sample number.

Query 6_15:

```
SELECT [Sample Number],
Max([Sample Value]) - Min([Sample Value]) AS Range,
Avg([Sample Value]) AS [Sample Mean]
FROM [Initial Samples]
WHERE [Sample Number] NOT IN (12)
GROUP BY [Sample Number]
ORDER BY [Sample Number];
```

If we rescan the 14 sample ranges in the original data, we see that all are within these new control limits, and there are no runs of points. Therefore, we conclude the process is under control according to its ranges, and proceed to the construction of the $\bar{X}$ chart, remembering that Sample 12 is no longer included.

The central line for the $\bar{X}$ chart (sometimes denoted $\bar{\bar{X}}$) is found by adding the 14 sample means and dividing by 14. We can do this because the sample sizes are all the same, whereas in the stock market example earlier, the sizes changed from 5 to 4 on a few samples, so we had to fudge a little. The sum of the means is 281.6, so the central line is

$$CL = 281.6/14 = 20.11 \, \text{hours}$$

Referring back to Table 6-3, we see that the control limit factor for the $\bar{X}$ chart is equal to 0.58 for a sample size of 5. We multiply this by the average range $\bar{R}$ and use the result to estimate the upper and lower control limits as follows:

$$LCL = CL - (A_2)(\bar{R}) = 20.11 - (0.58)(5.57) = 16.88 \, \text{hours}$$
$$UCL = CL + (A_2)(\bar{R}) = 20.11 + (0.58)(5.57) = 23.34 \, \text{hours}$$

The $\bar{X}$ chart is shown in Figure 6-7. Notice that the central line and control limit values have been rounded to one decimal for convenience.

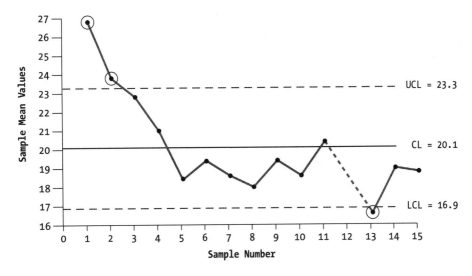

Figure 6-7. $\bar{X}$ chart for training hours (sample 12 omitted)

SQL/Query

Using Query 6_15, we can calculate the new $\bar{R}$ and the central line, CL, as shown in Query 6_16. Next, we calculate LCL and UCL by doing Query 6_17 that matches the Sample Size found previously by Query 6_11 with the Number of Observations in the Control Limits table. Query 6_18 gives the data that is needed for making the chart shown in Figure 6-7.

Query 6_16:

```
SELECT Avg(Range) AS Rbar, Avg([Sample Mean]) AS CL
FROM [Query 6_15];
```

Query 6_17:

```
SELECT [CL] - [Factor for Xbar Chart] * [Rbar] AS LCL,
[CL] + [Factor for Xbar Chart] * [Rbar] AS UCL
FROM [Query 6_11], [Query 6_16], [Control Limits]
WHERE [Sample Size] = [Number of Observations];
```

Query 6_18:

```
SELECT [Sample Number], [Sample Mean], UCL, CL, LCL
FROM [Query 6_15], [Query 6_16], [Query 6_17];
```

It's evident from the chart that the means of Samples 1 and 2 are above the upper control limit, and that of Sample 13 is below the lower control limit (these points are circled on the chart). Also, 9 of the last 10 means are below the central line, indicating a run or possible shift in the process average. The process is out of control and again needs to be investigated for special causes of variation.

The data miner dug deeper this time to ferret out the reasons for the instability. It was found that the software had been modified slightly after the second sample, requiring a revision in the training practice and a subsequent reduction in training time. No special cause could be found for Sample 13. It is decided to set up a new $\bar{X}$ chart for Samples 3 through 15 (less of course Sample 12 from before), since no other change in software was found during the current sampling period. The new central line will be the old sum of means (281.6) less the means for Samples 1 and 2, divided by 12: the number of remaining samples in the database. The result is shown as follows:

$$\text{New CL} = (281.6 - 26.8 - 23.8)/12 = 231.0/12 = 19.25 \text{ hours}$$

To find the new control limits, a new average range has to be calculated, because now Samples 1 and 2 are gone. The previous sum of sample ranges was 78 (after Sample 12 was eliminated), so we subtract from that the ranges of Samples 1 and 2, as follows:

$$\text{New } \bar{R} = (78 - 8 - 9)/12 = 61/12 = 5.08 \text{ hours}$$

This is now used with the control limit formulas for the $\bar{X}$ chart to obtain the new limits:

$$\text{New LCL} = 19.25 - (0.58)(5.08) = 16.30 \text{ hours}$$
$$\text{New UCL} = 19.25 + (0.58)(5.08) = 22.20 \text{ hours}$$

The new $\bar{X}$ chart is shown in Figure 6-8.

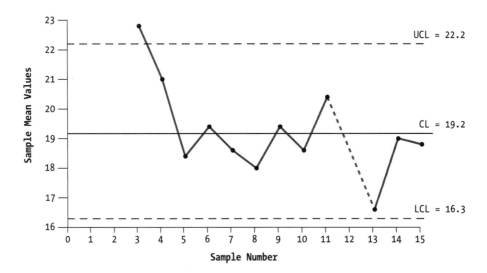

Figure 6-8. New $\bar{X}$ chart with Samples 1, 2, and 12 eliminated

SQL/Query

Query 6_19 eliminates samples 1, 2 and 12 and calculates the ranges and mean of the remaining samples. New values for $\bar{R}$ and CL ($\bar{\bar{X}}$) are calculated by Query 6_20. Query 6_21 determines the new values for LCL and UCL by looking up the control limit factor for the $\bar{X}$ chart in the Control Limits table. The data for the chart shown in Figure 6-8 is obtained by Query 6_22.

Query 6_19:

```
SELECT [Sample Number],
Max([Sample Value]) - Min([Sample Value]) AS Range,
Avg([Sample Value]) AS [Sample Mean]
FROM [Initial Samples]
WHERE [Sample Number] NOT IN (1, 2, 12)
GROUP BY [Sample Number]
ORDER BY [Sample Number];
```

Query 6_20:

```
SELECT Avg(Range) AS Rbar, Avg([Sample Mean]) AS CL
FROM [Query 6_19];
```

Query 6_21:

```
SELECT [CL] - [Factor for Xbar Chart] * [Rbar] AS LCL,
[CL] + [Factor for Xbar Chart] * [Rbar] AS UCL
FROM [Query 6_11], [Query 6_20], [Control Limits]
WHERE [Sample Size] = [Number of Observations];
```

Query 6_22:

```
SELECT [Sample Number], [Sample Mean], UCL, CL, LCL
FROM [Query 6_19], [Query 6_20], [Query 6_21];
```

Although it's evident that the mean of Sample 3 is slightly above the upper control limit, the process after that point is stabilizing. Therefore, it was decided to extend the present central line and control limit values into the next sampling period, so that sample means and ranges could be plotted as soon as they were known. Ten more samples were taken over the next 10 weeks, with the results shown in Table 6-4. Again the range and mean values are calculated for each sample and shown in the table.

Table 6-4. Training Times (in Hours) for the New Sampling Period

SAMPLE NUMBER	ITEM NUMBER					SAMPLE RANGE	SAMPLE MEAN
	1	2	3	4	5		
16	19	20	18	18	20	2	19.0
17	18	22	19	19	20	4	19.6
18	20	19	20	20	20	1	19.8
19	17	22	20	21	19	5	19.8
20	20	17	23	23	19	6	20.4
21	21	22	21	22	21	1	21.4
22	19	20	22	19	22	3	20.4
23	20	24	20	20	20	4	20.8
24	21	20	19	21	20	2	20.2
25	21	22	21	21	17	5	20.4

The $\bar{X}$ chart, using the previous limits and central line, is shown in Figure 6-9 (the *R* chart is not illustrated, although it would continue to be operated). Note from the chart that a shift upward in the process average is evident by the last 9 points being above the central line. New limits and central line should therefore be calculated for extension into the third sampling period, based on the last 10 samples. The sum of the 10 ranges is 33, and the sum of the means is 201.8. Dividing each by 10 yields the new central line values for the *R* and $\bar{X}$ charts:

CL (for *R* chart) = 33/10 = 3.3 hours, and CL (for $\bar{X}$ chart) = 201.8/10 = 20.18 hours

The new control limits for the *R* chart would be

LCL = 0 hours, and UCL = (2.11)(3.3) = 6.96 hours

For the $\bar{X}$ chart, the control limits are

LCL = 20.18 – (0.58)(3.3) = 18.27 hours, and UCL = 20.18 + (0.58)(3.3) = 22.09 hours

These new values for the $\bar{X}$ chart are illustrated in Figure 6-9 to the right of the chart, for comparison purposes.

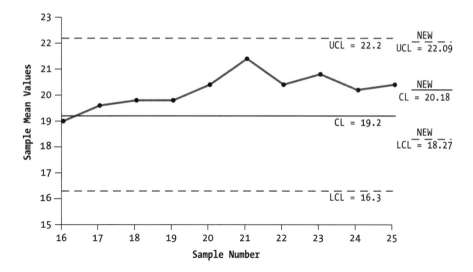

Figure 6-9. New $\bar{X}$ Chart for the Second Sampling Period, Showing New Control Limits on the Right Based on These 10 Samples

All points on the $\bar{X}$ chart for this second sampling period are within the new control limits, so the process is stabilizing at a mean training time of about 20.2 hours. This average may only be changed through an outside action, such as a simplification in the software package, or perhaps a qualifying questionnaire to all new customers to assess their relative experience levels before developing the training materials. It would be advisable to continue the use of the R and $\bar{X}$ charts as more data become available.

SQL/Query

Suppose the data for the new sampling period is in the database table called New Sampling Period that has three columns: Sample Number, Item Number, and Sample Value. Query 6_23 and Query 6_24 determine the ranges and sample means with respect to the Sample Number, and Query 6_25 determines the sample size. The control limits for both the R chart and $\bar{X}$ chart are calculated by Query 6_26. Notice how we used the WHERE clause to match the Sample Size to the Number of Observations in the control limits table, so that we can select the desired factor.

Query 6_23:

```
SELECT [Sample Number],
Max([Sample Value]) - Min([Sample Value]) AS Range,
Avg([Sample Value]) AS [Sample Mean]
FROM [New Sample Period]
GROUP BY [Sample Number]
ORDER BY [Sample Number];
```

Query 6_24:

```
SELECT Avg(Range) AS [CL for R chart],
Avg([Sample Mean]) AS [CL for X chart]
FROM [Query 6_23];
```

Query 6_25:

```
SELECT Max([Item Number]) AS [Sample Size]
FROM [New Sample Period];
```

Query 6_26:

```
SELECT [CL for R chart] * [LCL Factor for R Chart] AS [LCL R Chart],
[CL for R chart] * [UCL Factor for R Chart] AS [UCL R chart],
[CL for X chart] - [CL for R chart] * [Factor for Xbar Chart] AS [LCL X chart],
[CL for X chart] + [CL for R chart] * [Factor for Xbar Chart] AS [UCL X chart]
FROM [Query 6_24], [Query 6_25], [Control Limits]
WHERE [Sample Size] = [Number of Observations];
```

The range and mean charts are not the only control charts in use for variables data. Two other less frequently used charts are those that plot medians of samples and those for individual observations (rather than sample means). Median charts have proven easier for workers to plot, since the median of a sample may be determined without any calculations. Charts for individual items are only used when clustering into samples is not convenient or possible. Strictly speaking, they are not as powerful as charts for means of samples, but are somewhat easier to use. These types of charts are not presented here, but may be studied in most texts on quality assurance.

Control Chart for Fraction Nonconforming

As we discovered earlier, not all data is available as measured quantities. Attributes data may be in the form of "yes/no" or "defective/nondefective" classifications, for example. Or they may involve counts of the number of occurrences of some event in a sample space, such as defects in a square yard of cloth, or missing rivets in an airplane wing. In these situations, other control charts are available to analyze the data. Usually, the sample sizes for attributes charts must be larger than for variables charts, in order to assure the same degree of confidence in the results.

A frequently encountered situation involves inspecting a group or sample of items to see how many fail to conform to some standard. These might be number of defective circuit boards in manufactured batches, number of insect infested trees in forest plots, or number of sales completed per 100 customer calls. The number counted in each instance is termed the "number nonconforming." It has no "good" or "bad" connotation, but only depends on how it is defined for the data collection. Often the number is converted to a fraction or a percentage of the sample size, so it then becomes the "fraction nonconforming." The control chart used to analyze such data is called the *p* chart. The symbol *p* has been used for years to represent the fraction nonconforming. It is interesting to note that a *p* chart may be set up, with slight modifications, for a situation where the sample size *varies* from sample to sample. We haven't seen anything like that yet, but it is common in practical problems. For the moment, let's confine ourselves to an example where the sample size is constant. Then we'll muddy the water later.

Let's suppose we keep tabs on an automobile sales person for several weeks. Every time this person completes a sales pitch with 100 prospective car buyers, we record how many sales are actually completed, where a new car leaves the lot. We do this until the sales person has talked to 1,000 folks. For each 100 prospects, Table 6-5 shows the numbers of people who ended up buying a new car.

Table 6-5. Automobile Sales Results for Several Weeks

SAMPLE NUMBER	NUMBER OF CARS SOLD PER 100 PEOPLE	FRACTION OF PEOPLE BUYING A CAR
1	5	0.05
2	7	0.07
3	3	0.03
4	9	0.09
5	8	0.08
6	4	0.04
7	5	0.05
8	10	0.10
9	8	0.08
10	6	0.06

Now we ask, "Is the sales performance stable?" We can answer this question by setting up a little *p* chart. First we need a central line, called $\bar{p}$, the average fraction of sales per 100 people. This is found by adding the 10 numbers above and dividing by 1,000. Their sum is 65, so $\bar{p} = 0.065$ (or 6.5% "success" rate on average). Since a prospective buyer can be classified into one of two categories ("buys a car" or "doesn't buy a car"), the underlying statistical distribution that characterizes this situation is the binomial (see Appendix C for a brief description). We can calculate the control limits for our *p* chart by using an estimate for the standard deviation of the binomial distribution, then multiplying it by 3 as we did on the control charts earlier. We then add the result to $\bar{p}$ to get the upper control limit, and subtract it from $\bar{p}$ for the lower control limit. The expression we use, and the results, are shown below. Remember that n is again the sample size, in this case 100.

$$UCL = \bar{p} + 3\sqrt{\frac{\bar{p}(1-\bar{p})}{n}} = 0.065 + 3\sqrt{\frac{(0.065)(0.935)}{100}} = 0.139$$

$$LCL = \bar{p} - 3\sqrt{\frac{\bar{p}(1-\bar{p})}{n}} = 0.065 - 3\sqrt{\frac{(0.065)(0.935)}{100}} = -0.009 \text{ or } 0$$

Notice that the lower control limit value is actually less than zero. Since we can never have less than 0 sales per 100 prospects, we just set this limit at 0. Now we have a central line and control limits, so what do we need to plot on the chart? The points plotted are the fraction of completed sales per 100 prospects for each sample. For the first sample, the sales person had 5 sales, so the fraction completed is 0.05; for the second sample, it's 0.07, and so forth. The control chart is shown in Figure 6-10.

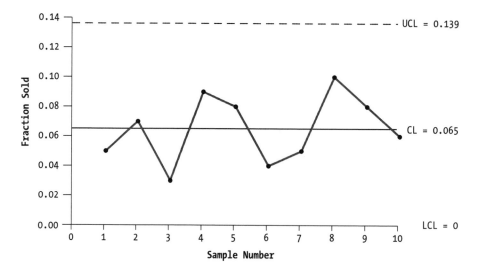

Figure 6-10. p *chart for fraction of car sales per 100 sales prospects*

SQL/Query

Suppose the automobile sales data is in the database table Automobile Sales that has two columns: Sample Number and Cars Sold. Query 6_27 calculates the fraction of people buying a car, and Query 6_28 calculates the average. The upper and lower control limits are determined by Query 6_29. The function Zero_Or_Bigger (given in Appendix D) returns zero if the value is negative; otherwise, it returns the value. The data for the chart shown in Figure 6-10 is provided through Query 6_30. Notice that Query 6_30 has no join condition because the result of Query 6_28 and Query 6_29 is a single row.

Query 6_27:

```
SELECT [Sample Number], [Cars Sold], [Cars Sold] / 100 AS [Fraction Buying]
FROM [Automobile Sales];
```

Query 6_28:

```
SELECT Avg([Fraction Buying]) AS [p bar]
FROM [Query 6_27];
```

Query 6_29:

```
SELECT [p bar] + 3 * Sqr([p bar] * (1 - [p bar]) / 100) AS UCL,
Zero_Or_Bigger([p bar] - 3 * Sqr([p bar] * (1 - [p bar]) / 100)) AS LCL
FROM [Query 6_28];
```

Query 6_30:

```
SELECT [Sample Number], [Fraction Buying], LCL, [p bar] AS CL, UCL
FROM [Query 6_27], [Query 6_28], [Query 6_29];
```

The sales per 100 prospects are in control. Assigning a quota to the sales person that exceeds 6 sales per 100 prospects will be counter-productive and frustrating, because half the time the quota can't be met. Yet managers do this sort of thing constantly, without any knowledge of the process or the circumstances under which the sales staff have to operate. The only way to improve the sales figures would be through some action by management, such as sales training, or improving the reputation of their automobiles.

You might be wondering why we didn't just plot the number of sales rather than the fraction per 100. This can easily be done. However, the control chart we presented is traditionally set up in the manner illustrated, for *fraction* nonconforming. If we want to plot actual numbers, the central line has to be 6.5, the average number of sales per 100 prospects. In general, the central line is equal to $n\bar{p}$, or in this case, (100)(0.065). As you may have guessed, this necessitates a little different approach for the control limits. They are calculated from the following formulas:

$$UCL = n\bar{p} + 3\sqrt{n\bar{p}(1-\bar{p})} = 6.5 + 3\sqrt{(6.5)(1-0.065)} = 13.9$$
$$LCL = n\bar{p} - 3\sqrt{n\bar{p}(1-\bar{p})} = 6.5 - 3\sqrt{(6.5)(1-0.065)} = -0.9 \text{ or } 0$$

The chart looks exactly the same as Figure 6-10, except the scale is changed to reflect the actual number of sales rather than fraction of sales.

As we mentioned earlier, the sample size may not be fixed as it was in the previous example. The next example illustrates how to modify the *p* chart again to accommodate the varying sample sizes. The resulting chart is called a *stabilized p* chart.

An electronics manufacturer was producing circuit boards for installation in personal computers. Problems with defective circuit boards were evident, so management decided to test all circuit boards each day and record the number defective. The results over a 15-day period are shown in Table 6-6.

Table 6-6. Results of Circuit Board Inspection Over a 15-Day Period

TEST DAY	1	2	3	4	5	6	7	8	9	10	11	12	13	14	15
NUMBER INSPECTED	442	138	481	593	530	488	503	596	476	568	492	373	636	633	438
NUMBER DEFECTIVE	26	38	93	105	81	124	72	72	72	73	83	31	63	93	50

We first need the average fraction defective for the sampling period, but we have to be careful how we calculate it. We won't get the correct answer if we find the 15 sample fraction defective values, add them, and divide by 15. This is because the sample sizes vary. The correct way is to find the total defectives for the period and divide it by the total items inspected. The calculation is shown next:

$$\bar{p} = 1,076/7,387 = 0.146 \text{ or } 14.6\% \text{ defective}$$

As we showed before, the standard deviation of the fractions defective may be estimated by the following formula:

$$\sqrt{\frac{\bar{p}(1-\bar{p})}{n}}$$

However, in this situation, *n* changes for every sample! This means that, conceivably, each sample should have its own set of control limits. But that can be alleviated by a standardization, or "stabilization," of each sample fraction defective value. The procedure is very similar to the way we standardized the normal distribution variate when we looked it up in the table. The trick involves shifting the observed value by subtracting the mean, then converting it to standard deviation units. The result may then be compared to a set of consistent control limits. For each sample we calculate

$$\frac{p-\bar{p}}{\sqrt{\dfrac{\bar{p}(1-\bar{p})}{n}}}$$

where p is the fraction defective for that particular sample, and plot it on the control chart. The central line will *always* be, for any stabilized p chart, equal to zero, and the control limits will *always* be at ±3. For the first sample, the value plotted will be

$$\frac{p-\bar{p}}{\sqrt{\dfrac{\bar{p}(1-\bar{p})}{n}}} = \frac{\dfrac{26}{442}-0.146}{\sqrt{\dfrac{(0.146)(1-0.146)}{442}}} = -5.18$$

(Your results may vary a little depending on the number of significant digits you carry in intermediate calculations.) This value is below the central line because it is negative, and in this case it is also below the lower control limit of –3. Table 6-7 shows all the stabilized p values for the data.

Table 6-7. Stabilized p *Values for the Circuit Board Data*

TEST DAY	1	2	3	4	5	6	7	8	9	10	11	12	13	14	15
NUMBER INSPECTED	442	138	481	593	530	488	503	596	476	568	492	373	636	633	438
NUMBER DEFECTIVE	26	38	93	105	81	124	72	72	72	73	83	31	63	93	50
STABILIZED P	−5.18	4.32	2.96	2.17	0.47	6.79	−0.16	−1.72	0.35	−1.16	1.45	−3.42	−3.33	0.09	−1.87

Now we can set up the stabilized p chart and plot these values (see Figure 6-11).

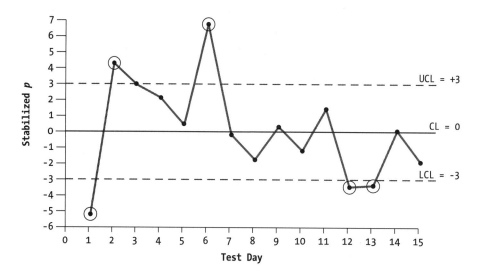

Figure 6-11. Stabilized p *chart for the circuit board data*

SQL/Query

Suppose the sampling data for the circuit board inspection is in a database table called Circuit Board Inspection that has three columns: Test Day, Number Inspected, and Number Defective. Query 6_31 calculates the average fraction defective, $\bar{p}$ (p bar). Next, Query 6_32 calculates the stabilized *p* values for each test day, and produces the values shown in Table 6-7.

Query 6_31:

```
SELECT Sum([Number Defective]) / Sum([Number Inspected]) AS [p bar]
FROM [Circuit Board Inspection];
```

Query 6_32:

```
SELECT [Test Day], [Number Inspected], [Number Defective],
((([Number Defective] / [Number Inspected]) - [p bar]) /
Sqr(([p bar] * (1 - [p bar])) / [Number Inspected]) AS [Stabilized p]
FROM [Circuit Board Inspection], [Query 6_31];
```

It is obvious from the chart that the process is out of control, because five points are outside the control limits. Before the average fraction defective can be reduced through continuous improvement, the process has to be stabilized. The special causes of variation that contribute to the out of control points need to be identified and corrected. These may be due to assembly problems, defective components, defective solder, poor working conditions, lack of adequate training, or any number of possibilities.

John's Jewels
Would You Want to Accept Bad Quality?

There is a paradox concerning control charts. It is possible to bring a process under statistical control generating 95% defectives! We have to realize that just because a process is stable, it isn't necessarily producing at a desirable level. By identifying and dealing with special causes of variation, we can usually systematically bring the process under control. At that point we have to look at its process *capability* and *average*, and take steps to improve them if possible. This normally requires action by management. In the past, managers in the United States came to accept a certain "percent defective" in their processes as inevitable. The story circulated about the U.S. manufacturer that ordered a batch of auto parts from Japan. The U.S. plant's specifications called for "3 percent defective," meaning their typical experience with this type of part was finding 3 bad ones in every lot of 100. As the story goes, the Japanese manufacturer sent the shipment in two boxes. One large box was marked "97 good parts." The other, very small box, was marked, "3 defective parts." In the little box was a note that read, "We are enclosing the three defective parts as specified. We don't know why they are needed, and they were very difficult to make, but we hope this pleases the customer."

Control Chart for Number of Nonconformities

We've saved the best for last. The control chart that plots nonconforming numbers (called the *c* chart), such as defects per square yard of fabric, hits per hour on a Web site, and the like, is the easiest to set up. However, it differs from the previous charts in that there is no classification of data into one of two categories. In this case, the data represent counts only. For example, we can count the number of hits we had on our Web site in an hour, but we don't know how many we *didn't* get. The statistical distribution that characterizes this situation is the Poisson. The Poisson distribution has a peculiar trait that we can use to advantage in control charting—its mean and variance are equal. Once we know the mean of our data, we can simply take its square root to get the standard deviation. Then we multiply the result by 3, and we have our control limits.

Let's revisit an example from Chapter 5. In Table 5-4, we presented the average number of hits per hour on a Web site. We fit a polynomial regression equation to this data. Now we can check it to see if the process generating the data is in control. We have reproduced the data as Table 6-8.

Table 6-8. Average Number of "Hits" per Hour on a Web Site

HOUR	HITS
0	120
1	105
2	100
3	80
4	70
5	80
6	75
7	85
8	90

The sum of the hits is 805. The mean of the 9 values is called $\bar{c}$, and is equal to

$$805/9 = 89.44$$

If the data are Poisson distributed, the standard deviation is $\sqrt{\bar{c}} = \sqrt{89.44} = 9.46$. The control limits for the c chart are

$$\text{UCL} = \bar{c} + 3\sqrt{\bar{c}} = 89.44 + 3(9.46) = 117.82$$
$$\text{LCL} = \bar{c} - 3\sqrt{\bar{c}} = 89.44 - 3(9.46) = 61.06$$

For convenience, we round $\bar{c}$ to 89 and the control limits to 118 and 61. The c chart is shown in Figure 6-12. We can see one point (120) just slightly above the upper control limit, so we might want to investigate that. It might be due to the initial opening of the Web site, with a larger than usual number of hits, then tapering off somewhat. It would be prudent to try to collect more data, since 9 points are not usually adequate to have a good feel for the process variation.

SQL/Query

Suppose we have a table in our database called Web Site Hits that is composed of two columns: Hour and Hits. Query 6_33 gives the mean and standard deviation for the hourly hits. In the query we did not use the StDev function because the hits are Poisson. Query 6_34 calculates the upper and lower control limits. The data for the chart given in Figure 6-12 is provided by Query 6_35.

Query 6_33:

```
SELECT Avg(Hits) AS Mean, Sqr(Avg([Hits])) AS SD
FROM [Web Site Hits];
```

Query 6_34:

```
SELECT [Mean] + 3 * [SD] AS UCL,
[Mean] - 3 * [SD] AS LCL
FROM [Query 6_33];
```

Query 6_35:

```
SELECT Hour, Hits, LCL, Mean AS CL, UCL
FROM [Web Site Hits], [Query 6_33], [Query 6_34];
```

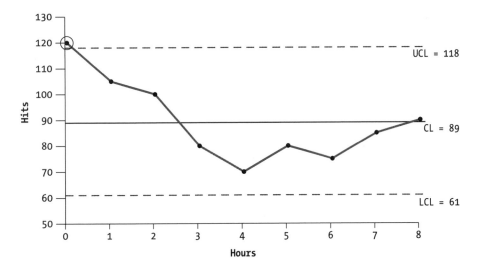

Figure 6-12. c *chart for hits per hour on a Web site*

Conclusion

We have presented four of the most useful types of control charts (R, $\bar{X}$, p, and c), but there are others. The important consideration is that the control chart is a powerful, yet simple tool for analyzing process variability and guiding the data miner toward control of the process generating the data.

T-SQL Source Code

Following are four T-SQL source code listings for generating the data necessary for plotting each of the control charts discussed in this chapter. The first procedure is called Sample_Range_and_Mean_Charts. The procedure creates two tables of data: one for plotting the R chart and one for plotting the $\bar{X}$ chart. The second and third procedures generate the data table for plotting the p chart. The last procedure is called the C_Chart, and it generates the data for plotting the c chart. The best technique for plotting these charts is to export the generated plot data tables to Microsoft's Excel and use Excel to draw the graphs. (We used Excel to create the charts shown in this chapter.) Following the procedure listings is a set of statements illustrating how to invoke each of the procedures.

Sample_Range_and_Mean_Charts

```
SET QUOTED_IDENTIFIER OFF
GO
SET ANSI_NULLS OFF
GO

ALTER PROCEDURE [Sample_Range_and_Mean_Charts]
@SrcTable Varchar(50) = 'Initial Samples',
@Sample_Num Varchar(50) = 'Sample Number',
@Item_Num Varchar(50) = 'Item Number',
@Sample_Value Varchar(50) = 'Sample Value',
@Plot_R_Chart Varchar(50) = 'Plot_R_Chart',
@Plot_X_Chart Varchar(50) = 'Plot_X_Bar_Chart'
AS

/***********************************************************/
/*                                                       */
/*           Sample_Range_And_Mean_Charts                */
/*                                                       */
/*    This procedure creates a table of data for plotting */
/*  the R chart and the X-bar chart. For these charts, each */
/*  sample has a sample ID number and is composed of a    */
/*  fixed number of observations (items). Each item in the */
/*  sample is also numbered (ID-ed) and each item has a   */
/*  value of type float. The R chart is determined first  */
/*  and then the X-bar chart is determined.               */
/*                                                       */
```

```
/* INPUTS:                                                    */
/*    SrcTable - name of table containing sample data         */
/*    Sample_Num - a column in SrcTable, and is the ID        */
/*       number of each sample                                */
/*    Item_Num - a column in SrcTable, ID number of           */
/*       observation or item with the sample                  */
/*    Sample_Value - a column in SrcTable, and is the         */
/*       observed value of data type float                    */
/*    Plot_R_Chart - name of table to receive plotting data   */
/*       for R chart                                           */
/*    Plot_X_Chart - name of table to receive plotting data   */
/*       for X-bar chart                                       */
/* STATISTICAL TABLES USED:                                    */
/*    Control Limits                                           */
/*                                                             */
/***************************************************************/


/* Local Variables */
Declare @Q varchar(5000)      /* Query string */
Declare @SampSize Int         /* Sample size */
Declare @Rbar Float           /* R bar */


/*  R CHART */

/* Determine sample size */
SET @Q = 'SELECT Max([' + @Item_Num + ']) AS [SampSize] ' +
     'INTO ##TempSRMC1 '      +
     'FROM [' + @SrcTable + ']'
EXEC(@Q)
SELECT @SampSize = (SELECT SampSize FROM ##TempSRMC1)

/* Determine the range and mean of each sample */
SET @Q = 'SELECT [' + @Sample_Num + '] AS [Sample_ID_Num], ' +
     'Max([' + @Sample_Value + '])-Min([' + @Sample_Value + ']) AS Range, ' +
     'Avg([' + @Sample_Value + ']) AS [Sample Mean] ' +
     'INTO ##TempSRMC2 '      +
     'FROM [' + @SrcTable + '] ' +
     'GROUP BY [' + @Sample_Num + ']'
EXEC(@Q)

/* Determine average range */
SET @Q = 'SELECT Avg(Range) AS Rbar ' +
     'INTO ##TempSRMC3 '      +
     'FROM ##TempSRMC2'
EXEC(@Q)
SELECT @Rbar = Rbar FROM ##TempSRMC3
```

217

```
/* Determine LCL, CL, and UCL */
SET @Q = 'SELECT [Rbar]*[LCL Factor for R Chart] AS LCL, ' +
    'Rbar AS CL, ' +
    '[Rbar]*[UCL Factor for R Chart] AS UCL ' +
    'INTO ##TempSRMC4 '       +
    'FROM [##TempSRMC1], [##TempSRMC3], [Control Limits] ' +
    'WHERE [SampSize] = ' +
    '[Control Limits].[Number of Observations]'
EXEC(@Q)


/* If plot for R-chart exist, then delete it */
SET @Q = 'SELECT * ' +
    'INTO ##TempSRMC5 ' +
    'FROM ..sysobjects ' +
    'WHERE Name = "' + @Plot_R_Chart + '"'
EXEC(@Q)
IF Exists (SELECT * FROM ##TempSRMC5)
Begin
    /* Delete table */
    SET @Q = 'DROP TABLE [' + @Plot_R_Chart + ']'
    EXEC(@Q)
End



/* R Chart Data */
SET @Q = 'SELECT [Sample_ID_Num], Range, LCL, CL, UCL ' +
    'INTO [' + @Plot_R_Chart + ']' +
    'FROM [##TempSRMC1], [##TempSRMC2], ##TempSRMC4'
EXEC(@Q)
PRINT 'Data for plotting the R-chart is in the table called ' +
    @Plot_R_Chart + '.'



/* X-BAR CHART */

/* Determine sample mean */
SET @Q = 'SELECT Avg(Range) AS Rbar, ' +
    'Avg([Sample Mean]) AS CL ' +
    'INTO ##TempSRMC6 ' +
    'FROM ##TempSRMC2'
print @Q
EXEC(@Q)
```

```
/* Determine LCL and UCL */
SET @Q = 'SELECT [CL] - [Factor for Xbar Chart] * [Rbar] AS LCL, ' +
    '[CL] + [Factor for Xbar Chart] * [Rbar] AS UCL ' +
    'INTO ##TempSRMC7 ' +
    'FROM ##TempSRMC1, ##TempSRMC6, [Control Limits] ' +
    'WHERE [SampSize] = [Number of Observations]'
EXEC(@Q)


/* If plot for X-bar chart exists then delete it */
SET @Q = 'SELECT * ' +
    'INTO ##TempSRMC8 ' +
    'FROM ..sysobjects ' +
    'WHERE Name = "' + @Plot_X_Chart + '"'
EXEC(@Q)
IF Exists (SELECT * FROM ##TempSRMC8)
Begin
    /* Delete table */
    SET @Q = 'DROP TABLE [' + @Plot_X_Chart + ']'
    EXEC(@Q)
End


/* X-bar Chart Data */
SET @Q = 'SELECT [Sample_ID_Num], [Sample Mean], UCL, CL, LCL ' +
    'INTO [' + @Plot_X_Chart + '] ' +
    'FROM ##TempSRMC2, ##TempSRMC6, ##TempSRMC7'
EXEC(@Q)
PRINT 'Data for plotting X-bar chart is in the table called ' +
    @Plot_X_Chart + '.'


GO
SET QUOTED_IDENTIFIER OFF
GO
SET ANSI_NULLS ON
GO
```

Standard_P_Chart

```
SET QUOTED_IDENTIFIER OFF
GO
SET ANSI_NULLS OFF
GO
```

```
ALTER    PROCEDURE [Standard_P_Chart]
@SrcTable Varchar(50) = 'Automobile Sales',
@Sample_Size Int = 100,
@Sample_Num Varchar(50) = 'Sample Number',
@Non_Conf Varchar(50) = 'Cars Sold',
@Plot_P_Chart Varchar(50) = 'Plot_Standard_P_Chart'
AS


/***********************************************************/
/*                                                         */
/*                   Standard_P_Chart                      */
/*                                                         */
/*    This procedure creates a table of data for plotting  */
/*  the standard p chart. For the p chart, the sample size */
/*  has a fixed (constant) number of observations per      */
/*  sample, and each sample is numbered. The number of     */
/*  'nonconforming' instances is counted, and becomes      */
/*  (or is used to determine) the value of the sample. The */
/*  sample value must be of data type float to prevent     */
/*  round-off errors during calculations.                  */
/*                                                         */
/* INPUTS:                                                 */
/*    SrcTable - name of table containing sample data      */
/*    Sample_Size - fixed number of observations           */
/*    Sample_Num - a column in SrcTable, and is the ID     */
/*      number of each fixed-size sample                   */
/*    Non_Conf - a column in SrcTable, and is the number of */
/*      nonconforming in each sample of type float         */
/*    Plot_P_Chart - name of table to receive plotting data */
/*                                                         */
/***********************************************************/


/* Local Variables */
Declare @Q varchar(5000)     /* query string */
Declare @Pbar Float          /* P-bar value (also CL) */
Declare @LCL Float           /* LCL value */
Declare @CL Float            /* CL value (also P-bar) */
Declare @UCL Float           /* UCL value */
```

```
/* Determine fraction nonconforming */
SET @Q = 'SELECT [' + @Sample_Num + '] AS [Sample_Number], ' +
    '[' + @Non_Conf + '] AS [Num_Non_Conforming], ' +
    '[' + @Non_Conf + '] / ' + convert(varchar(10), @Sample_Size) + ' ' +
    'AS [Fraction_Non_Conforming] ' +
    'INTO ##TempStanPC1 ' +
    'FROM [' + @SrcTable + ']'
EXEC(@Q)


/* Determine P-Bar */
SET @Q = 'SELECT Avg([Fraction_Non_Conforming]) AS P_bar ' +
    'INTO ##TempStanPC2 ' +
    'FROM ##TempStanPC1'
EXEC(@Q)
SELECT @Pbar = P_bar FROM ##TempStanPC2
SELECT @CL = @Pbar


/* Determine LCL */
SET @LCL = @Pbar - 3 * Sqrt(@Pbar * (1.0 - @Pbar) / @Sample_Size)
IF @LCL < 0 SET @LCL = 0


/* Determine UCL */
SET @UCL = @Pbar + 3 * Sqrt(@Pbar * (1.0 - @Pbar) / @Sample_Size)


/* If the table for plotting the p-chart exist, then delete it */
SET @Q = 'SELECT * ' +
    'INTO ##TempStanPC3 ' +
    'FROM ..sysobjects ' +
    'WHERE Name = "' + @Plot_P_Chart + '"'
EXEC(@Q)
IF exists (SELECT * FROM ##TempStanPC3)
Begin
    /* Delete table */
    SET @Q = 'DROP TABLE [' + @Plot_P_Chart + ']'
    EXEC(@Q)
End
```

```
/* P Chart Data */
SET @Q = 'SELECT [Sample_Number], [Fraction_Non_Conforming], ' +
    convert(varchar(10), @LCL) + ' AS [LCL], ' +
    convert(varchar(10), @CL) + ' AS [CL], ' +
    convert(varchar(10), @UCL) + ' AS [UCL] ' +
    'INTO [' + @Plot_P_Chart + '] ' +
    'FROM ##TempStanPC1'
EXEC(@Q)
PRINT 'Data for plotting the Standard p-chart is in the table called ' +
    @Plot_P_Chart + '.'


GO
SET QUOTED_IDENTIFIER OFF
GO
SET ANSI_NULLS ON
GO
```

Stabilized_P_Chart

```
SET QUOTED_IDENTIFIER OFF
GO
SET ANSI_NULLS OFF
GO


ALTER PROCEDURE [Stabilized_P_Chart]
@SrcTable Varchar(50) = 'Circuit Board Inspection',
@Sample_ID Varchar(50) = 'Test Day',
@Num_Sample Varchar(50) = 'Number Inspected',
@Num_Non_Conf Varchar(50) = 'Number Defective',
@Plot_P_Chart Varchar(50) = 'Plot_Stablized_P_Chart'
AS


/***********************************************************/
/*                                                       */
/*                 Stabilized_P_Chart                    */
/*                                                       */
/*    This procedure creates a table of data for plotting */
/*  the stabilized p chart. For the p chart, the sample  */
/*  size has a variable number of observations per sample, */
/*  each sample is assigned an ID sequence number, and the */
/*  number of observations within each sample is recorded. */
/*  The number of 'nonconforming' instances is counted,  */
/*  and becomes (or is used to determine) the value of the */
```

```
/*  sample. The sample value must be of data type float to  */
/*  prevent round-off errors during calculations.          */
/*                                                          */
/*  INPUTS:                                                 */
/*    SrcTable - name of table containing sample data       */
/*    Sample_ID - a column in SrcTable, and is the ID       */
/*       number of each sample                              */
/*    Num_Sample - a column in SrcTable, number of          */
/*       observations per sample                            */
/*    Num_Non_Conf - a column in SrcTable, and is the number */
/*       of nonconforming in each sample of data type float */
/*    Plot_P_Chart - name of table to receive plotting data */
/*                                                          */
/************************************************************/

/* Local Variables */
Declare @Q varchar(5000)        /* query string */
Declare @LCL Float              /* LCL value = -3 */
Declare @CL Float               /* CL value = 0 */
Declare @UCL Float              /* UCL value = 3 */
Declare @Pbar Float             /* p-bar value */
Declare @PbarTxt varchar(50)    /* p-bar value as text */

/* Assign UCL, CL, and LCL values */
SET @UCL = 3
SET @CL = 0
SET @LCL = -3

/* Determine P-Bar */
SET @Q = 'SELECT Sum([' + @Num_Non_Conf + ']) / ' +
     'Sum([' + @Num_Sample + ']) AS Pbar ' +
     'INTO ##TempStabPC1 ' +
     'FROM [' + @SrcTable + ']'
EXEC(@Q)
SELECT @Pbar = Pbar FROM ##TempStabPC1
SET @PbarTxt = convert(varchar(20), @Pbar)

/* If the table for plot p chart exists, then delete it */
SET @Q = 'SELECT * ' +
     'INTO ##TempStabPC2 ' +
     'FROM ..sysobjects ' +
     'WHERE name = ''' + @Plot_P_Chart + ''''
print @Q
EXEC(@Q)
```

```
IF exists (SELECT * FROM ##TempStabPC2)
Begin
    /* Delete table */
    SET @Q = 'DROP TABLE [' + @Plot_P_Chart + ']'
    EXEC(@Q)
End


/* P Chart Data */
SET @Q = 'SELECT [' + @Sample_ID + '] AS [Sample ID], ' +
    '[' + @Num_Sample + '] AS [Number Samples], ' +
    '[' + @Num_Non_Conf + '] AS [Number Non-Conforming], ' +

    '((([' + @Num_Non_Conf + '] / [' + @Num_Sample + ']) - ' +
    @PbarTxt + ') / Sqrt((' + @PbarTxt + '* (1.0 - ' + @PbarTxt + ')) / ' +
    '[' + @Num_Sample + ']) AS [Stablized p], ' +

    convert(varchar(10), @LCL) + ' AS [LCL], ' +
    convert(varchar(10), @CL) + ' AS [CL], ' +
    convert(varchar(10), @UCL) + ' AS [UCL] ' +
    'INTO [' + @Plot_P_Chart + '] ' +
    'FROM [' + @SrcTable + ']'
EXEC(@Q)
PRINT 'Data for plotting the Stabalized p-chart is in the table called ' +
    @Plot_P_Chart + '.'

GO
SET QUOTED_IDENTIFIER OFF
GO
SET ANSI_NULLS ON
GO
```

C_Chart

```
SET QUOTED_IDENTIFIER ON
GO
SET ANSI_NULLS OFF
GO

ALTER PROCEDURE [C_Chart]
@SrcTable Varchar(50) = 'Web Site Hits',
@Sample_ID Varchar(50) = 'Hour',
@Sample_Value Varchar(50) = 'Hits',
@Plot_C_Chart Varchar(50) = 'Plot_C_Chart'

AS
```

```
/*********************************************************/
/*                                                       */
/*                      C_Chart                          */
/*                                                       */
/*     This procedure creates a table of data for plotting */
/*  the c chart. For the c chart, each sample is numbered, */
/*  and each sample has a value of data type float to    */
/*  prevent round-off errors during calculations.        */
/*                                                       */
/* INPUTS:                                               */
/*    SrcTable - name of table containing sample data    */
/*    Sample_ID - a column in SrcTable, and is the ID    */
/*       number of each sample                           */
/*    Sample_Value - a column in SrcTable, and is the value */
/*       of each sample and has a data type of float     */
/*    Plot_C_Chart - name of table to receive plotting data */
/*                                                       */
/*********************************************************/


/* Local Variables */
Declare @Q varchar(5000)      /* query string */
Declare @LCL Float            /* LCL value */
Declare @CL Float             /* CL value */
Declare @UCL Float            /* UCL value */
Declare @Mean Float           /* Mean */
Declare @SD Float             /* Standard deviation */

/* Determine the mean */
SET @Q = 'SELECT Avg([' + @Sample_Value + ']) AS Mean ' +
     'INTO ##TempCC1 ' +
     'FROM [' + @SrcTable + ']'
EXEC(@Q)
SELECT @Mean = Mean FROM ##TempCC1
print '@Mean = ' + convert(varchar(10), @Mean)

/* Determine the standard deviation */
/* Since the data is Poisson, the standard */
/* deviation is the square root of the mean. */
SET @SD = Sqrt(@Mean)
print '@SD = ' + convert(varchar(10), @SD)
```

```
/* Calculate LCL, CL, and UCL */
SET @UCL = @Mean + 3 * @SD
SET @CL = @Mean
SET @LCL = @Mean - 3 * @SD

/* If the table for plot c chart exists, then delete it */
SET @Q = 'SELECT * ' +
    'INTO ##TempCC2 ' +
    'FROM ..sysobjects ' +
    'WHERE name = ''' + @Plot_C_Chart + ''''
EXEC(@Q)
IF exists (SELECT * FROM ##TempCC2)
Begin
    /* Delete table */
    SET @Q = 'DROP TABLE [' + @Plot_C_Chart + ']'
    EXEC(@Q)
End

/* C Chart Data */
SET @Q = 'SELECT [' + @Sample_ID + '] AS [Sample ID], ' +
    '[' + @Sample_Value + '] AS [Sample Value], ' +
    convert(varchar(20), @LCL) + ' AS [LCL], ' +
    convert(varchar(20), @CL) + ' AS [CL], ' +
    convert(varchar(20), @UCL) + ' AS [UCL] ' +
    'INTO [' + @Plot_C_Chart + '] ' +
    'FROM [' + @SrcTable + ']'
EXEC(@Q)
PRINT 'Data for plotting the c-chart is in the table called ' +
    @Plot_C_Chart + '.'

GO
SET QUOTED_IDENTIFIER OFF
GO
SET ANSI_NULLS ON
GO
```

Procedure Calls

The call statements for calling the procedures Sample_Range_and_Mean_Charts, Standard_P_Chart, Stabilized_P_Chart, and C_Chart are as follows:

```
DECLARE @Item_Num varchar(50)
DECLARE @Sample_Value varchar(50)
DECLARE @Plot_R_Chart varchar(50)
DECLARE @Plot_X_Chart varchar(50)

SET @SrcTable = 'Initial Samples'
SET @Sample_Num = 'Sample Number'
SET @Item_Num = 'Item Number'
SET @Sample_Value  = 'Sample Value'
SET @Plot_R_Chart = 'Plot_R_Chart'
SET @Plot_X_Chart = 'Plot_X_Bar_Chart'

EXEC [CH6].[dbo].[Sample_Range_and_Mean_Charts]
     @SrcTable, @Sample_Num, @Item_Num,
@Sample_Value, @Plot_R_Chart, @Plot_X_Chart


DECLARE @SrcTable varchar(50)
DECLARE @Sample_Size int
DECLARE @Sample_Num varchar(50)
DECLARE @Non_Conf varchar(50)
DECLARE @Plot_P_Chart varchar(50)

SET @SrcTable = 'Automobile Sales'
SET @Sample_Size = 100
SET @Sample_Num = 'Sample Number'
SET @Non_Conf = 'Cars Sold'
SET @Plot_P_Chart = 'Plot_Standard_P_Chart'

EXEC [CH6]..[Standard_P_Chart] @SrcTable, @Sample_Size, @Sample_Num,
     @Non_Conf, @Plot_P_Chart


DECLARE @SrcTable varchar(50)
DECLARE @Sample_ID varchar(50)
DECLARE @Num_Sample varchar(50)
DECLARE @Num_Non_Conf varchar(50)
DECLARE @Plot_P_Chart varchar(50)
```

```
SET @SrcTable = 'Circuit Board Inspection'
SET @Sample_ID = 'Test Day'
SET @Num_Sample = 'Number Inspected'
SET @Num_Non_Conf = 'Number Defective'
SET @Plot_P_Chart = 'Plot_Stablized_P_Chart'

EXEC [CH6]..[Stabilized_P_Chart] @SrcTable, @Sample_ID,
    @Num_Sample, @Num_Non_Conf, @Plot_P_Chart


DECLARE @SrcTable varchar(50)
DECLARE @Sample_ID varchar(50)
DECLARE @Sample_Value varchar(50)
DECLARE @Plot_C_Chart varchar(50)

SET @SrcTable = 'Web Site Hits'
SET @Sample_ID = 'Hour'
SET @Sample_Value = 'Hits'
SET @Plot_C_Chart = 'Plot_C_Chart'

EXEC [CH6]..[C_Chart] @SrcTable, @Sample_ID,
    @Sample_Value, @Plot_C_Chart
```

Analysis of Experimental Designs

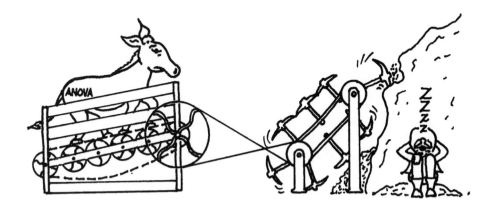

Successful Design

IN CHAPTER 4 WE SHOWED how to compare the variances of two samples to determine whether the corresponding data sets appeared to represent the same population. We calculated a variance estimate for each sample, and then compared their ratio to the *F* distribution. In this chapter, we extend that concept to the *analysis of variance* (ANOVA) of several (more than two) samples. In Figure 1-3, our diagnostic tree directs us to ANOVA when we want to analyze relationships among factors involved in an experimental design.

If you recall, the comparison of the two variances in Chapter 4 was a preliminary step in comparing the means of the two samples (using the *t* test). In actuality, ANOVA is also a technique where variances are compared to draw a conclusion about the means of samples. Clear as mud? Well, let's proceed.

Unfortunately, the presentation of analysis of variance in most statistical texts is convoluted and confusing to say the least. It is really a very simple concept, made complicated by too much theory and not enough application. There is nothing magical about the calculations required to conduct an analysis of variance. However, it sometimes requires a sorcerer to structure a data collection effort so that an appropriate ANOVA may be applied. This body of knowledge is called "experimental design." The situation that requires a design for the data collection commonly goes something like this: The analyst wants to test the reaction of something to one or more factors whose levels are known or can be controlled. We've seen these databases in previous chapters. For example, in the vehicle mileage data in Chapter 5, we initially recorded mileage data as we changed payload. We looked for an equation (or regression) that would relate these two variables. Then we threw in another factor in the form of speed. We not only changed payload, but speed also, and again checked the resulting mileage values.

In ANOVA lingo, the independent variable we are controlling and changing (say payload) is called the *treatment* (also sometimes called the *factor*). This term comes from the earliest applications of ANOVA to agricultural growth plots, where the treatment might be a fertilizer type. When only one treatment is of interest in the experiment, the results are analyzed using a *one-way* analysis of variance. If there are two treatments, we use a two-way ANOVA, and so forth. In the one-way design, it is standard practice to set one level of the treatment and run the test several times, then move to the next treatment level, and run the test again. It is not necessary to collect the same number of results each time the treatment level is changed, but it does make the analysis a little easier.

John's Jewels
The Danger in Enthusiasm

I always tried to spend an adequate amount of time when I taught experimental design and analysis of variance. One day I bumped into one of my former industrial engineering students, who now had a good job with a local textile company. He told me how he was applying these design principles in his work with good success. He related, however, how close he came to being fired the first week on the job. He said he was eager to please his boss, and looked in earnest for any way to improve the production process and save the company money. He noticed one production line where the yarn was wrapped onto cardboard cones, dipped into a dye tank, off-wound, and the cones were thrown away. This looked wasteful, he thought, so he re-designed the process to eliminate the cones. He enthusiastically burst into his boss' office with his projected cost savings, which were in the neighborhood of $750,000 per year. His boss turned pale, then said, "I'm glad you came to me first. Our president owns the company that makes those cones!"

One-Way ANOVA

Let's look at the technique first with a one-way ANOVA, to try and understand the inner workings of the method. It would be nice to be able to use the same mileage example and the data from Table 5-1, but in that case we only have one observation of mileage per treatment level (or payload value). This won't help us much in a one-way analysis of variance, because we don't have any sample means to compare (only individual values). Let's extend the experiment so that we operate the vehicle four times for each payload level. Maybe we obtain the results shown in Table 7-1.

Table 7-1. Results of Testing Vehicle Mileage for Various Payload Levels

	PAYLOAD (IN POUNDS)				
	0	500	1,000	1,500	2,000
	19.6	18.8	18.5	17.6	16.9
MILEAGE (MILES PER GALLON)	19.1	18.9	18.5	17.9	16.6
	19.3	19.0	18.6	17.4	16.7
	19.1	18.7	18.4	17.7	16.8

The hypothesis we are testing is this: Is there any significant difference among the means of the five samples as a result of varying payloads, or are the differences only due to chance? In other words,

H_0: $\mu_1 = \mu_2 = \mu_3 = \mu_4 = \mu_5$ and

H_1: $\mu_i \neq \mu_j$ for some i and j (in other words, at least two

of the means are significantly different).

In the actual experiment, we would estimate μ_1, for example, from the corresponding sample mean $\bar{x}_1$. If our ANOVA reveals no significant difference among sample means (and we accept H_0), then we conclude that payload changes don't affect mileage to any degree beyond random variation.

If this is the case, each sample is assumed to have been drawn from a population with variance equal to σ^2. Therefore, we could estimate σ^2 by finding the variance of any one of the five samples. Using our old variance formula, we find the variances of the five samples to be $s_1^2 = 0.05583$, $s_2^2 = 0.01667$, $s_3^2 = 0.00667$, $s_4^2 = 0.04333$, and $s_5^2 = 0.01667$. We could take their average as our estimate of σ^2:

$$\sigma^2 \cong \left(s_1^2 + s_2^2 + s_3^2 + s_4^2 + s_5^2 \right)/5 = 0.02783$$

This is called the "within-sample" variance estimate.

SQL/Query

We can use SQL to calculate the within-sample variance estimate by first executing Query 7_1 to determine the variances of the samples. Then we can use Query 7_2 to finish the job by computing the average.

Query 7_1:

```
SELECT "s" & (([ID]-1)\4)+1 AS Varance,
Var([Table 7_1].Mileage) AS VarOfMileage
FROM [Table 7_1]
GROUP BY "s" & (([ID]-1)\4)+1, [Table 7_1].Payload;Query 7_2:
SELECT Avg([Query 7_1].VarOfMileage) AS AvgOfVarOfMileage
FROM [Query 7_1];
```

We could also estimate σ^2 another way. We could estimate the variance of the five sample means ($\sigma_{\bar{x}}^2$) and remember from the Central Limit Theorem (see Chapter 6) that

$$\sigma_{\bar{x}} \cong \sigma/\sqrt{n}$$

where n is the sample size of each sample (in this case 4) and σ is the population standard deviation. We are of course estimating σ from the sample data. If we square both sides of the above equation and solve for σ^2, we obtain the equivalent relationship

$$\sigma^2 \cong (n)\left(\sigma_{\bar{x}}^2\right)$$

in terms of the variances. This says we can also estimate the population variance by finding the variance of the five sample means and multiplying the result by $n = 4$. The five sample means are $\bar{x}_1 = 19.275$, $\bar{x}_2 = 18.850$, $\bar{x}_3 = 18.500$, $\bar{x}_4 = 17.650$, and $\bar{x}_5 = 16.750$. The variance of these five values is equal to 1.01825, and this is our estimate of $\sigma_{\bar{x}}^2$. We can then multiply this number by $n = 4$ to obtain the estimate of $\sigma^2 = 4.073$. This estimate is called the "between-sample" variance.

SQL/Query

We use three SQL queries to determine the between-sample variance. The first query, Query 7_3, determines the sample means. Query 7_4 counts the number of means. The estimated between-sample variance is calculated by Query 7_5.

Query 7_3:

```
SELECT [Table 7_1].Payload, Avg([Table 7_1].Mileage) AS Xbar
FROM [Table 7_1]
GROUP BY [Table 7_1].Payload;
```

Query 7_4:

```
SELECT [Table 7_1].Payload, Count([Table 7_1].Payload) AS CountOfPayload
FROM [Table 7_1]
GROUP BY [Table 7_1].Payload;
```

Query 7_5:

```
SELECT Var([Query 7_3].[Xbar]) * Avg([Query 7_4].[CountOfPayload])
AS [Between-sample variance]
FROM [Query 7_3], [Query 7_4]
WHERE [Query 7_3].Payload = [Query 7_4].Payload;
```

> **NOTE** *We could have combined Queries 7_3 and 7_4 into one query, but we separated them to make it consistent with the discussion of the calculations.*

If the two estimates of σ^2 thus obtained (the within-sample and between-sample variance estimates) were equal, we could expect no difference whatsoever in the means of the five samples. This is not the case in a practical situation, due to inherent variation and experimental error. However, if the ratio of the two variance estimates (the between-sample variance over the within-sample variance) is less than the table F statistic (see Appendix B) for the appropriate numbers of degrees of freedom, we accept the null hypothesis H_0 and conclude there is no significant difference among the five sample means, at the chosen significance level. If, on the other hand, the ratio exceeds the table F value, we reject H_0 and conclude that at least two of the means are significantly different. In other words, payload does influence gas mileage significantly.

We can compare our two variance estimates, but first we need to know how to determine the degrees of freedom for the test. You may recall our discussion of the concept of degrees of freedom from Chapter 3. Another way of presenting this concept is to think of a point in three-dimensional space. The point can be located by its coordinates, say, x, y, and z. Unless we impose a constraint on the point, it can move freely within the three dimensions, so it has three degrees of freedom. However, let's constrain it to move only within a plane. Maybe the equation of a plane is

$$ax + by + cz = d$$

Now the point has only two degrees of freedom, since it can only move within two dimensions. In other words, there are three variables (x, y, and z) and one constraint (the equation of the plane), so the number of degrees of freedom is equal to 3–1=2. In general, then, the number of degrees of freedom is the number of independent variables in a problem less the number of constraints. Now we can apply this concept to the ANOVA example.

Since we've let n represent the number of observations for each sample collected at each treatment (payload) level, let's use k to denote the number of treatment levels ($k = 5$ in our example). The degrees of freedom for the between-sample variance in the numerator of the ratio are $(k-1) = 4$, since we only used five values to estimate the variance. However, for the within-sample variance in the denominator, the degrees of freedom are $k(n-1) = (5)(4-1) = 15$. This is because four values were used to estimate each of five sample variances. If we choose a 5% significance level for our test, the ratio is equal to

$$4.07300/0.02783 = 146.35$$

This value is clearly greater than the table F statistic for 4 and 15 degrees of freedom (again see the table in Appendix B), which is 3.06, so we reject the null hypothesis and conclude that payload significantly influences mileage in our experiment.

This seems like a lot of calculating to reach this simple conclusion. Fortunately, over the years a simplified approach for calculating the required variances has developed, so we present it now and use it in subsequent examples. It may be accomplished by using the original data table to sum the columns, then using these results to develop a little one-way ANOVA table that summarizes the results. First, let's reproduce the mileage results in Table 7-2, and add the column totals and the grand total.

Table 7-2. Mileage Results (from Table 7-1) with Totals Added

	PAYLOAD (IN POUNDS)					
	0	500	1,000	1,500	2,000	
	19.6	18.8	18.5	17.6	16.9	
MILEAGE (MILES PER GALLON) WITHIN TABLE	19.1	18.9	18.5	17.9	16.6	
	19.3	19.0	18.6	17.4	16.7	
	19.1	18.7	18.4	17.7	16.8	
TOTALS	77.1	75.4	74.0	70.6	67.0	364.1

We first calculate a "correction term" that is used to determine quantities referred to as "sum of squares" (abbreviated SS) for estimating the total variance and the between-sample variance. This correction term is the square of the overall total (364.1) divided by the number of observations that contributed to the total (20), or

$$(364.1)^2 / 20 = 6,628.4405$$

SQL/Query

The following is an SQL query to find the correction term.

Query 7_6:

```
SELECT ((Sum([Mileage]))^2) / Count([Mileage]) AS [correction term]
FROM [Table 7_1];
```

The sum of squares for the total is found by squaring each of the 20 values, adding their squares together, and then subtracting the correction term, as follows:

$$\begin{aligned}
SS_{total} &= (19.6)^2 + (19.1)^2 + (19.3)^2 + L + (16.8)^2 - 6,628.4405 \\
&= 6,645.1500 - 6,628.4405 \\
&= 16.7095
\end{aligned}$$

SQL/Query

Query 7_7 finds the sum of squares for the total. Notice we used no join condition in Query 7_7 because the result of Query 7_6 is a single value.

Query 7_7:

```
SELECT (Sum([Mileage]^2))-Avg([correction term]) AS SS_total
FROM [Table 7_1], [Query 7_6];
```

The sum of squares for the between-sample variance (also called the treatment sum of squares, or in our case, the payload sum of squares) is found by adding the squares of each of the column totals from Table 7-2, dividing this result by 4 (the number of observations that contributed to each sum), and finally subtracting the correction term.

$$\begin{aligned} SS_{\text{between samples}} &= \left[(77.1)^2 + (75.4)^2 + (74.0)^2 + (70.6)^2 + (67.0)^2 \right] / 4 - 6,628.4405 \\ &= 6,644.7325 - 6,628.4405 \\ &= 16.2920 \end{aligned}$$

SQL/Query

Before we can calculate the sum of squares for the between-sample variance, we must first run Query 7_8 to sum the mileage as show in the Totals row of Table 7-2. Once we have these totals, we run Query 7_9 to do the calculations.

Query 7_8:

```
SELECT [Table 7_1].Payload, Sum([Table 7_1].Mileage) AS SumOfMileage
FROM [Table 7_1]
GROUP BY [Table 7_1].Payload;
```

Query 7_9:

```
SELECT (Sum([SumOfMileage]^2) / Avg([Query 7_4.[CountOfPayload])) -

Avg([correction term]) AS SS_between_samples
FROM [Query 7_6], [Query 7_8], [Query 7_4]
WHERE [Query 7_8].Payload = [Query 7_4].Payload;
```

The within-sample (or error) sum of squares is easily found by subtracting the between-sample sum of squares from the total sum of squares, as follows:

$$\begin{aligned} SS_{\text{within samples}} &= 16.7095 - 16.2920 \\ &= 0.4175 \end{aligned}$$

SQL/Query

The within-sample sum of squares is given by Query 7_10.

Query 7_10:

```
SELECT [SS_total] - [SS_between_samples] AS [SS within samples]
FROM [Query 7_7], [Query 7_9];
```

All that we need to do now is divide each of the last two sum of squares values by their respective degrees of freedom to obtain our variance estimates. Remember that the number of degrees of freedom is 4 for the between-sample variance and 15 for the within-sample variance. The variance estimates thus obtained are called in the ANOVA table the "mean square" results, abbreviated MS. They are

$$\begin{aligned} MS_{between\ samples} &= SS_{between\ samples} / 4 \\ &= 16.2920 / 4 \\ &= 4.07300 \end{aligned}$$

$$\begin{aligned} MS_{within\ samples} &= SS_{within\ samples} / 15 \\ &= 0.4175 / 15 \\ &= 0.02783 \end{aligned}$$

SQL/Query

Before we can compute the mean square values we must first determine the degrees of freedom by executing Query 7_11. Query 7_12 calculates the between samples mean square and Query 7_13 calculates the within samples mean square.

Query 7_11:

```
SELECT Sum([CountOfPayload]) - 1 AS [DF total],

Count([Payload]) - 1 AS [DF sample],
 (Sum([CountOfPayload]) - 1) - (Count([Payload]) - 1) AS [DF error]
FROM [Query 7_4];
```

Query 7_12:

```
SELECT [SS_between_samples] / [DF sample] AS [MS between samples]
FROM [Query 7_11], [Query 7_9];
```

Query 7_13:

```
SELECT [SS within samples] / [DF error] AS [MS within samples]
FROM [Query 7_11], [Query 7_10];
```

These results should look familiar. They are about the same (except for a little round off error) as the two variance estimates we obtained previously. Their ratio is again compared to the F statistic. All these results are summarized in a compact table called the ANOVA table, as shown in Table 7-3.

Table 7-3. ANOVA Table of Results for the Mileage Data Example

SOURCE OF VARIANCE	SS	DEGREES OF FREEDOM	MS	F CALCULATED	F TABLE (5%)	CONCLUSION
Between samples (Treatments)	16.2920	4	4.07300	146.35	3.06	Significant Difference (Reject H_0)
Within samples (Error)	0.4175	15	0.02783			
Total	16.7095	19				

Two-Way ANOVA

Wasn't that easy? Now we look at a little more complicated example. This is a two-way analysis of variance, with two treatments rather than just one. The data for a situation such as this are collected as one observation per unique combination of treatment levels. If multiple observations are collected per combination of treatment levels, we may introduce another source of variation. We look at one of these later in the chapter.

Remember how, in Chapter 5, we added the factor of speed to the mileage experiment? The data appear again in Table 7-4, with the totals of both the rows and columns shown, since there are now two treatments (payload and speed).

Table 7-4. Mileage Values Collected for Combinations of Payload and Speed

PAYLOAD (POUNDS)	SPEED (MILES/HR)			
	40	50	60	TOTAL
0	19.4	19.3	19.1	57.8
500	18.4	18.8	18.9	56.1
1,000	18.2	18.6	18.6	55.4
1,500	16.0	17.9	17.8	51.7
2,000	15.7	16.8	16.6	49.1
Total	87.7	91.4	91.0	270.1

The analysis of variance proceeds in much the same way as before, except now we have the two treatments to contend with. This means we'll have *three* mean square estimates before we're finished (one for the payloads, one for the speeds, and one for the error). The trick is remembering for each sum of squares calculation the appropriate number to divide by. It is always the number of observations that contributed to each term in the numerator of the fraction. For the correction term, this would be 15; for the payloads, it would be 3; for the speeds, it would be 5. The payload sum of squares is found from the row totals in the table, and the speed sum of squares from the column totals. The sum of squares calculations are as follows:

$$\text{Correction term} = (270.1)^2 / 15$$
$$= 4,863.6007$$
$$\text{SS}_{total} = (19.4)^2 + (18.4)^2 + (18.2)^2 + \text{L} + (16.6)^2 - 4,863.6007$$
$$= 4,883.5700 - 4,863.6007$$
$$= 19.9693$$
$$\text{SS}_{payload} = \left[(57.8)^2 + (56.1)^2 + (55.4)^2 + (51.7)^2 + (49.1)^2 \right] / 3 - 4,863.6007$$
$$= 4,880.3033 - 4,863.6007$$
$$= 16.7026$$
$$\text{SS}_{speed} = \left[(87.7)^2 + (91.4)^2 + (91.4)^2 \right] / 5 - 4,863.6007$$
$$= 4,865.2500 - 4,863.6007$$
$$= 1.6493$$
$$\text{SS}_{error} = \text{SS}_{total} - \text{SS}_{payload} - \text{SS}_{speed}$$
$$= 19.9693 - 16.7020 - 1.6493$$
$$= 1.6173$$

SQL/Query

The following set of queries performs all the necessary calculations for determining the sum of squares values. Each query is preceded with a statement denoting what calculation is being performed.

Query 7_14 gives the Totals column shown in Table 7-4.

Query 7_14:

```
SELECT [Table 7_4].Payload, Sum([Table 7_4].Mileage) AS Total_Payload_Mileage
FROM [Table 7_4]
GROUP BY [Table 7_4].Payload;
```

Query 7_15 gives the Totals row shown in Table 7_4.

Query 7_15:

```
SELECT [Table 7_4].Speed, Sum([Table 7_4].Mileage) AS Total_Speed_Mileage
FROM [Table 7_4]
GROUP BY [Table 7_4].Speed;
```

Query 7_16 gives the count, total, and degrees of freedom.

Query 7_16:

```
SELECT Count([Table 7_4].[Mileage]) AS Number_Of_Mileages,
Sum([Table 7_4].[Mileage]) AS Total_Mileage,
Count([Mileage]) - 1 AS [Degrees of Freedom]
FROM [Table 7_4];
```

Query 7_17 calculates the correction term.

Query 7_17:

```
SELECT ([Total_Mileage]^2) / [Number_Of_Mileages] AS [Correction term]
FROM [Query 7_16];
```

Query 7_18 calculates the sum of squares total.

Query 7_18:

```
SELECT Sum([Mileage]^2) - Avg([Correction term]) AS [SS total]
FROM [Table 7_4], [Query 7_17];
```

Query 7_19 determines the number of payloads and the degrees of freedom with respect to payload.

Query 7_19:

```
SELECT Count([Query 7_14].[Payload]) AS [Number Of Payloads],
Count([Payload]) - 1 AS [Payload Degrees of Freedom]
FROM [Query 7_14];
```

Query 7_20 determines the number of speeds and the degrees of freedom with respect to the speeds.

Query 7_20:

```
SELECT Count([Speed]) AS [Number of Speeds],
Count([Speed]) - 1 AS [Speed Degrees of Freedom]
FROM [Query 7_15];
```

Query 7_21 calculates the sum of squares for payload.

Query 7_21:

```
SELECT Sum([Total_Payload_Mileage]^2) / (Avg([Number of Speeds])) -
Avg([Correction term]) AS [SS payload]
FROM [Query 7_14], [Query 7_17], [Query 7_20];
```

Query 7_22 calculates the sum of squares for speed.

Query 7_22:

```
SELECT Sum([Total_Speed_Mileage]^2) / Avg([Number Of Payloads]) -
Avg([Correction term]) AS [SS speed]
FROM [Query 7_17], [Query 7_15], [Query 7_19];
```

Query 7_23 calculates the sum of squares error term.

Query 7_23:

```
SELECT [SS total] - [SS payload] - [SS speed] AS [SS error]
FROM [Query 7_18], [Query 7_21], [Query 7_22];
```

The ANOVA table is shown in Table 7-5.

Table 7-5. Two-way ANOVA Table for Mileage Example Source of Variance

SOURCE OF VARIANCE	SS	DEGREES OF FREEDOM	MS	F CALCULATED	F TABLE (5%)	CONCLUSION
Payload	16.7026	4	4.17565	20.65	3.84	Significant
Speed	1.6493	2	0.82465	4.08	4.46	Not Significant
Error	1.6173	8	0.20216			
Total	19.9693	14				

A couple of things should be pointed out. First, the number of degrees of freedom for each treatment is one less than the number of levels of that treatment, and the error degrees of freedom are the product of the two treatment degrees of freedom. These three degrees of freedom values should always sum to the total number of degrees of freedom, which is one less than the total data values collected. Second, the F statistic for each treatment is calculated by dividing the mean square value for the treatment by the error mean square. The table F values depend on the degrees of freedom for the numerator and denominator of the respective treatment F statistics, so will not necessarily be the same numbers.

SQL/Query

The following set of queries creates the two-way ANOVA table shown in Table 7-5. Query 7_24 and Query 7_25 calculate the degrees of freedom.

Query 7_24:

```
SELECT [Degrees of Freedom] - [Speed Degrees of Freedom] -
[Payload Degrees of Freedom] AS [Error Degrees of Freedom]
FROM [Query 7_16], [Query 7_19], [Query 7_20];
```

Query 7_25:

```
SELECT "Payload" as [Source],
[Query 7_19].[Payload Degrees of Freedom] AS [Degrees of Freedom]
FROM [Query 7_19]
UNION ALL
SELECT "Speed" as [Source],
[Query 7_20].[Speed Degrees of Freedom] AS [Degrees of Freedom]
FROM [Query 7_20]
UNION ALL
SELECT "Error" as [Source],
[Query 7_24].[Error Degrees of Freedom] AS [Degrees of Freedom]
FROM [Query 7_24]
UNION ALL
SELECT "Total" as [Source],
[Query 7_16].[Degrees of Freedom]
FROM [Query 7_16];
```

Query 7_26 and Query 7_27 obtain the MS values.

Query 7_26:

```
SELECT "Error" AS MS, [SS error] / [Error Degrees of Freedom] AS [MS error]
FROM [Query 7_23], [Query 7_24];
```

Query 7_27:

```
SELECT "Payload" As [MS],
[SS payload] / [Payload Degrees of Freedom] AS [MS value],
[Payload Degrees of Freedom] AS [Degrees of Freedom]
FROM [Query 7_21], [Query 7_19]
UNION ALL
SELECT "Speed" As [MS],
[SS speed] / [Speed Degrees of Freedom] AS [MS value],
[Speed Degrees of Freedom] AS [Degrees of Freedom]
FROM  [Query 7_22], [Query 7_20]
UNION ALL
SELECT "Error" As [MS],
[SS error] / [Error Degrees of Freedom] AS [MS value],
[Error Degrees of Freedom] AS [Degrees of Freedom]
FROM [Query 7_23], [Query 7_24]
UNION ALL
SELECT "Total" As [MS], "" AS [MS value], [Degrees of Freedom]
FROM [Query 7_16];
```

Query 7_28 gets all the SS values.

Query 7_28:

```
SELECT "Payload" as [SS], [Query 7_21].[SS payload] As [SS value]
FROM [Query 7_21]
UNION ALL
SELECT "Speed" as [SS], [Query 7_22].[SS speed] As [SS value]
FROM [Query 7_22]
UNION ALL
SELECT "Error" as [SS], [Query 7_23].[SS error] As [SS value]
FROM [Query 7_23]
UNION ALL
SELECT "Total" as [SS], [Query 7_18].[SS total] As [SS value]
FROM [Query 7_18];
```

Query 7_29 computes the value for *F* calculated.

Query 7_29:

```
SELECT [Query 7_27].MS, [MS value] / [MS error] AS [F Calculated]
FROM [Query 7_27], [Query 7_26]
WHERE [Query 7_27].MS NOT IN ("Error","Total");
```

Query 7_30 looks up the *F* values from the *F* table for a significance level of 5%.

Query 7_30:

```
SELECT [Query 7_25].Source, [F Table].F
FROM [Query 7_25], [F Table], [Query 7_24]
WHERE [F Table].[Denomiator Degrees of Freedom] =
[Query 7_24].[Error Degrees of Freedom]
AND [Query 7_25].[Degrees of Freedom] = [F Table].[Numerator Degrees of Freedom])
AND [Query 7_25].Source NOT IN ("Error","Total")
AND [F Table].[Significance level] = 0.05;
```

Query 7_31 puts all the information together into a single result.

Query 7_31:

```
SELECT [Query 7_28].SS, [Query 7_28].[SS value], [Query 7_27].[Degrees of Freedom],
[Query 7_27].[MS value], [Query 7_29].[F Calculated], [Query 7_30].F AS [F table 5%]
FROM (([Query 7_28]
INNER JOIN [Query 7_27] N [Query 7_28].SS = [Query 7_27].MS)
LEFT JOIN [Query 7_29] ON [Query 7_27].MS = [Query 7_29].MS)
LEFT JOIN [Query 7_30] ON [Query 7_27].MS = [Query 7_30].Source;
```

The conclusion to this example is the same as that found when the data were used to conduct a regression analysis in Chapter 5. That is, payload has a significant influence on mileage, while speed does not.

Robert's Rubies
The Father of ANOVA

Sir Ronald Aylmer Fisher was born in London in 1890. He began his career as a teacher of mathematics, but in 1919 left that profession to work as a biologist at the Rothamsted Agricultural Experiment Station. His many contributions were primarily in the fields of genetics and statistics. He studied the design of experiments, and introduced the concepts of randomization and analysis of variance. He received many awards throughout his life for his significant contributions to modern statistics. He died in Adelaide, Australia, in 1962. The following quote is attributed to him at the Indian Statistical Congress in Sankhya about 1938: "To call in the statistician after the experiment is done may be no more than asking him to perform a postmortem examination: he may be able to say what the experiment died of."

ANOVA Involving Three Factors

A researcher has become interested in factors that might influence life expectancy in males. Although this is a very complex topic for study, a mortality database has been used, along with demographic and related information on individuals, to narrow the problem to three areas of interest. These factors are heredity, personal life/habits, and work environment. We'll simplify them considerably for our example. Let's suppose that a measure of heredity is the age of one's parents, the hypothesis being that, if your parents live long, your chances are also good of living to an advanced age. The researcher decides to take the average age of the parents at death and place this number into one of four categories: young (less than 40), relatively young (between 40 and 60), average (between 60 and 80), and advanced (above 80). We'll designate this factor by the letter A. Personal life and habits might include a wide range, such as diet, smoking, drinking, whether the subject married, whether he had children, whether he was law-abiding, and so forth. The researcher decides to establish three categories to assess this factor: low-risk, medium risk, and high risk. A model might be developed to place an individual into one of these categories, depending on a combination of lifestyle features. Let's call this factor B. Finally, work environment is classified as low stress or high stress, depending on the job and the individual's personality traits. This will be factor C.

The researcher used the large database to draw representative samples. She decided to randomly select three individuals for each combination of factors A, B, and C. In ANOVA terminology, we would say that each cell of the data table contains three *replicates*. The three data values in each cell of the table are the ages at death of the three individuals selected at random from the database. The resulting data are shown in Table 7-6.

Notice the way the table is constructed. There are two major columns divided by a double line. The left column represents all the individuals' ages that were classified under "low stress" jobs (factor C). The right column is, of course, the ages that fell into the "high stress" job category. Within each of these major columns are three more columns. These represent the ages of individuals who were considered in either low, medium, or high risk lifestyle/personality traits (factor B). The four rows of the table represent the four age brackets selected for the parents. This is the heredity factor (factor A).

The analysis of variance for this example is a little different than before, because the researcher wants to know more than just the individual influences of the three factors on life expectancy. She also would like to determine if there is *interaction* among the factors as well. For example, perhaps a person's genes tend to influence his or her personality and lifestyle choices, so there might be an interaction between these two factors. ANOVA can test for this. It can even test for a *three-way* interaction among factors A, B, and C. This is good information, but it requires a lot more calculations, as we'll see shortly.

Table 7-6. Three-Factor Data Table for the Life Expectancy Example

AVERAGE AGE OF PARENTS (A)	JOB STRESS (C)					
	LOW			HIGH		
	LIFESTYLE RISK (B)			LIFESTYLE RISK (B)		
	Low	Medium	High	Low	Medium	High
< 40	64	62	67	65	63	73
	65	61	70	73	67	70
	68	63	72	69	62	66
40 - 60	69	61	64	59	76	65
	71	65	63	68	45	61
	69	63	61	65	59	69
60 - 80	82	72	67	64	61	71
	85	70	66	70	69	70
	87	69	69	76	65	73
> 80	86	79	74	78	80	73
	92	81	74	84	77	71
	88	76	72	86	79	74

Before we tackle the number crunching, however, we'd like to mention a few important points about this experimental design. It goes by the name of a *factorial* design, because there is more than one observation per cell in the data table, and multiple factors involved. There are methods (which we won't go into in our discussion) in the field of experimental design that allow the analyst to obtain about the same degree of information from a collection of experimental data by reducing the required number of combinations of factors. This is especially useful when the testing or data collection is expensive, time-consuming, or destructive. The procedures are contained in virtually all advanced statistics texts.

Another point relates to the methods in which the sum of squares and degrees of freedom values are obtained for the ANOVA table. As we see in the following, there are seven tests we can conduct for this design. We can test the significance of each of the three main factors (A, B, and C), we can test the interactions among each pair of factors (denoted A×B, A×C, and B×C), and we can test the three-way interaction (A×B×C).

To do this, we also need the total sum of squares and the error sum of squares. These are calculated in the usual manner. However, it becomes a little more tricky to find the other sum of squares values, particularly for the interactions. This is partly a result of presenting what amounts to a three-dimensional collection of data in a two-dimensional table. We have to really keep our heads clear when we start extracting the appropriate sums from the table.

To get us off on the right foot, let's first enter into the data table all the sums we can conveniently write down, given the format of the table. These would be the cell totals, the totals for the six columns associated with factor B, the totals for the two major columns of the table, and the row totals (as well as the overall total of all data values). These sums are shown in Table 7-7. Note that the cell totals are shown within each cell.

Table 7-7. Data Table from Table 7-6 with Totals Shown

AVERAGE AGE OF PARENTS (A)	JOB STRESS (C)						TOTAL
	LOW LIFESTYLE RISK (B)			HIGH LIFESTYLE RISK (B)			
	Low	Medium	High	Low	Medium	High	
< 40	64 65 68 → 197	62 61 63 → 186	67 70 72 → 209	65 73 69 → 207	63 67 62 → 192	73 70 66 → 209	**1,200**
40 - 60	69 71 69 → 209	61 65 63 → 189	64 63 61 → 188	59 68 65 → 192	76 45 59 → 180	65 61 69 → 195	**1,153**
60 - 80	82 85 87 → 254	72 70 69 → 211	67 66 69 → 202	64 70 76 → 210	61 69 65 → 195	71 70 73 → 214	**1,286**
> 80	86 92 88 → 266	79 81 76 → 236	74 74 72 → 220	78 84 86 → 248	80 77 79 → 236	73 71 74 → 218	**1,424**
Subtotal	926	822	819	857	803	836	
Total	**2,567**			**2,496**			**5,063**

SQL/Query

The sums shown in Table 7-7 can be obatined by executing the queries that follow. Query 7_32 creates the table. Query 7_33 obtains the data to go into the table. Query 7_34 populates the table with data.

Query 7_32:

```
CREATE TABLE [Table 7_7]
(SeqNum integer,
Factor text(10),
[Factor Name] text(20),
Total  long,
[Count] long);
```

Query 7_33:

```
SELECT "1" as SeqNum, "A" AS Factor,
[Table 7_6].[Average Age of Parents] AS [Factor Name],
Sum([Table 7_6].[Life Expectancy]) AS [Total],
Count([Table 7_6].[Life Expectancy]) AS [Count]
FROM [Table 7_6]
GROUP BY [Table 7_6].[Average Age of Parents]

UNION ALL
SELECT "2" as SeqNum, "B" AS Factor,
[Table 7_6].[Lifestyle] AS [Factor Name],
Sum([Table 7_6].[Life Expectancy]) AS [Total],
Count([Table 7_6].[Life Expectancy]) AS [Count]
FROM [Table 7_6]
GROUP BY [Table 7_6].[Lifestyle]

UNION ALL
SELECT "3" as SeqNum, "C" AS Factor,
[Table 7_6].[Job Stress] AS [Factor Name],
Sum([Table 7_6].[Life Expectancy]) AS [Total],
Count([Table 7_6].[Life Expectancy]) AS [Count]
FROM [Table 7_6]
GROUP BY [Table 7_6].[Job Stress]
```

```
UNION ALL
SELECT "4" as SeqNum, "AxB" AS Factor,
[Table 7_6].[Average Age of Parents] & " - " & [Table 7_6].[Lifestyle]
AS [Factor Name],
Sum([Table 7_6].[Life Expectancy]) AS [Total],
Count([Table 7_6].[Life Expectancy]) AS [Count]
FROM [Table 7_6]
GROUP BY [Table 7_6].[Average Age of Parents], [Table 7_6].[Lifestyle]

UNION ALL
SELECT "5" as SeqNum, "AxC" AS Factor,
[Table 7_6].[Average Age of Parents] & " - " & [Table 7_6].[Job Stress]
AS [Factor Name],
Sum([Table 7_6].[Life Expectancy]) AS [Total],
Count([Table 7_6].[Life Expectancy]) AS [Count]
FROM [Table 7_6]
GROUP BY [Table 7_6].[Average Age of Parents], [Table 7_6].[Job Stress]

UNION ALL
SELECT "6" as SeqNum, "BxC" AS Factor,
[Table 7_6].[Job Stress] & " - " & [Table 7_6].[Lifestyle] AS [Factor Name],
Sum([Table 7_6].[Life Expectancy]) AS [Total],
Count([Table 7_6].[Life Expectancy]) AS [Count]
FROM [Table 7_6]
GROUP BY [Table 7_6].[Lifestyle], [Table 7_6].[Job Stress]

UNION ALL
SELECT "7" as SeqNum, "AxBxC" AS Factor,
[Table 7_6].[Average Age of Parents] & " - " & [Table 7_6].[Job Stress] &
" - " & [Table 7_6].[Lifestyle] AS [Factor Name],
Sum([Table 7_6].[Life Expectancy]) AS [Total],
Count([Table 7_6].[Life Expectancy]) AS [Count]
FROM [Table 7_6]
GROUP BY [Table 7_6].[Average Age of Parents], [Table 7_6].[Lifestyle],
[Table 7_6].[Job Stress];
```

Query 7_34:

```
INSERT INTO [Table 7_7]
SELECT [Query 7_33].*
FROM [Query 7_33];
```

Now we can proceed to the analysis of variance. First we need the total sum of squares. This is the most painful calculation, because we must take each of the 72 values in the table, square it, then add all the squares together. After that's done, we still have to subtract the correction term, as before.

$$SS_{total} = (64)^2 + (65)^2 + (68)^2 + L + (74)^2 - (5,063)^2/72$$
$$= 360,825 - 356,027.3472$$
$$= 4,797.6528$$

For each of the three main factors, we have to pull out the appropriate sums from the table. For example, when we want to isolate factor A (age of parents), we work with the squares of the row totals. There are 18 values in the table that contribute to each of these row totals, so we have to divide the sum of squares by 18 before we subtract the correction term. This is shown next.

$$SS_A = \left[(1,200)^2 + (1,153)^2 + (1,286)^2 + (1,424)^2\right]/18 - (5,063)^2/72$$
$$= 2,360.4861$$

The sum of squares for factor B is a little trickier. Here we have to remember that the three levels of factor B are divided in the table. In other words, to determine the sum for the low level of factor B, we must work with the two column sums in columns one and four of the table. We do likewise for the medium and high levels. Since there are 24 values in the table contributing to each of these three sums, we divide the first term in the sum of squares expression by 24, as shown.

$$SS_B = \left[(926+857)^2 + (822+803)^2 + (819+836)^2\right]/24 - (5,063)^2/72$$
$$= 586.7778$$

Factor C is a little easier. Since it has only two levels, we use the total for each half of the table, and divide the sum of their squares by 36 (half the values), again subtracting the correction term.

$$SS_C = \left[(2,567)^2 + (2,496)^2\right]/36 - (5,063)^2/72$$
$$= 70.0139$$

Now come the interaction sum of squares values. These are a bit strange. For A×B, we start out by scanning the table for the cells whose sums represent a unique combination of factors A and B. These would be the cells in the first and fourth columns of each row of the table. There are six values that contribute to each of these sums, so we divide the first term of the sum of squares expression by 6. From this term, we not only subtract the correction term, but the sum of squares values we just found for factor A and factor B.

$$SS_{A \times B} = \left[(197+207)^2 + (186+192)^2 + (209+209)^2 + L + (220+218)^2 \right] / 6$$
$$-SS_A - SS_B - (5,063)^2 / 72$$
$$= 439.8889$$

For the A×C interaction sum of squares, we have to look for the cells in the table representing unique combinations of factors A and C. The first row and first three columns would yield the first sum we need, since these are the only cells where we see low job stress and parents ages less than 40. There are nine values in these cells. So we would add $197 + 186 + 209$ and square the sum for our first term. The rest of the expression is shown below. Note that this time, SS_A and SS_C (and the correction term) are subtracted.

$$SS_{A \times C} = \left[(197+186+209)^2 + (209+189+188)^2 + L + (248+236+218)^2 \right] / 9$$
$$-SS_A - SS_C - (5,063)^2 / 72$$
$$= 114.4861$$

The six column totals in the table should be used to find the B×C interaction sum of squares. There are 12 values that make up each of these totals. We also remember to subtract SS_B and SS_C (and the correction term) this time.

$$SS_{B \times C} = \left[(926)^2 + (822)^2 + L + (836)^2 \right] / 12 - SS_B - SS_C - (5,063)^2 / 12$$
$$= 155.4444$$

The final interaction sum of squares is for A×B×C. Here we use every cell total (three values contributing to each total). Once we have their sum of squares, we subtract every sum of squares we've calculated so far, even the interactions, and also our old correction term. The result is shown next.

$$SS_{A \times B \times C} = \left[(197)^2 + (186)^2 + L + (218)^2 \right] / 3 - SS_A - SS_B - SS_C$$
$$-SS_{A \times B} - SS_{A \times C} - SS_{B \times C} - (5,063)^2 / 72$$
$$= 196.5555$$

We only have the error sum of squares left, and it's easy. We just subtract every sum of squares we've previously calculated from the total sum of squares.

$$SS_{error} = SS_{total} - SS_A - SS_B - SS_C - SS_{A \times B} - SS_{A \times C} - SS_{B \times C} - SS_{A \times B \times C}$$
$$= 874.0000$$

Now we have all the information we need for the ANOVA table. The degrees of freedom for each main factor are equal to one less than the number of levels of the factor. The interaction degrees of freedom are the products of the appropriate main factor degrees of freedom. For example, the A×B interaction degrees of freedom are the product of the degrees of freedom for factor A and those for factor B (3 × 2 = 6). The degrees of freedom for A×B×C are the product of all three main factor degrees of freedom. The error degrees of freedom are equal to the product of the number of levels of the three factors and one less than the number of observations (or replicates) per cell. In our example, this would be 4 × 3 × 2 × 2 = 48. The total degrees of freedom are always one less than the total number of observations, or in this case 71. The table *F* values are found for the appropriate pairs of degrees of freedom, as before. The entire ANOVA table is shown in Table 7-8.

Table 7-8. ANOVA Table for the Life Expectancy ExampleSource of Variance

SOURCE OF VARIANCE	SS	DEGREES OF FREEDOM	MS	F CALCULATED	F TABLE (5%)	CONCLUSION
Heredity (A)	2,360.4861	3	786.8287	43.21	>2.80	Significant
Lifestyle (B)	586.7778	2	293.3889	16.11	>3.19	Significant
Job Stress (C)	70.0139	1	70.0139	3.85	<4.04	Not Significant
yAxB Interaction	439.8889	6	73.3148	4.03	>2.29	Significant
AxC Interaction	114.4861	3	38.1620	2.10	<2.80	Not Significant
BxC Interaction	155.4444	2	77.7222	4.27	>3.19	Significant
AxBxC Interaction	196.5555	6	32.7593	1.80	<2.29	Not Significant
Error	874.0000	48	18.2083			
Total	4,797.6528	71				

From reading the table results, the researcher can conclude that heredity appears to be highly significant as an indicator of life expectancy, with lifestyle also important. Job stress, however, does not seem to be a significant predictor. There is a significant interaction between heredity and lifestyle, which might indicate that genes somewhat influenced lifestyle choices in the population. There is also a significant interaction between lifestyle and job stress, although job stress alone is not significant.

SQL/Query

When executed in sequence, the following set of SQL queries generates the ANOVA table shown in Table 7-8. We first create the table and then we proceed to make the needed calculations and to populate the table. You will notice some of the queries have the same query number but with an "a" or "b" suffix. Those queries with the "a" suffix are selection queries, and the result of these queries is used in later queries. The "b" suffixed queries are insertion queries that populate the ANOVA table with values. The "a" and "b" queries perform the same numerical calculations. The reason we made two queries is to reduce the complexity of the queries and to show the calculations along the way.

Query 7_35 creates an empty ANOVA table that is to receive the values shown in Table 7-8.

Query 7_35:

```
CREATE TABLE [Three Way ANOVA]
(Source text(20),
SS  single,
[Degrees of Freedom] integer,
MS single,
[F Calculated] single,
[F Table 5%]  single);
```

The correction term is calculated by Query 7_36 and Query 7_37 calculates the sum of squares for total. Query 7_37 also determines the overall degrees of freedom, which are needed later.

Query 7_36:

```
SELECT (Sum([Life Expectancy])^2) / Count([Life Expectancy]) AS [Correction Term]
FROM [Table 7_6];
```

Query 7_37:

```
SELECT "Total" AS Source, Sum([Life Expectancy]^2) -
Avg([Correction Term]) AS [SS Total],
Count([Life Expectancy]) - 1 AS [Degrees of Freedom]
FROM [Table 7_6], [Query 7_36];
```

The next pair of queries calculates the SS_A.

Query 7_38a:

```
SELECT [Table 7_7].[Factor],
Sum([Total]^2) / Avg([count]) - Avg([Correction Term]) AS [SS A],
Count([Count]) - 1 AS [A Degrees of Freedom]
FROM [Query 7_36], [Table 7_7]
GROUP BY [Table 7_7].[Factor]
HAVING [Table 7_7].[Factor] = "A";
```

Query 7_38b:

```
INSERT INTO [Three Way ANOVA] ( Source, SS, [Degrees of Freedom] )
SELECT [Table 7_7].[Factor],
Sum([Total]^2) / Avg([count]) - Avg([Correction Term]) AS [SS A],
Count([Count]) - 1 AS [Degrees of Freedom]
FROM [Query 7_36], [Table 7_7]
GROUP BY [Table 7_7].[Factor]
HAVING [Table 7_7].[Factor] = "A";
```

The next pair of queries calculates the SS_B.

Query 7_39a:

```
SELECT [Table 7_7].[Factor], Sum([Total]^2) / Avg([count]) -
Avg([Correction Term]) AS [SS B],
Count([Count]) - 1 AS [B Degrees of Freedom]
FROM [Query 7_36], [Table 7_7]
GROUP BY [Table 7_7].[Factor]
HAVING [Table 7_7].[Factor] = "B";
```

Query 7_39b:

```
INSERT INTO [Three Way ANOVA] (Source, SS, [Degrees of Freedom])
SELECT [Table 7_7].[Factor],
Sum([Total]^2) / Avg([count]) - Avg([Correction Term]) AS [SS B],
Count([Count]) - 1 AS [Degrees of Freedom]
FROM [Query 7_36], [Table 7_7]
GROUP BY [Table 7_7].[Factor]
HAVING [Table 7_7].[Factor] = "B";
```

The next pair of queries calculates the SS_C.

Query 7_40a:

```
SELECT [Table 7_7].[Factor], Sum([Total]^2) / Avg([count]) -
Avg([Correction Term]) AS [SS C],
Count([Count]) - 1 AS [C Degrees of Freedom]
FROM [Query 7_36], [Table 7_7]
GROUP BY [Table 7_7].[Factor]
HAVING [Table 7_7].[Factor] = "C";
```

Query 7_40b:

```
INSERT INTO [Three Way ANOVA] (Source, SS, [Degrees of Freedom])
SELECT [Table 7_7].[Factor],
Sum([Total]^2) / Avg([count]) - Avg([Correction Term]) AS [SS C],
Count([Count]) - 1 AS [Degrees of Freedom]
FROM [Query 7_36], [Table 7_7]
GROUP BY [Table 7_7].[Factor]
HAVING [Table 7_7].[Factor] = "C";
```

The next pair of queries calculates the SS_{AxB}.

Query 7_41a:

```
SELECT [Table 7_7].Factor, Sum([Total]^2) / Avg([count]) -
Avg([SS A]) - Avg([SS B]) -Avg([Correction Term]) AS [SS AxB],
Avg([A Degrees of Freedom]) * Avg([B Degrees of Freedom])
AS [AxB Degrees of Freedom]
FROM [Query 7_36], [Query 7_38a], [Query 7_39a], [Table 7_7]
GROUP BY [Table 7_7].Factor
HAVING [Table 7_7].Factor = "AxB";
```

Query 7_41b:

```
INSERT INTO [Three Way ANOVA] (Source, SS, [Degrees of Freedom])
SELECT [Table 7_7].Factor,
Sum([Total]^2) / Avg([count]) - Avg([SS A]) - Avg([SS B]) - Avg([Correction Term])
AS [SS AxB],
Avg([A Degrees of Freedom]) * Avg([B Degrees of Freedom])
AS [AxB Degrees of Freedom]
FROM [Query 7_36], [Query 7_38a], [Query 7_39a], [Table 7_7]
GROUP BY [Table 7_7].Factor
HAVING [Table 7_7].[Factor] = "AxB";
```

The next pair of queries calculates the SS_{AxC}.

Query 7_42a:

```
SELECT [Table 7_7].Factor,
Sum([Total]^2) / Avg([count]) - Avg([SS A]) - Avg([SS C]) - Avg([Correction Term])
AS [SS AxC],
Avg([A Degrees of Freedom]) * Avg([C Degrees of Freedom])
AS [AxC Degrees of Freedom]
FROM [Query 7_36], [Query 7_38a], [Query 7_40a], [Table 7_7]
GROUP BY [Table 7_7].Factor
HAVING [Table 7_7].Factor = "AxC";
```

Query 7_42b:

```
INSERT INTO [Three Way ANOVA] (Source, SS, [Degrees of Freedom])
SELECT [Table 7_7].Factor,
Sum([Total]^2) / Avg([count]) - Avg([SS A]) - Avg([SS C]) - Avg([Correction Term])
AS [SS AxC],
Avg([A Degrees of Freedom]) * Avg([C Degrees of Freedom])
AS [AxC Degrees of Freedom]
FROM [Query 7_36], [Query 7_38a], [Query 7_40a], [Table 7_7]
GROUP BY [Table 7_7].Factor
HAVING [Table 7_7].[Factor] = "AxC";
```

The next pair of queries calculates the SS_{BxC}.

Query 7_43a:

```
SELECT [Table 7_7].[Factor],
Sum([Total]^2) / Avg([count]) - Avg([SS B]) - Avg([SS C]) - Avg([Correction Term])
AS [SS BxC],
Avg([B Degrees of Freedom]) * Avg([C Degrees of Freedom])
AS [BxC Degrees of Freedom]
FROM [Query 7_36], [Query 7_40a], [Query 7_39a], [Table 7_7]
GROUP BY [Table 7_7].[Factor]
HAVING [Table 7_7].[Factor] = "BxC";
```

Query 7_43b:

```
INSERT INTO [Three Way ANOVA] (Source, SS, [Degrees of Freedom])
SELECT [Table 7_7].Factor,
Sum([Total]^2) / Avg([count]) - Avg([SS B]) - Avg([SS C]) - Avg([Correction Term])
AS [SS BxC],
Avg([B Degrees of Freedom]) * Avg([C Degrees of Freedom])
AS [BxC Degrees of Freedom]
FROM [Query 7_36], [Query 7_40a], [Query 7_39a], [Table 7_7]
GROUP BY [Table 7_7].Factor
HAVING [Table 7_7].[Factor] ="BxC";
```

The next pair of queries calculates the SS_{AxBxC}.

Query 7_44a:

```
SELECT [Table 7_7].[Factor],
Sum([Total]^2) / Avg([count]) - Avg([SS A]) - Avg([SS B]) - Avg([SS C]) -
Avg([SS AxB]) – Avg([SS AxC]) - Avg([SS BxC]) -
Avg([Correction Term]) AS [SS AxBxC],
Avg([A Degrees of Freedom]) * Avg([B Degrees of Freedom]) *
Avg([C Degrees of Freedom]) AS [AxBxC Degrees of Freedom]
FROM [Query 7_36], [Query 7_41a], [Query 7_42a], [Query 7_43a],
[Query 7_38a], [Query 7_39a], [Query 7_40a], [Table 7_7]
GROUP BY [Table 7_7].[Factor]
HAVING [Table 7_7].[Factor] = "AxBxC";
```

Query 7_44b:

```
INSERT INTO [Three Way ANOVA] (Source, SS, [Degrees of Freedom])
SELECT [Table 7_7].Factor,
Sum([Total]^2) / Avg([count]) - Avg([SS A]) - Avg([SS B]) - Avg([SS C]) –
Avg([SS AxB]) - Avg([SS AxC]) - Avg([SS BxC]) - Avg([Correction Term])
AS [SS AxBxC],
Avg([A Degrees of Freedom]) * Avg([B Degrees of Freedom]) *
Avg([C Degrees of Freedom]) AS [AxBxC Degrees of Freedom]
FROM [Query 7_36], [Query 7_41a], [Query 7_42a], [Query 7_43a],
[Query 7_38a], [Query 7_39a], [Query 7_40a], [Table 7_7]
GROUP BY [Table 7_7].Factor
HAVING [Table 7_7].[Factor] = "AxBxC";
```

The next pair of queries calculates the SS$_{error}$.

Query 7_45a:

```
SELECT [SS Total] - [SS A] - [SS B] - [SS C] -
[SS AxB] - [SS AxC] - [SS BxC] - [SS AxBxC] AS [SS error],
[Query 7_37].[Degrees of Freedom] -
[A Degrees of Freedom] - [B Degrees of Freedom] - [C Degrees of Freedom] -
[AxB Degrees of Freedom] - [AxC Degrees of Freedom] - [BxC Degrees of Freedom] -
[AxBxC Degrees of Freedom] AS [Error Degrees of Freedom]
FROM [Query 7_37], [Query 7_38a], [Query 7_39a], [Query 7_40a],
[Query 7_41a], [Query 7_42a], [Query 7_43a], [Query 7_44a];
```

Query 7_45b:

```
INSERT INTO [Three Way ANOVA] (Source, SS, [Degrees of Freedom])
SELECT "Error" AS Source,
[SS Total] - [SS A] - [SS B] - [SS C] -
[SS AxB] - [SS AxC] - [SS BxC] - [SS AxBxC] AS [SS error],
[Query 7_37].[Degrees of Freedom] -
[A Degrees of Freedom] - [B Degrees of Freedom] - [C Degrees of Freedom] -
[AxB Degrees of Freedom] - [AxC Degrees of Freedom] - [BxC Degrees of Freedom] -
[AxBxC Degrees of Freedom] AS [Error Degrees of Freedom]
FROM [Query 7_37], [Query 7_38a], [Query 7_39a], [Query 7_40a],
[Query 7_41a], [Query 7_42a], [Query 7_43a], [Query 7_44a];
```

Now we calculate the MS values.

Query 7_46:

```
UPDATE [Three Way ANOVA]
SET [Three Way ANOVA].MS = [SS] / [Degrees of Freedom]
WHERE [Three Way ANOVA].Source <> "Total";
```

Then we calculate the error SS and degrees of freedom.

Query 7_47:

```
SELECT [Three Way ANOVA].Source,

[Three Way ANOVA].[Degrees of Freedom] AS [Error Degrees of Freedom],

[Three Way ANOVA].MS AS [Error MS]

FROM [Three Way ANOVA]
WHERE [Three Way ANOVA].Source = "Error";
```

Next we calculate the *F* statistics.

Query 7_48:

```
UPDATE [Three Way ANOVA], [Query 7_47]
SET [Three Way ANOVA].[F Calculated] = [MS] / [Error MS]
WHERE [Three Way ANOVA].Source NOT IN ("Error","Total");
```

Now we retrieve the *F* table values.

Query 7_49:

```
UPDATE [F Table], [Query 7_47], [Three Way ANOVA]
SET [Three Way ANOVA].[F Table 5%] = [F Table].[F]
WHERE [Three Way ANOVA].Source NOT IN ("Error","Total")
AND [F Table].[Significance level] = 0.05
AND [F Table].[Numerator Degrees of Freedom] =
[Three Way ANOVA].[Degrees of Freedom]
AND [F Table].[Denomiator Degrees of Freedom] =
[Query 7_47].[Error Degrees of Freedom];
```

Finally we append the totals.

Query 7_50:

```
INSERT INTO [Three Way ANOVA] (Source, SS, [Degrees of Freedom])
SELECT "Total" AS Source,
Sum([Life Expectancy]^2) - Avg([Correction Term]) AS [SS Total],
Count([Life Expectancy]) - 1 AS [Degrees of Freedom]
FROM [Table 7_6], [Query 7_36];
```

Don't Knock It If It Works

Analysis of variance has been the traditional approach to assessing experimental designs. One of its drawbacks, however, is the requirement of a great deal of expensive experimental trials to obtain the data necessary for certain analyses. Dr. Genichi Taguchi of Japan, beginning primarily in the 1980s, developed a new approach that was simpler to implement in practice and produced dramatic improvements. It has come to be known as "parameter design" or "robust design" in the United States. Basically, Taguchi's approach includes several related concepts. First, he felt that customer satisfaction with a product or service could be characterized by a quadratic *loss function*, demonstrating that dissatisfaction increased significantly as quality deviated more and more from the customer's expectations. Dr. Taguchi attempted to reduce the sensitivity of engineering designs to factors that were uncontrollable. He called these "noise" factors, and they could include uncontrollable environmental conditions, variation in materials, tolerances that are unnecessarily tight, and the like. His design approach set about to reduce the influence of these noise factors by more robust designs, and he achieved this goal through simpler experimental designs and fewer trials and data gathering. Although condemned by some theoretical statisticians in the United States, Taguchi's methods have worked well in practice in Japan.

Conclusion

In this brief discussion, we have tried to highlight the basic solution techniques for ANOVA problems of various designs. The body of literature on this subject, however, is extensive. Many variations of the models and designs we have presented occur in practice, although the basic ANOVA techniques are comparable. In addition, an innovative approach to experimental design was developed several years ago by Dr. Genichi Taguchi and his associates (see the sidebar). It incorporates some of the same theory as analysis of variance, but has the advantage of extracting a great deal of information from smaller designs, thereby providing a more economical set of tools for the researcher or data miner.

T-SQL Source Code

Below is the T-SQL source code listing for performing the ANOVA as discussed in this chapter. The procedure is called ANOVA and will perform either an *n*-way ANOVA or an *n*-factor ANOVA for *n* = 1, 2, 3, … factors. Following the procedure listing is a set of statements illustrating how to invoke the ANOVA procedure.

ANOVA

```
SET QUOTED_IDENTIFIER OFF
GO
SET ANSI_NULLS OFF
GO

CREATE PROCEDURE [ANOVA]
@SrcTable Varchar(50) = 'Table 7_6',
@Num_Factors Int = 3,
@First_Factor_Position Int = 2,
@RsltTable Varchar(50) = 'ANOVA RESULT'
AS


/**********************************************************/
/*                                                      */
/*                      ANOVA                           */
/*                                                      */
/*    This procedure performs an n-way or n-factor (n=1,2,…) */
/* ANOVA on a set of data given in the table denoted by   */
/* SrcTable, and creates a ANOVA result table with the name */
/* supplied by RstlTable. An n-way ANOVA is performed if   */
/* there is only one single observation per cell; otherwise, */
/* an n-factor ANOVA is performed if there are replications. */
/* The SrcTable must contain n consecutive columns, one for */
/* each factor, and the column name serves as an identifier */
/* for the factor. The next column after the n-th factor is */
/* the sample data value. For n > 1, combinations of factors */
/* are considered and these combinations are generated by   */
/* this procedure. After combinations are generated the    */
/* procedure proceeds to calculate the sums, sum of squares, */
/* and the degrees of freedom for each factor and for each  */
/* combination of factors. Next the mean squares are        */
/* determined. The F statistic is calculated and the        */
/* respective F table value is retrieved. The calculations  */
/* of the ANOVA are summarized in a result table denoted by */
/* RsltTable.                                              */
```

```
/*                                                    */
/* INPUTS:                                            */
/*   SrcTable - name of table containing sample data  */
/*   Num_Factors - the number of factors participating in */
/*     the ANOVA                                      */
/*   First_Factor_Position - the column number of the */
/*     column in SrcTable that contains the first factor */
/*                                                    */
/* OUTPUTS:                                           */
/*   RsltTable - a name for the ANOVA table that is created */
/*     by this procedure                              */
/*                                                    */
/******************************************************/


/* Local Variables */
DECLARE @Q Varchar(5000)                    /* Query string */
DECLARE @Do_nWay Int                        /* Flag denoting n-way ANOVA */
DECLARE @Tid Int                            /* Table ID number */
DECLARE @ci Int                             /* Column ID number */
DECLARE @si Int                             /* Column ID number */
DECLARE @fi Int                             /* Factor ID number */
DECLARE @ColName Varchar(50)                /* Column name */
DECLARE @ValColName Varchar(50)             /* Value column name */
DECLARE @AllColNames Varchar(500)           /* All the column names */
DECLARE @Fid Varchar(10)                    /* Factor ID as text */
DECLARE @Last_Cid Int                       /* ID of last combination */
DECLARE @Fname Varchar(50)                  /* Factor name */
DECLARE @Comb Varchar(500)                  /* Combination of factors */
DECLARE @SComb Varchar(500)                 /* Combination of factors */
DECLARE @CombLast Varchar(500)              /* Combination of factors */
DECLARE @CombGroup Varchar(500)             /* Combination of factors */
DECLARE @CombNameGroup Varchar(500)         /* Combination of factors */
DECLARE @CorrectionTerm Float               /* Correction Term */
DECLARE @Csums Float                        /* A sum of factor values */
DECLARE @Sums Float                         /* A sum of factor values */
DECLARE @CombPat Varchar(500)               /* Search pattern */
DECLARE @TotalAll Float                     /* Total of all values */
DECLARE @DF1 Int                            /* Degrees of freedom */
DECLARE @DF2 Int                            /* Degrees of freedom */
DECLARE @TotalDF Int                        /* Total deg. of freedom */
DECLARE @SumSS Float                        /* Sum of Sum of Squares */
DECLARE @SumDF Int                          /* Sum of deg. of freedom */
```

```
DECLARE @ErrorSS Float                    /* Error Sum of Squares */
DECLARE @ErrorDF Int                      /* Error deg. of freedom */
DECLARE @ErrorMS Float                    /* Error MS */


/*****************/
/* OBTAIN FACTORS */
/*****************/


/* Get source table ID number */
SELECT @Tid = (SELECT id FROM ..sysobjects
      WHERE name = @SrcTable)


/* Create a work table containing the factor names and */
/* the respective column names in the source table */
CREATE TABLE ##TempFactors(Factor varchar(10),
      Factor_Name varchar(50))


/* Initialize */
SET @fi = 1
SET @ci = @First_Factor_Position
SET @AllColNames = ''


/* Get column names from source table and */
/* save them in the factors table. Also */
/* collect all the column names as a text string */
WHILE @ci <= @Num_Factors + @First_Factor_Position - 1
Begin

     /*Get source table column name */
SELECT @ColName = (SELECT name FROM ..syscolumns
          WHERE id = @Tid And Colid = @ci)

     /* Generate factor id */
     SET @Fid = 'F' + convert(varchar(10), @fi)

     /* Save info in factors table */
     INSERT INTO ##TempFactors
          VALUES(@Fid, @ColName)

     /* Accumulate the column names */
     SET @AllColNames = @AllColNames + ', [' + @ColName + ']'
```

```
        /* Increment column index pointer */
        SET @fi = @fi + 1
        SET @ci = @ci + 1
End

    /* Get the column name of the data values column */
    SELECT @ValColName = (SELECT name FROM ..syscolumns
        WHERE id = @Tid And Colid = @ci)

    /* Trim off the ', ' from the accumulated column names */
    SET @AllColNames = substring(@AllColNames, 3, len(@AllColNames)-2)

    /* Determine if this is an n-way or n-factor ANOVA */
    /* Assume it is n_way ANOVA by setting flag to one. */
    SET @Do_nWay = 1
    SET @Q = 'SELECT ' + @AllColNAmes +
        'INTO ##TempNway ' +
        'FROM [' + @SrcTable + ']' +
        'GROUP BY ' + @AllColNames + 'HAVING Count(*) > 1'
    EXEC(@Q)

    /* If there are more than one set of values, */
    /* then it is an n-factor ANOVA. So, set the */
    /* flag to zero to denote not n-Way ANOVA. */
    IF exists (SELECT * FROM ##TempNway)
    Begin
        SET @Do_nWay = 0
    end


    /****************************/
    /* BUILD FACTOR COMBINATIONS */
    /****************************/


    /* Establish cursor pointing to factor names */
    DECLARE cr_F INSENSITIVE SCROLL CURSOR
        FOR SELECT Factor, Factor_Name FROM ##TempFactors
```

```
/* Create a work table containing all the factor */
/* combinations and insert a dummy first record */
CREATE TABLE ##TempCombs
    (Cid varchar(10),
    Comb_Name varchar(100),
    Comb_Name_Group varchar(100),
    Comb_Group Varchar(500),
    Comb_Name_Len Int,
    Sums Float,
    DegFreedom Int,
    MS Float,
    [F Calculated] Float,
    [F Table 5%] Float)
INSERT INTO ##TempCombs VALUES('0', '1', '1', '1', 0, 0, 0, 0, 0, 0)


/* Establish cursor pointing to factor combinations */
/* Note: 'INSENSITIVE' is necessary because the */
/* cursor must reference the preceding combinations */
/* and not the combinations currently being added. */
DECLARE cr_C INSENSITIVE SCROLL CURSOR
    FOR SELECT Comb_Name, Comb_Name_Group, Comb_Group FROM ##TempCombs


/* Set up cursor for factor */
OPEN cr_F
FETCH FIRST FROM cr_F INTO @Fid, @Fname


/* Loop through each factor and build the */
/* combinations by appending the factor */
/* to all previous combinations */
WHILE @@FETCH_STATUS = 0
Begin

    /* Get each preceding combination and append */
    /* the newly selected factor to yield a new */
    /* combination. Afterwards, add the newly */
    /* formed combination to list of combinations */

    OPEN cr_C
    FETCH FIRST FROM cr_C INTO @Comb, @CombNameGroup, @CombGroup
```

```
            WHILE @@FETCH_STATUS = 0
         Begin
             INSERT INTO ##TempCombs
                 VALUES('0', @Comb + @Fid,
                     @CombNameGroup + ', ' + @Fid,
                     @CombGroup + ', [' + @Fname + ']',
                     0, 0, 0, 0, 0, 0)
             FETCH NEXT FROM cr_C INTO @Comb, @CombNameGroup, @CombGroup
         End

         /* Prepare to do next factor */
         CLOSE cr_C
         FETCH NEXT FROM cr_F INTO @Fid, @Fname

    End

    /* Close cursor */
    CLOSE cr_F

    /* Remove dummy row */
    DELETE FROM ##TempCombs WHERE Comb_Name = '1'

    /* Remove the dummy prefix '1' */
    UPDATE ##TempCombs
        SET Comb_Group = substring(Comb_Group, 4, len(Comb_Group)-3)
    UPDATE ##TempCombs
        SET Comb_Name = substring(Comb_Name, 2, Len(Comb_Name)-1),
            Comb_Name_Group = substring(Comb_Name_group, 4, len(Comb_Name_Group)-3)

    /* Check for n-Way ANOVA */
    IF @Do_nWay = 1
    Begin
        /* Doing an n-way ANOVA, so remove the combination */
        /* record that has all the factors in it. */
        SET @Q = 'DELETE FROM ##TempCombs ' +
            'WHERE Comb_Group = ''' + @AllColNames + ''''
        EXEC(@Q)
    End

    /* Assign lengths to combinations */
    UPDATE ##TempCombs
        SET Comb_Name_Len = Len(Comb_Name)
```

```
/* Next, we need to create an ID number */
/* (sequence number) for each combination */

/* Set up cursor to go through each combination of */
/* factors in increasing order by combination complexity */
DECLARE cr_S INSENSITIVE SCROLL CURSOR
     FOR SELECT Comb_Name FROM ##TempCombs
          ORDER BY Comb_Name_Len, Comb_Name

/* Initialize ID number, open cursor, */
/* and get the first record */
SET @ci = 0
OPEN cr_S
FETCH FIRST FROM cr_S INTO @Comb

/* For each factor combination, calculate the */
/* ID number and save it in the table */
WHILE @@FETCH_STATUS = 0
Begin
     SET @ci = @ci + 1
     SET @Q = 'UPDATE ##TempCombs '+
          'SET Cid = ' + convert(varchar(10), @ci) + ' ' +
          'WHERE Comb_Name = ''' + @Comb + ''''
     EXEC(@Q)
     FETCH NEXT FROM cr_S INTO @Comb
End

/* Remember last ID number */
SET @Last_Cid = @ci

CLOSE cr_S
DEALLOCATE cr_S

/***************************/
/* GROUP TALLIES AND COUNTS */
/***************************/

/* Establish a table to receive the totals and */
/* counts for each combination of factors */
CREATE TABLE ##TempTotals
     (Comb Varchar(100), [Total] Float, [Count] Int)
```

```
/* Set up cursor to go through */
/* each combination of factors */
OPEN cr_C
FETCH FIRST FROM cr_C INTO @Comb, @CombNameGroup, @CombGroup


/* For each factor combination, calculate the */
/* total and count the number of data values */
WHILE @@FETCH_STATUS = 0
Begin
    SET @Q = 'INSERT INTO ##TempTotals '+
        'SELECT ''' + @Comb + ''', ' +
        'Sum([' + @ValColName + ']) AS [Total], ' +
        'Count([' + @ValColName + ']) AS [Count] ' +
        'FROM [' + @SrcTable + '] ' +
        'GROUP BY ' + @CombGroup
    EXEC(@Q)
    FETCH NEXT FROM cr_C INTO @Comb, @CombNameGroup, @CombGroup
End
CLOSE cr_C


/********************************/
/* CALCULATE THE SUM OF SQUARES */
/********************************/


/* Get Correction term */
SET @Q = 'SELECT (Sum([' + @ValColName + ']) * ' +
    'Sum([' + @ValColName + '])) / ' +
    'Count([' + @ValColName + ']) ' +
    'AS [Correction Term] ' +
    'INTO ##TempCorrectTerm ' +
    'FROM [' + @SrcTable + ']'
EXEC(@Q)
SELECT @CorrectionTerm = [Correction Term] FROM ##TempCorrectTerm


/* Calculate the sums for each combination of factors */


/* Set up cursor to go through */
/* each combination of factors */
OPEN cr_C
FETCH FIRST FROM cr_C INTO @Comb, @CombNameGroup, @CombGroup
```

```
/* For each factor combination, calculate the sum of */
/* squares minus the correction term and the degrees */
/* of freedom for the single factor combinations */
WHILE @@FETCH_STATUS = 0
Begin

SET @Q = 'UPDATE ##TempCombs ' +
    'SET Sums = ' +
        '(SELECT Sum([Total] * [Total]) / Avg([Count]) - ' +
            Str(@CorrectionTerm, 12, 4) +
            'FROM ##TempTotals ' +
            'WHERE ##TempTotals.Comb = ''' + @comb + '''), ' +
    'DegFreedom = ' +
        '(SELECT Count([Total]) - 1 ' +
            'FROM ##TempTotals ' +
            'WHERE ##TempTotals.Comb = ''' + @comb + ''') ' +
    'WHERE ##TempCombs.Comb_Name = ''' + @comb + ''''
    EXEC(@Q)
    FETCH NEXT FROM cr_C INTO @Comb, @CombNameGroup, @CombGroup
End

/* Close cursor */
Close cr_C

/* To complete the sum of squares, the overlapping */
/* sub-combinations need to be subtracted off */

DECLARE cr_CB SCROLL CURSOR
    FOR SELECT Cid, Comb_Name, Sums FROM ##TempCombs
        ORDER BY Cid

DECLARE cr_OVLP SCROLL CURSOR
    FOR SELECT Cid, Comb_Name, Sums FROM ##TempCombs
        ORDER BY Cid

/* Open cursor and get first combination */
OPEN cr_CB
FETCH FIRST FROM cr_CB INTO @ci, @Comb, @Csums

/* For each factor combination, subtract */
/* the overlapping combinations */
WHILE @@FETCH_STATUS = 0
Begin
```

```
/* Open cursor and get first candidate sub-combination */
OPEN cr_OVLP
FETCH FIRST FROM cr_OVLP INTO @si, @Scomb, @Sums

/* For each factor combination, subtract */
/* the overlapping sub-combinations */
WHILE @@FETCH_STATUS = 0 and @si < @ci
Begin
    /* Build search pattern for sub-combination */
    SET @CombPat = REPLACE(@Scomb,'F','%F') + '%'

    /* Check to see if sub-combination is in the combination */
    IF PatIndex(@CombPat, @Comb) > 0 and @Scomb <> @Comb
    Begin
        /* sub-combination is in the combination, */
        /* so subtract out the overlap */
        SET @Csums = @Csums - @Sums
    End

    /* Get next candidate sub-combination */
    FETCH NEXT FROM cr_OVLP INTO @si, @Scomb, @Sums
End

/* Close cursor */
CLOSE cr_OVLP

/* Save the sum of squares */
SET @Q = 'UPDATE ##TempCombs '+
    'SET Sums = ' + convert(varchar(20), @Csums) + ' ' +
    'WHERE Cid = ' + convert(varchar(10), @ci)
EXEC(@Q)

/* Get next combination */
FETCH NEXT FROM cr_CB INTO @ci, @Comb, @Csums
End

/* Close and release cursors */
CLOSE cr_CB
DEALLOCATE cr_OVLP
DEALLOCATE cr_CB
```

```
/******************************/
/* CALCULATE DEGREES OF FREEDOM */
/******************************/


/* Calculate the degrees of freedom for those */
/* combinations with more than one factor */


/* Establish a work table for computing degrees of freedom */
CREATE TABLE ##TempDF(DF Int)
INSERT INTO ##TempDF VALUES (1)


/* Define cursor for obtaining degrees of freedom */
DECLARE cr_DF SCROLL CURSOR
     FOR SELECT DF FROM ##TempDF


/* Set up cursor for looping through each combination */
Open cr_C
FETCH FIRST FROM cr_C INTO @Comb, @CombNameGroup, @CombGroup


/* For each combination calculate its degrees of freedom */
/* as the product of the degrees of freedom of the factors */
/* in the combination. */
WHILE @@FETCH_STATUS = 0
Begin

     /* Does the combination have more than one factor? */
     IF CharIndex(',', @CombNameGroup) <> 0
     Begin

          SET @CombNameGroup = Replace(@CombNameGroup, 'F', '"F')
          SET @CombNameGroup = Replace(@CombNameGroup, ',', '","')
          SET @CombNameGroup = @CombNameGroup + '"'

          /* Get degrees of freedom for each factor */
          /* in the multi-factor combination */
          SET @Q = 'INSERT INTO ##TempDF (DF) ' +
               'SELECT DegFreedom AS DF ' +
               'FROM ##TempCombs ' +
               'WHERE Comb_Name IN (' + @CombNameGroup + ')'
          EXEC(@Q)
```

```
                    /* Obtain the product of the degrees of freedom */
                    /* of those factors in the combination */
                    SET @DF2 = 1
                    Open cr_DF
                    FETCH FIRST FROM cr_DF INTO @DF1
                    WHILE @@FETCH_STATUS = 0
                    Begin
                        Set @DF2 = @DF2 * @DF1
                        FETCH NEXT FROM cr_DF INTO @DF1
                    End

                    /* Clean up temp table so it can be used */
                    /* for the next combination. */
                     Close cr_DF
                    delete from ##TempDF

                    /* Save the combinations degrees of freedom */
                    UPDATE ##TempCombs
                    SET DegFreedom = @DF2
                    WHERE Comb_Name = @comb

            End        /* IF */

            /* Get next combination record */
            FETCH NEXT FROM cr_C INTO @Comb, @CombNameGroup, @CombGroup

    End        /* WHILE */

Close cr_C

/*****************************/
/*  CALCULATE TOTAL AND ERROR  */
/*****************************/

/* Calculate the sum of squares and the degrees of freedom */
/* for the 'Total' */
SET @Q = 'SELECT Sum([' + @ValColName + '] * [' +
    @ValColName + ']) - ' +
    Str(@CorrectionTerm, 12, 4) + ' AS TTall, ' +
    'Count([' + @ValColName + ']) - 1 AS DF ' +
    'INTO ##TempTTall ' +
    'FROM [' + @SrcTable + ']'
EXEC(@Q)
SELECT @TotalAll = (SELECT TTall FROM ##TempTTall)
SELECT @TotalDF = (SELECT DF FROM ##TempTTall)
```

```
/* Save the 'TOTAL' by appending it to ##TempCombs */
INSERT INTO ##TempCombs
    (Cid, Comb_Name, Comb_Group, Comb_Name_Len, Sums, DegFreedom)
    Values(convert(Varchar(20), (@Last_cid + 2)), 'Total', 'Total', 5,
    convert(Varchar(20), @TotalAll), convert(Varchar(20), @TotalDF))

/* Calculate the 'Error' sum of squares and degrees of freedom */
SELECT @SumSS = (SELECT Sum([Sums]) FROM ##TempCombs WHERE Cid <= @Last_Cid)
SELECT @SumDF = (SELECT Sum(DegFreedom) FROM ##TempCombs WHERE Cid <= @Last_Cid)
SET @ErrorSS = @TotalAll - @SumSS
SET @ErrorDF = @TotalDF - @SumDF

/* Save the 'Error' by appending it to ##TempCombs */
INSERT INTO ##TempCombs
    (Cid, Comb_Name, Comb_Group, Comb_Name_Len, Sums, DegFreedom)
    Values(convert(Varchar(20), (@Last_cid + 1)), 'Error', 'Error', 5,
    convert(Varchar(20), @ErrorSS), convert(Varchar(20), @ErrorDF))

/*******************************/
/*  CALCULATE MEAN SQUARE (MS)  */
/*******************************/

/* Set the MS values */
UPDATE ##TempCombs
    SET MS = Sums / DegFreedom
    WHERE Cid <= (@Last_Cid + 1)

/**************************************/
/*  DETERMINE F CALCULATED AND F TABLE  */
/**************************************/

/* Get the value of MS for Error */
SELECT @ErrorMS = (SELECT MS FROM ##TempCombs WHERE Comb_Name = 'Error')

/* Set the F calculated values */
UPDATE ##TempCombs
    SET [F Calculated] = MS / @ErrorMS
    WHERE Cid <= @Last_Cid
```

```
/* Set the F Table values */
UPDATE ##TempCombs
    SET [F Table 5%] =
        (SELECT F FROM [F Table]
            WHERE [Numerator Degrees of Freedom] = ##TempCombs.DegFreedom
            AND [Denomiator Degrees of Freedom] = convert(Varchar(10), @ErrorDF)
            AND [Significance level] = .05)
    WHERE Cid <= @Last_Cid


/*********************/
/*  SAVE ANOVA TABLE  */
/*********************/

/* Remove previous result table if it exists */
IF Exists (SELECT id FROM ..sysobjects
    WHERE name = @RsltTable)
Begin
    SET @Q = 'DROP TABLE [' + @RsltTable + ']'
    EXEC(@Q)
End

/* Set up result table */
SET @Q = 'CREATE TABLE [' + @RsltTable + '] ' +
    '(Source varchar(500), ' +
    'SS Float, ' +
    '[Degrees of Freedom] Int, ' +
    'MS Float, ' +
    '[F Calculated] Float, ' +
    '[F Table 5%] Float)'
EXEC(@Q)

/* Full result table */
SET @Q = 'INSERT INTO [' + @RsltTable + '] ' +
    '(Source, SS, [Degrees of Freedom], MS, [F Calculated], [F Table 5%]) ' +
    'SELECT Comb_Group, Sums, DegFreedom, MS, [F Calculated], [F Table 5%] ' +
    'FROM ##TempCombs ORDER BY Cid'
EXEC(@Q)

GO
SET QUOTED_IDENTIFIER OFF
GO
SET ANSI_NULLS ON
GO
```

Procedure Calls

Below is an example of the call statement for invoking the ANOVA procedure.

```
DECLARE @RC int
DECLARE @SrcTable varchar(50)
DECLARE @Num_Factors int
DECLARE @First_Factor_Position int
DECLARE @RsltTable varchar(50)

SET @SrcTable = 'Table 7_6'
SET @Num_Factors = 3
SET @First_Factor_Position = 2
SET @RsltTable = 'ANOVA RESULT'

EXEC [CH7]..[ANOVA] @SrcTable, @Num_Factors,
    @First_Factor_Position, @RsltTable
```

Time Series Analysis

The Importance of Time Series

IN MOST ORGANIZATIONS, it is helpful, even vital, to forecast activities into the future. We would like to predict, as accurately as possible, information such as demand for goods and services, personnel requirements, sales, expenses, inventory requirements, profits, market share, and many other aspects of our operations. This knowledge obviously helps us plan for the future and stay competitive. If the data changes over time, and we are primarily interested in using past performance to predict the future, our diagnostic tree in Figure 1-3 directs us to time series analysis, or forecasting.

Most forecasting techniques are based on historical *time series* data and the trends the data reveal. These trends manifest themselves in various ways. For example, a company's quarterly sales figures may be increasing over the last five years. There may be seasonal variations as well. For tourist-related organizations, services or sales may be high in the summer and lower in winter, depending on the area of the country and the type of attraction. Cyclical variations may also occur, spread over several years, as consumer tastes change or the economic climate varies. And, of course, random (or unpredictable) variation is always present.

A time series is simply a collection of data that varies over time, much like the data we presented for control chart analysis in Chapter 6. As you would expect, numerous forecasting techniques have been developed for analyzing time series data and making predictions. Some are very simple to employ, and some involve highly sophisticated statistical models and techniques. Most are designed to predict general trends, and some offer adjustments for seasonal variation as well. Generally speaking, the predictor relationships are developed from historical data and modified as current data are collected and analyzed.

For our discussion, we look at two of the most common techniques for forecasts involving time series. These are known as *moving averages* and *exponential smoothing*. We also show you how to adjust the forecast to accommodate seasonal changes, when this is appropriate. Our examples illustrate both simple and weighted moving averages, single and double exponential smoothing, and a multiplicative seasonal method. Then we look at measures of forecast errors and criteria for selection of a technique.

Simple Moving Average

In Table 8-1, the gross sales figures for a small business are presented by month over a 10-year period, from 1991 through 2000. Sales come from several sources, including retail products, wholesale products, mail order shipments, and services.

Table 8-1. Monthly Gross Sales Figures for a Small Business for 10 Years

	1991	1992	1993	1994	1995	1996	1997	1998	1999	2000
Jan.	1,511.98	2,503.91	3,175.11	2,264.75	2,912.40	3,220.08	2,425.53	3,233.90	3,605.65	5,544.77
Feb.	2,109.43	2,779.49	2,614.33	2,700.92	2,809.92	2,846.85	3,193.74	3,363.78	3,888.04	5,594.75
Mar.	3,243.90	3,202.68	3,345.64	3,615.84	3,943.03	3,417.77	4,674.19	4,407.05	5,548.25	6,056.26
Apr.	2,839.26	4,819.20	3,994.10	4,348.88	5,693.04	5,311.95	5,056.87	5,675.45	5,952.36	6,580.86
May	4,690.78	5,787.95	6,341.94	6,030.03	5,499.29	6,245.12	6,291.26	6,797.99	5,526.09	11,771.86
Jun.	4,785.68	4,635.35	5,096.10	5,811.04	5,733.06	5,422.64	7,127.70	6,344.44	8,128.48	7,226.25
Jul.	4,551.25	6,165.84	6,254.09	7,209.43	7,586.10	5,372.56	6,615.50	6,340.09	7,168.99	9,104.72
Aug.	4,656.89	5,438.25	6,428.15	4,981.24	5,585.09	4,724.46	5,734.26	6,545.85	6,742.05	8,093.08
Sep.	3,967.72	6,373.47	5,244.76	6,948.26	4,905.27	5,120.06	6,913.97	5,143.82	7,205.62	8,473.97
Oct.	6,532.07	7,674.18	8,482.82	7,884.53	10,190.89	10,648.20	9,017.47	9,194.57	9,058.43	12,643.28
Nov.	4,342.33	6,482.80	6,099.76	4,657.16	6,573.18	6,479.60	7,315.96	8,660.43	14,100.88	9,164.20
Dec.	3,476.78	4,694.05	3,493.90	4,001.13	4,281.98	4,723.02	4,928.34	4,866.15	8,289.78	5,276.09

The numbers by themselves don't mean much until we graph them. Figure 8-1 represents the plot of the 120 sales figures. We can glean some bits of information from this graph. First, notice that over time the general trend in gross sales reveals an increase. Also, it appears there is seasonal variation in sales as well. The sales figures reach a low point in the winter months, then increase through the summer. In fact, the particular business is tourist-oriented to some degree, and this shows up primarily as an increase in retail sales during high visitor periods (especially the summer). Other sales and services, however, are steadier throughout the year. The seasonal variation shows up more dramatically if the quarterly (i.e., seasonal) sales figures are graphed. This is shown in Figure 8-2.

Of course, the seasonal variation is obscured when only the annual gross sales figures are plotted, as shown in Figure 8-3. However, these totals will be helpful to us a little later when we need a predictor equation for annual sales.

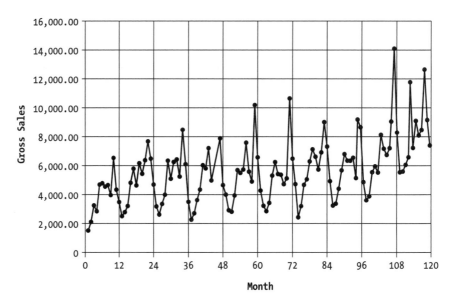

Figure 8-1. Graph of monthly gross sales figures for the small business

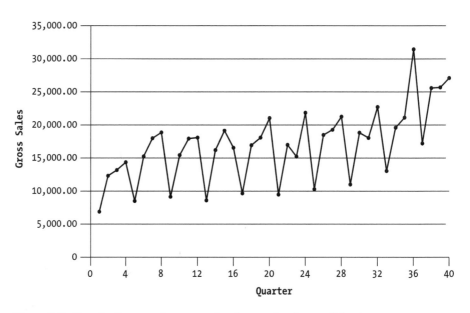

Figure 8-2. Graph of quarterly gross sales figures for the small business

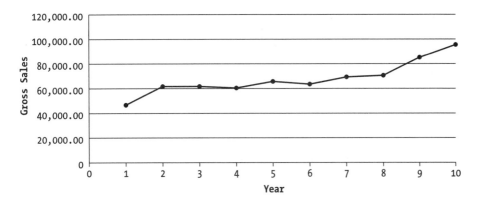

Figure 8-3. Graph of annual gross sales for the small business

An Example Close to Home

The example presented in this chapter is developed from the accounting books of author John Lovett's small business, Falls Mill & Country Store, in Belvidere, Tennessee. He and his wife have operated the 1873 water-powered mill and museum since 1984. The mill primarily operates as a tourist attraction, but also produces flour and meal products, includes a country store, and has a bed-and-breakfast log cabin on the grounds. Sales have increased through word of mouth, numerous national and regional articles, and e-commerce through the Web site at www.fallsmill.com. Although the sales figures shown here may appear pretty good, the expenses always equal the income, so the business is at best a break even proposition (an expensive hobby, in other words!).

How could we predict the gross sales for this business for the year 2001? One way would be to revert back to the discussion in Chapter 5 and use regression analysis. We might try a linear regression of gross sales relative to months, quarters, or years. We could then run a correlation test to check the accuracy of the line we fit to the data. If it is poor, perhaps a nonlinear model is more appropriate. (In fact, we employ regression a little later in the chapter). There are simpler techniques for forecasting available. The easiest is the simple moving average method. To get us started, it is more convenient to list the data in columnar format to show how the method works. Referring to Table 8.2, we list the month in the left column, in this case sequentially as Month 1 through Month 9 (for brevity in illustrating the technique). The actual gross sales for the appropriate month are then listed in the second column. Now we have to decide how many months we want to average at a time. This is a difficult decision without any criteria for selection (which we present at the end of the chapter). If we choose, for example, a three-month moving average, we would take the first three months' sales figures and calculate the mean, or average. For Months 1, 2, and 3, this is the average of 1511.98, 2109.43, and 3243.90, or 2288.44. This becomes our forecast for Month 4. To obtain the forecast for Month 5, we average the sales figures for Months 2, 3, and 4. This average is 2730.86, so it becomes the forecast for Month 5. Next, we obtain the Month 6 forecast by averaging the sales figures for Months 3, 4, and 5. Now you see why the method is termed "moving average." We simply move along one month as we average each set of three months' sales figures, dropping the first month of the previous calculation as we proceed. The general formula for these calculations is

$$F_{i+n} = \left(X_i + X_{i+1} + X_{i+2} + \text{L} + X_{i+n-1}\right)/n$$

where n = time period for moving average (e.g., 3 months)
F_{i+n} = forecast for time period $i+n$
X_i = data value for time period i (e.g., Month 1)
$i = 1, 2, \ldots$

The error column in the table is a measure of how much our forecast missed the actual sales figure, once it was known. It is the difference between the actual sales and the forecast sales, so may be either positive or negative. We'll need the error measurements a little later.

Table 8-2. Three-Month Moving Average for the First 9 Months Sales Data

MONTH	ACTUAL GROSS SALES S	3-MONTH AVERAGE A	FORECAST F	ERROR S - F
1	1511.98			
2	2109.43			
3	3243.90	2288.44		
4	2839.26	2730.86	2288.44	550.82
5	4690.78	3591.31	2730.86	1959.92
6	4785.68	4105.24	3591.31	1194.37
7	4551.25	4675.90	4105.24	446.01
8	4656.89	4664.61	4675.90	-19.01
9	3967.72	4391.95	4664.61	-696.89

SQL/Query

The following two queries create the data for Table 8-2. Notice the FROM clause in Query 8_1 works with three copies of Table 8_1. This is necessary in order to accomplish the three-month moving average. If we happen to be doing a five-month moving average, the FROM clause would have five copies of Table 8_1.

Query 8_1:

```
SELECT [table 8_1].[Month] + 1 AS [Forecasted Month],
([Table 8_1].[Actual Gross Sales] + [Table 8_1_1].[Actual Gross Sales] +
[Table 8_1_2].[Actual Gross Sales]) / 3 AS [3 Month Average]
FROM [Table 8_1], [Table 8_1] AS [Table 8_1_1], [Table 8_1] AS [Table 8_1_2]
WHERE ((([Table 8_1].Month) = [Table 8_1_1].[Month] + 1
AND ([Table 8_1].Month) = [Table 8_1_2].[Month] + 2));
```

Query 8_2:

```
SELECT [Table 8_1].Month, [Table 8_1].Date, [Table 8_1].[Actual Gross Sales],
[Query 8_1].[3 Month Average], [Actual Gross Sales] - [3 Month Average] AS Error
INTO [Table 8_2]
FROM [Query 8_1], [Table 8_1]
WHERE [Query 8_1].[Forecasted Month] = [Table 8_1].[Month];
```

We could have selected a 6-month moving average as another option, or even a 12-month moving average. However, these tend to be less sensitive to large swings in the data, such as the seasonal changes we see in this example. But they sometimes produce better forecasts over the long haul!

Figure 8-4 shows the moving average forecasts relative to the actual sales figures. Of course the first average occurs in Month 4, since we needed the first three months' data to generate the Month 4 forecast.

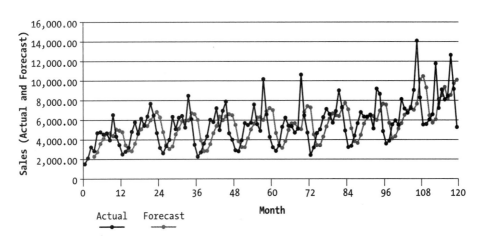

Figure 8-4. Graph of simple moving average forecasts versus actual sales

Some analysts prefer to place more weight on recent sales data. The *simple* moving average method cannot handle this, so we resort to a modification called *weighted* moving average. For example, we could allow the most recent month's sales figure to carry 50% of the weight for the forecast, the previous month's figure to carry 30%, and the month before that to carry 20% of the weight. The first 3-month weighted moving average would be calculated as follows:

$$(0.2)(1511.98)+(0.3)(2109.43)+(0.5)(3243.90)=2557.18$$

This becomes the forecast for Month 4. The next weighted moving average uses the weighting factors and sales figures from Months 2, 3, and 4. The general formula for these calculations is

$$F_{i+n} = \left(W_1 X_i + W_2 X_{i+1} + W_3 X_{i+2} + \text{L} + W_n X_{i+n-1} \right)/n$$

where n = time period for moving average (e.g., 3 months)

F_{i+n} = forecast for time period $i+n$

X_i = data value for i (e.g., Month 1)

W_j = weighting factor for time period $j = 1, 2, \text{K}, n$

$i = 1, 2, \ldots$

Often these forecast results produce smaller errors than those of the simple moving average method. The weighted 3-month moving average results for Months 1 through 9 are shown in Table 8-3.

Table 8-3. Weighted Moving Average Results for the First 9 Months Sales

MONTH	ACTUAL GROSS SALES	3-MONTH AVERAGE	FORECAST	ERROR
	S	A	F	S-F
1	1,511.98			
2	2,109.43			
3	3,243.90	2,557.18		
4	2,839.26	2,814.69	2,557.18	282.09
5	4,690.78	3,845.95	2,814.69	1,876.09
6	4,785.68	4,367.93	3,845.95	939.73
7	4,551.25	4,649.49	4,367.93	183.32
8	4,656.89	4,650.96	4,649.49	7.41
9	3,967.72	4,291.18	4,650.96	- 683.24

SQL/Query

Query 8_3 and Query 8_4 generate the data for the weighted moving average shown in Table 8-3.

Query 8_3:

```
SELECT [Table 8_1].[Month] + 1 AS [Forecased Month],
0.2 * [Table 8_1_2].[Actual Gross Sales] + 0.3*[Table 8_1_1].[Actual Gross Sales] +
0.5 * [Table 8_1].[Actual Gross Sales] AS [3 Month Average]
FROM [Table 8_1], [Table 8_1] AS [Table 8_1_1], [Table 8_1] AS [Table 8_1_2]
WHERE [Table 8_1].Month = [Table 8_1_1].[Month] + 1
AND [Table 8_1].Month = [Table 8_1_2].[Month] + 2;
```

Query 8_4:

```
SELECT [Table 8_1].Month, [Table 8_1].Date, [Table 8_1].[Actual Gross Sales],

[Query 8_3].[3 Month Average], [Actual Gross Sales] - [3 Month Average] AS Error

INTO [Table 8_3]
FROM [Table 8_1], [Query 8_3]
WHERE [Query 8_3].[Forecased Month] = [Table 8_1].[Month];
```

The graph of the 3-month weighted moving average forecasts is shown relative to the actual sales in Figure 8-5. The results are a little more accurate than the simple moving average method. Notice how the peaks and valleys are predicted a bit better.

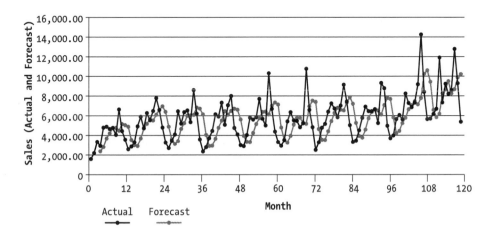

Figure 8-5. Graph of the weighted moving average forecasts and the actual sales

John's Jewels
Forecasting Grits

At Falls Mill, my wife and I have about 80 restaurant accounts to which we ship old-fashioned stone-ground grits. In order to meet their demands, we have to contract a year ahead with our local farmer to insure enough white corn to last through the season. The corn is harvested in our county in September and stored in silos. By January or so, most of the white corn is sold, and about all has left the bins by May. It becomes impossible to find any between May and the new crop, so we have to plan ahead for the farmer to save enough for our customers. I kept track of our corn demands over a three-year period and set up a simple little fore-casting technique. It has helped us avoid running short in the summer, and allowed us to service our grits customers without problems.

Single Exponential Smoothing

The moving average methods are sensitive to fairly wide fluctuations in the time series data. There is another group of techniques available to the data miner that offers a better ability to smooth out these variations and provide potentially more accurate forecasts. Although the tradeoff lies in the greater complexity of the method versus moving aver-ages, it is still perhaps the most commonly employed forecasting tool. The name of this family of techniques is *exponential smoothing*. The first type we examine is termed *single* exponential smoothing.

One drawback of the method is that we have to use a great deal of our data to cal-culate a starting point value for the technique. This value is usually the mean or average of the data over some period. In our sales example, we could calculate the average gross sales over the first 5 years, for example. If we take the 60 monthly sales amounts from 1991 through 1995 and find the mean, the result is $4,900.04. This becomes the starting point for the January 1996 forecast. But we also need a second value. This is the monthly sales figure just prior to our first forecast, or the December 1995 value ($4,281.98). Now the data miner has a decision to make. A smoothing factor denoted by α must be chosen to jump-start the forecast calculations. The smoothing factor must be between 0 and 1. The smaller the factor, the more weight will be given to (initially) the 60-month average we just found. In subsequent forecast calculations, the previous month's forecast will take the place of this 60-month average, as we'll see momentarily. The factor is applied directly to the previous month's actual sales value, when it becomes known. Therefore, if the factor is small, we are putting less emphasis on the immediately preceding sales value, and more emphasis on a number that represents the average of a greater amount of data. The decision is up to the data miner. Does that make any sense? Well, if not, an example will probably clear things up.

Let's pick $\alpha = 0.1$ for our smoothing factor. Using the December 1995 sales of $4,281.98 and the 60-month average of $4,900.04, we calculate the January 1996 sales forecast in the following way:

$$(\alpha)(4,281.98)+(1-\alpha)(4,900.04)=(0.1)(4,281.98)+(0.9)(4,900.04)=4,838.23$$

Now we know the actual sales for January 1996 were $3,220.08, so we use this figure and the January 1996 forecast to calculate the February 1996 forecast, as follows:

$$(0.1)(3,220.08)+(0.9)(4,838.23)=4,676.42$$

Proceeding in this same manner, we find the actual sales for February 1996 were $2,846.85, so the forecast for March 1996 becomes

$$(0.1)(2,846.85)+(0.9)(4,676.42)=4,493.46$$

The general formula for these calculations is

$$F_i = (\alpha)(X_{i-1})+(1-\alpha)(F_{i-1})$$

where $F_0 = $ starting average

$F_i = $ forecast for time period $i = 1, 2, \ldots$

$X_0 = $ previous time period's data value

$X_i = $ data value for time period $i = 1, 2, \ldots$

$\alpha = $ smoothing factor, $0 < \alpha < 1$

The results for the first 9 months of 1996 are shown in Table 8-4. The graph in Figure 8-6 shows the single exponential smoothing results compared to the actual sales for 1996 through 2000. Notice from the graph that the performance of the exponential smoothing forecasting technique with $\alpha = 0.1$ is poor, since the predictions do not pick up the peaks and valleys of the actual sales.

Table 8-4. Single Exponential Smoothing Results for the First 9 Months of the 1996-2000 Sales Data

MONTH	ACTUAL SALES S	FORECAST SALESS F	ERROR S-F
61	3,220.08	4838.23	1,618.15
62	2,846.85	4676.42	1,829.57
63	3,417.77	4493.46	1,075.69
64	5,311.95	4385.89	-926.06
65	6,245.12	4478.50	-1,766.62
66	5,422.64	4655.16	-767.48
67	5,372.56	4731.91	-640.65
68	4,724.46	4795.97	71.51
69	5,120.06	4788.82	-331.24

SQL/Query

It is impractical to do the calculations for the exponential smoothing as a set of SQL queries because SQL does not provide any means for remembering a calculation in a preceding row so it can be applied in the succeeding row. Thus, we resort to a Visual Basic function for calculating exponential smoothing.

The function begins by setting alpha equal to 0.1 and then obtains the average actual sales for the first 60 months (5 years). The average is used as the initial sales forecast. Next, the function sets up a query to extract the monthly sales for all those months after the 5th year (60th month). Notice that the query includes the 60th month. We use the sales of the 60th month to predict the sales for the 61st month. As a setup step, the function clears Table 8_4 of all previous data. Next, the function enters a Do While loop that calculates the error, inserts the data into Table 8_4, and forecasts the next month's sales. After processing all the months, the function ends.

TIP *One way to run a Visual Basic function is to simply make a single line macro as follows:*

```
RunCode Single_Exp_Forecast()
```

Save the macro and give it a name such as "Do Single exp forecast." To run (execute) the macro, select the Tools menu and choose Macro. From the Macro list select "Run Macro..." Select the macro named "Do Single exp forecast" and click OK. Once you click OK, Access will run the macro that in turn invokes the Visual Basic function Single_Exp_Forecast.

```
Public Function Single_Exp_Forecast()

    ' This procedure performs the single exponential smoothing forecasting

    Dim Sales_Forecast      As Currency                 ' sales forecast
    Dim Forecast_Error      As Currency                 ' forecasting error
    Dim Alpha1              As Single                   ' Alpha factor
    Dim Alpha2              As Single                   ' one minus Alpha factor
    Dim Q1                  As String                   ' A query
    Dim Q2                  As String                   ' A query
    Dim Q3                  As String                   ' A query
    Dim rs_in               As New ADODB.Recordset      ' input records

    ' Initialize
    Alpha1 = 0.1
    Alpha2 = 1# - Alpha1

    ' OBTAIN THE FIRST 5 YEARS' (60 MONTH) AVERAGE

    ' Build query to obtain average
    Q1 = "SELECT Avg([Actual Gross Sales]) AS [60-month average] "
    Q1 = Q1 & "FROM [Table 8_1] "
    Q1 = Q1 & "WHERE [Table 8_1].Month <= 60; "

    ' Process the query
    rs_in.Open Q1, Application.CurrentProject.Connection

    ' Get the average  that serves as first forecast sales
    rs_in.MoveFirst
    Sales_Forecast = rs_in("60-month average")

    ' Close out query
    rs_in.Close
    Set rs_in = Nothing


    ' DO THE FORECAST OVER THE LAST 5 YEARS

    ' Build query to obtain month and actual gross sales
    Q2 = "SELECT Month, [Actual Gross Sales] "
    Q2 = Q2 & "FROM [Table 8_1] "
    Q2 = Q2 & "WHERE Month >= 60; "
```

```
' Process the query
rs_in.Open Q2, Application.CurrentProject.Connection

' Do the first forecast for the 61st month using
' the 60th month actual sales and the 60-month average
rs_in.MoveFirst
Sales_Forecast = Alpha1 * rs_in("Actual Gross Sales") + Alpha2 * Sales_Forecast

' Move to the first month (61st month) of the forecast period
rs_in.MoveNext

' Turn off records modification confirmation message so that
' it does not appear while populating Table 8_4
DoCmd.SetWarnings False

' Clear table of old records
DoCmd.RunSQL "DELETE * FROM [Table 8_4]; "

' Calculate the error and save current month's information,
' and then calculate next month's sales forecast
Do While Not rs_in.EOF

                ' Determine this month's error
                Forecast_Error = Sales_Forecast - rs_in("Actual Gross Sales")

                ' Append record to Table 8_4
                Q3 = "INSERT INTO [Table 8_4] "
                Q3 = Q3 & "([Month], [Actual Gross Sales]
                Q3 = Q3 & "[Forecast Sales], [Forecast Error])
                Q3 = Q3 & "VALUES (" & rs_in("Month") & ", "
                Q3 = Q3 & rs_in("Actual Gross Sales") & ", "
                Q3 = Q3 & Sales_Forecast & ", "
                Q3 = Q3 & Forecast_Error & "); "
                DoCmd.RunSQL Q3

                ' Do forecast for next month
                Sales_Forecast = Alpha1 * rs_in("Actual Gross Sales") + _
                        Alpha2 * Sales_Forecast

                ' Get next month
                rs_in.MoveNext

Loop
```

```
' Turn on confirmation message
DoCmd.SetWarnings True

' Close out query
rs_in.Close
Set rs_in = Nothing


End Function
```

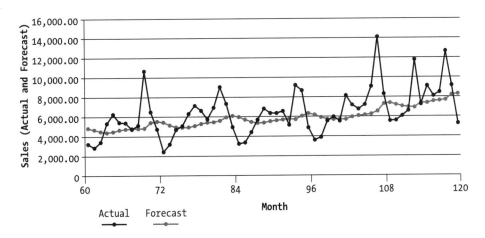

Figure 8-6. Graph of single exponential smoothing results versus actual sales

It is interesting to examine the differences in the results if the smoothing factor is changed to, let's say, 0.2 or 0.5. The graphs of these two results compared to $\alpha = 0.1$ are shown in Figure 8-7, along with the actual sales.

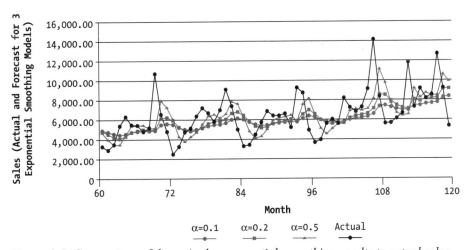

Figure 8-7. Comparison of three single exponential smoothing results to actual sales

Notice from the graph that the performance of the forecasting method has been significantly improved as the smoothing factor increases from 0.1 to 0.5. With $\alpha = 0.5$, the highs and lows of the actual sales data are predicted with much greater accuracy. This will not necessarily be the case in general. For this reason, the data miner may want to run the model with varying α values and assess the results based on some measures we present later in the chapter.

Double Exponential Smoothing

The limitation of the single exponential smoothing method is that it cannot incorporate a trend very well. In our example, the annual sales are increasing, so it would be helpful to factor this knowledge into our forecasting model. *Double* exponential smoothing provides a way to do just that.

The first step is to determine a measure for the trend for some period. Since we used the first five years of sales data in the previous illustration, we'll keep that 60-month period for our trend analysis. One measure of trend would be the slope of the line of best fit for the first 60 months of sales data. We could perform a linear regression as we illustrated in Chapter 5 and obtain the following regression line, where y represents the monthly sales and x represents the month (e.g., 1, 2, 3, etc.):

$$y = 3,778.054 + 36.786x$$

SQL/Query

A linear regression is performed by Query 8_5 and Query 8_6. These two queries are similar to Query 5_1 and Query 5_2 in Chapter 5, which contains a discussion on how to do linear regression. The table called Forecast_Parameters contains the year for which forecasting is being done. As we'll see later, we can change the forecast year to the next year and reexecute the queries to obtain subsequent years' forecasts. However, for now we can think of it as being the constant 6.

Query 8_5:

```
SELECT Count([Table 8_1].Month) AS n,
Sum([Table 8_1].Month) AS Sx,
Sum([Table 8_1].[Actual Gross Sales]) AS Sy,
Sum([Month] ^ 2) AS Sx2,
Sum([Actual Gross Sales] ^ 2) AS Sy2,
Sum([Month]*[Actual Gross Sales]) AS Sxy
FROM [Table 8_1], Forecast_Parmeters
WHERE [Table 8_1].Month <= 12 * ([Forecast_Parameters].[forecast year] - 1);
```

Query 8_6:

```
SELECT ([Sy] * [Sx2] - [Sx] * [Sxy]) / ([n] * [Sx2] - [Sx] ^ 2) AS a,
([n] * [Sxy] - [Sx] * [Sy]) / ([n] * [Sx2] - [Sx] ^ 2) AS b
FROM [Query 8_5];
```

The slope or trend is therefore about \$36.79 per month on average for the 60-month period. You may recall that the average sales figure for the same period was \$4,900.04. We need a starting point for the method, so we again choose the December 1995 sales of \$4,281.98. This figure is weighted against the exponential smoothing factor that we want to use. In this example we pick $\alpha = 0.3$. From the single exponential smoothing formula, we also need some value to multiply by $(1 - \alpha) = 0.7$. We begin by using the average 6-months sales of \$4,900.04 plus the average trend amount of \$36.79. The calculation is therefore

$$(0.3)(4,281.98) + (0.7)(4,900.04 + 36.79) = 4,740.38$$

From this we subtract the average sales for 60 months, as follows:

$$4,740.38 - 4,900.04 = -159.66$$

In this case the result is negative, and represents a component of trend. In order to make use of it, we must select another smoothing factor called β. Now you see why the method is termed "double" exponential smoothing. For illustration, we'll pick $\beta = 0.2$. We multiply β by the trend figure above (–159.66), then add the result to $(1 - \beta)$ times the average trend (36.79), as shown below, to obtain a weighted trend.

$$(0.2)(-159.66) + (0.8)(36.79) = -2.50$$

Finally, this result is added to 4,740.38 to obtain the forecast for January 1996:

$$4,740.38 - 2.50 = 4,737.88$$

This is our kickoff forecast. Then we find the actual sales for January 1996 are 3,220.08. The forecast for February 1996 then proceeds as follows:

$$(\alpha)(\text{actual January 1996 sales}) + (1 - \alpha)(\text{January 1996 forecast}) =$$
$$(0.3)(3,220.08) + (0.7)(4,737.88) = 4,282.54$$

Then we would have

$$4,282.54 - 4,740.38 = -457.84$$

Notice that the 4,740.38 came from the first calculation of the January forecast. Again the trend component is negative, so we use it in the trend forecast that follows.

$$(\beta)(-457.84) + (1 - \beta)(-2.50) = (0.2)(-457.84) + (0.8)(-2.50) = -93.57$$

Notice here that the –2.50 came from the January trend forecast. Finally, the February 1996 forecast is obtained as follows:

$$4,282.54 - 93.57 = 4,188.97$$

Maybe we should do one more. The actual February 1996 sales are now known to be \$2,846.85. Then the forecast for March would be found as follows (notice where the numbers come from):

$$(0.3)(2,846.85) + (0.7)(4,188.97) = 3,786.33$$
$$3,786.33 - 4,282.54 = -496.21$$
$$(0.2)(-496.21) + (0.8)(-93.57) = -174.10$$
$$3,786.33 - 174.10 = 3,612.23$$

The March 1996 forecast is thus \$3,612.23. The general formula for these calculations is

Let X_i = actual data value in time period i
S_i = Smoothed sales in time period i
TF_i = Trend factor in time period i
WT_i = Weighted trend in time period i
F_i = Forecast in time period i

F_0 = starting average + average trend
$S_0 = (\alpha)(X_0) + (1-\alpha)(F_0)$
$TF_0 = S_0$ – starting average
$WT_0 = (\beta)(TF_0) + (1-\beta)(\text{average trend})$
$F_1 = S_0 + WT_0$
$S_1 = (\alpha)(X_1) + (1-\alpha)(F_1)$
$TF_1 = S_1 - S_0$
$WT_1 = (\beta)(TF_1) + (1-\beta)(WT_0)$

In general,

$F_i = S_{i-1} + WT_{i-1}$
$S_i = (\alpha)(X_i) + (1-\alpha)(F_i)$
$TF_i = S_i - S_{i-1}$
$WT_i = (\beta)(TF_i) + (1-\beta)(WT_{i-1})$

The results of the first 9 months' forecasts (for 1996 – 2000) for the double exponential smoothing method are shown in Table 8-5.

Table 8-5. Double Exponential Smoothing Results for the First 9 Months of the 1996–2000 Sales Data

MONTH	ACTUAL SALES	SMOOTHED SALES	TREND	WEIGHTED TREND	FORECAST	ERROR
61	3,220.08	4,740.38	-159.67	-2.50	4,737.87	-1,517.79
62	2,846.85	4,282.54	-457.84	-93.57	4,188.97	-1,342.12
63	3,417.77	3,786.33	-496.20	-174.10	3,612.23	-194.46
64	5,311.95	3,553.89	-232.44	-185.77	3,368.12	1,943.82
65	6,245.12	3,951.27	397.38	-69.14	3,882.14	2,362.98
66	5,422.64	4,591.03	639.76	72.64	4,663.67	758.97
67	5,372.56	4,891.36	300.33	118.18	5,009.54	363.02
68	4,724.46	5,118.45	227.09	139.96	5,258.41	-533.95
69	5,120.06	5,098.23	-20.22	107.92	5,206.15	-86.09

SQL/Query

The same impracticality also exists for the double exponential smoothing method as it did in the single exponential case. So, we developed a Visual Basic function for calculating the double exponential smoothing results. The Function begins by setting the parameter alpha to 0.3 and the parameter beta to 0.2. The initial trend, which is the slope of the linear regression line, is obtained from the execution of Query 8_6. Next, the initial forecast is taken from the results of a query that computes the average sales of the first 5 years (60 months). With these values in place, the function is ready to determine the sales forecast for the remaining 5 years. We first set up a query that can provide each month and its actual sales. Notice that the query begins with the 60th month. To forecast the 61st month sales, we need the 60th month sales. The function calculates the 61st month smoothed sales, determines the trend and weighted trend, and then makes the forecast for the 61st month. Next, the function clears Table 8_5 of any previous data values. A Do While loop is used to perform the calculations for each month. Within the loop the forecast error is determined for the current month, the values are appended to Table 8_5, and the next month's smoothed sales, trends, and weighted trends, are computed. On completion of the loop, the function ends.

> **TIP**   *One way to run a Visual Basic function is to simply make a single line macro as follows:*
>
> ```
> RunCode Double _Exp_Forecast()
> ```
>
> *Save the macro and give it a name such as "Do Double exp forecast." To run (execute) the macro, select the Tools menu and choose Macro. From the Macro list select "Run Macro…" Select the macro named "Do Double exp forecast" and click OK. Once you click OK, Access will run the macro that in turn invokes the Visual Basic function Double _Exp_Forecast.*

```
Public Function Double_Exp_Forecast()

    ' This procedure performs the double exponential smoothing forecasting

        Dim Initial_Forecast     As Double     ' the 60 month average
        Dim Smoothed_Sales       As Double     ' sales forecast from 1st exponential
        Dim Previous_Sales       As Double     ' previous sales forecast
        Dim Forecast             As Double     ' the forecasted sales
        Dim Forecast_Error       As Double     ' forecasting error
        Dim Alpha1               As Double     ' Alpha factor
        Dim Alpha2               As Double     ' one minus Alpha factor
        Dim Beta1                As Double     ' Beta factor
        Dim Beta2                As Double     ' one minus Beta factor
        Dim Trend                As Double     ' trend of 1st exponential
        Dim Previous_Trend       As Double     ' previous trend of 1st exponential
        Dim Wtd_Trend            As Single     ' weighted trend from 2nd exponential
        Dim Q1                   As String     ' A query
        Dim Q2                   As String     ' A query
        Dim Q3                   As String     ' A query
        Dim rs_in                As New ADODB.Recordset     ' input records
        ' Initialize
        Alpha1 = 0.3
        Alpha2 = 1# - Alpha1
        Beta1 = 0.2
        Beta2 = 1# - Beta1
```

```
' OBTAIN THE INITIAL TREND (SLOPE)


rs_in.Open "[Query 8_6]", _
                    Application.CurrentProject.Connection
Previous_Trend = rs_in("b")
rs_in.Close



' OBTAIN THE FIRST 5 YEARS' (60 MONTH) AVERAGE


' Build query to obtain average
Q1 = "SELECT Avg([Actual Gross Sales]) AS [60-month average] "
Q1 = Q1 & "FROM [Table 8_1] "
Q1 = Q1 & "WHERE [Table 8_1].Month <= 60; "


' Process the query
rs_in.Open Q1, Application.CurrentProject.Connection


' Get the average that serves as first forecast sales

rs_in.MoveFirst
Initial_Forecast = rs_in("60-month average")


' Close out query
rs_in.Close
Set rs_in = Nothing



' DO THE FORECAST OVER THE LAST 5 YEARS


' Build query to obtain month and actual gross sales
Q2 = "SELECT Month, [Actual Gross Sales] "
Q2 = Q2 & "FROM [Table 8_1] "
Q2 = Q2 & "WHERE Month >= 60; "


' Process the query
rs_in.Open Q2, Application.CurrentProject.Connection


' Do the first forecast for the 61st month using the
'_60th month actual sales and the 60 month average and trend
rs_in.MoveFirst
Smoothed_Sales = Alpha1 * rs_in("Actual Gross Sales") + _
                    Alpha2 * (Initial_Forecast + Previous_Trend)
```

```
' Determine the trend and do second exponential
Trend = Smoothed_Sales - Initial_Forecast
Wtd_Trend = Beta1 * Trend + Beta2 * Previous_Trend

' Make the forecast for next month
Forecast = Smoothed_Sales + Wtd_Trend

' Move to the first month (61st month) of the forecast period
rs_in.MoveNext

' Turn off records modification confirmation message so that
' it does not appear while populating Table 8_5
DoCmd.SetWarnings False

' Clear table of old records
DoCmd.RunSQL "DELETE * FROM [Table 8_5]; "

' Calculate the error and save current month's information,
' and then calculate next month's sales forecast
Do While Not rs_in.EOF

    ' Determine this month's error
    Forecast_Error = rs_in("Actual Gross Sales") - Forecast

    ' Append record to Table 8_5
    Q3 = "INSERT INTO [Table 8_5] "
    Q3 = Q3 & "([Month], [Actual Sales], [Smoothed Sales], "
    Q3 = Q3 & "[Trend], [Wtd Trend], [Forecast], [Forecast Error]) "
    Q3 = Q3 & "VALUES (" & rs_in("Month") & ", "
    Q3 = Q3 & rs_in("Actual Gross Sales") & ", "
    Q3 = Q3 & Smoothed_Sales & ", "
    Q3 = Q3 & Trend & ", "
    Q3 = Q3 & Wtd_Trend & ", "
    Q3 = Q3 & Forecast & ", "
    Q3 = Q3 & Forecast_Error & "); "
    DoCmd.RunSQL Q3

    ' Calculate smoothed sales for next month
    Previous_Sales = Smoothed_Sales
    Smoothed_Sales = Alpha1 * rs_in("Actual Gross Sales") + Alpha2 * Forecast

    ' Calculate trends
    Trend = Smoothed_Sales - Previous_Sales
    Wtd_Trend = Beta1 * Trend + Beta2 * Wtd_Trend
```

```
    ' Determine the forecast for next month
    Forecast = Smoothed_Sales + Wtd_Trend

    ' Get next month
    rs_in.MoveNext

Loop

' Turn on confirmation message
DoCmd.SetWarnings True

' Close out query
rs_in.Close
Set rs_in = Nothing

End Function
```

The graph of the double exponential smoothing results compared to the original monthly sales figures appears in Figure 8-8. If we look closely and compare this graph to those of the simple and weighted moving averages, and to the single exponential smoothing graph, we detect a little better set of predictions. In fact, we show a bit later how we can compare the performances of the various methods.

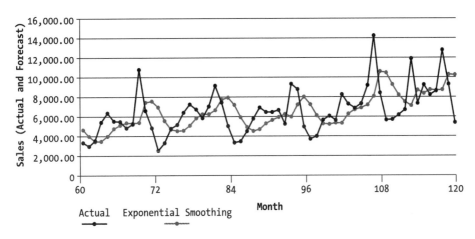

Figure 8-8. Graph of double exponential smoothing results versus actual sales

John's Jewels
It Got Here Just in Time

I suppose there are few forecasting applications more critical than "just-in-time" inventory control. This is a term applied to the receipt of raw materials and sub-assemblies from suppliers at precisely the time they are needed in the manufacturing process. The advantages are the savings in storage, warehousing, and the accompanying paperwork. However, a manager has to know exactly the demands for these materials in the production process, and must schedule shipments with adequate lead-time for the arrival of materials at the critical points in production. As a result, the need for accurate forecasting techniques is of primary importance to the smooth operation of the plant. Of course the demand for raw materials is governed by the customers' demands and the diligence of the marketing department, so these become factors for input to the forecasting model. Fortunately, there is considerable software available today for forecasting, and this has relieved the burden of analyzing vast amounts of past data.

Incorporating Seasonal Influences

Earlier we stated that the sales data exhibited a definite seasonal change, with sales increasing during the warmer months and decreasing in the winter. The previous forecasting methods have not been able to help us predict these seasonal ups and downs with much accuracy. There are techniques available, however, for incorporating seasonal influences. The one we'll use is called *multiplicative seasonal adjustment.*

First, we need to decide on a time period for the model. It seems logical to use quarters of the year, since these correspond roughly to the seasons in which we are interested. What we want to do is look again at a period of historical data in order to arrive at a starting point for our forecasts. Let's stay consistent and choose the first 5 years of sales data. This time, however, we want to look at the total sales figures *per quarter.* Then we'll calculate the percent (or decimal fraction) that each quarter sales figure represents of the year's *average* quarterly sales. These will become our *seasonal factors.* Once we calculate these factors for the first 5 years (1991-1995), we can average the results by quarters. These average seasonal factors get us started toward the forecast for 1996. Then when the actual 1996 sales figures become available, we can average the factors from 1991–1996, if we want to, or drop the 1991 factors and keep using 5-year intervals. In our example, we've used all the factors from 1991 on through the year in which we are forecasting seasonal sales.

In addition to the seasonal factors, we need a projected sales figure for 1996 on which to base our seasonal forecasts. This figure could be determined by regression or one of the other forecasting techniques we presented. For our example, we fit a linear regression equation to the first 5 years *annual* sales figures (in other words, there will be 5 points used to determine the regression line). The result is

$$y = 47,428.56 + 3,790.64x$$

where *x* represents the year (1, 2, 3, etc.) and *y* represents the annual sales. We can now use this equation to predict the sales figures in Year 6 (1996), by substituting $x = 6$ into the equation. The result is shown below.

$$y = 47,428.56 + (3,790.64)(6) = 70,172.40$$

Now we can begin to develop a table. We first sum the actual monthly sales figures for each quarter of 1991. Then we total these to get the annual sales. The total annual sales for 1991 were $46,708.07. If we divide this figure by 4, we obtain the average quarterly sales for 1991, which are equal to $11,677.02. For the first quarter, the total sales were actually $6,865.31. If we divide the actual first quarter sales by the average quarterly sales, we arrive at the fraction of average quarterly sales the first quarter represents, as follows:

$$6,865.31/11,677.02 = 0.587933 \text{ (or about 59\%)}$$

A similar procedure will produce the seasonal factors shown in Table 8-6. In the last column of the table, we see the average seasonal factors for each quarter.

Table 8-6. Seasonal Factors for the First 5 Years of Sales Data

QUARTER	1991		1992		1993		1994		1995		AVERAGE SEASONAL FACTORS
	SALES	SEASONAL FACTORS	SALES	SEASONAL FACTORS	SALES	SEASONAL FACTORS	SALES	SEASONAL FACTORS	SALES	SEASONAL FACTORS	
1	6,865.31	0.587933	8,486.08	0.560534	9,135.08	0.603267	8,581.51	0.567812	9,665.35	0.588335	0.581576
2	12,315.72	1.054697	15,242.50	1.006817	15,432.14	1.019116	16,189.95	1.071239	16,925.39	1.030258	1.036425
3	13,175.86	1.128358	17,977.56	1.187477	17,927.00	1.183872	19,138.93	1.266363	18,076.46	1.100324	1.173279
4	14,351.18	1.229010	18,851.03	1.245173	18,076.48	1.193744	16,542.82	1.094587	21,046.05	1.281084	1.208720
Total	46,708.07		60,557.17		60,570.70		60,453.21		65,713.25		
Quarterly Average sales	11,677.02		15,139.29		15,142.68		15,113.30		16,428.31		Projected Qtr. Sales for 1996 17,543.10

SQL/Query

The following queries determine the seasonal factors as shown in Table 8-6.

Query 8_7 expands the date field of Table 8-1 by adding columns for months, years, and quarters.

Query 8_7:

```
SELECT Int((([Month] - 1) / 3) + 1 AS Quarters,
Format(Mid([date], 1, 2) & "/01/" & Mid([date], 3), "q") AS [Quarter ID],
[Table 8_1].Month, [Table 8_1].Date,
Int((([Month] - 1) / 12) + 1 AS [Year],
[Table 8_1].[Actual Gross Sales]
FROM [Table 8_1]
ORDER BY [Table 8_1].Month;
```

In the next two queries, Query 8_8 finds the total sales for each quarter, and Query 8_9 finds the total sales for each year.

Query 8_8:

```
SELECT [Query 8_7].Quarters, Avg([Query 8_7].[Quarter ID]) AS [Quarter ID],
Avg([Query 8_7].Year) AS [Year],
Sum([Query 8_7].[Actual Gross Sales]) AS Sales
FROM [Query 8_7]
GROUP BY [Query 8_7].Quarters;
```

Query 8_9:

```
SELECT [Query 8_7].Year,
Sum([Query 8_7].[Actual Gross Sales]) AS [Yearly Sales]
FROM [Query 8_7]
GROUP BY [Query 8_7].Year;
```

Using the quarterly sales, Query 8_10 finds the yearly average of the quarterly sales.

Query 8_10:

```
SELECT [Query 8_8].Year,
Avg([Query 8_8].Sales) AS [Quarterly Average Sales]
FROM [Query 8_8]
GROUP BY [Query 8_8].Year;
```

Next, Query 8_11 determines the seasonal factor by dividing each quarterly sales total by the quarterly average sales.

Query 8_11:

```
SELECT [Query 8_8].Quarters, [Query 8_8].[Quarter ID], [Query 8_8].Year,
[Query 8_8].[Sales] / [Quarterly Average Sales] AS [Seasonal Factor]
FROM [Query 8_8], [Query 8_10]
WHERE [Query 8_8].Year = [Query 8_10].Year;
```

The results obtained from Queries 8_8, 8_10, and 8_11 are combined into a single result when Query 8_12 is executed. Since we desire to keep these results, Query 8_12a is the same as Query 8_12 except the results are saved in Table 8_6. Notice how Query 8_12 makes use of the table called Forecast_Parameters that is used to identify the year being forecasted.

Query 8_12:

```
SELECT [Query 8_10].Year, [Query 8_8].[Quarter ID],
[Query 8_10].[Quarterly Average Sales], [Query 8_8].Sales,
[Query 8_11].[Seasonal Factor]
FROM Forecast_Parameters, [Query 8_10], [Query 8_8], [Query 8_11]
WHERE [Query 8_10].Year = [Query 8_8].Year
AND [Query 8_10].Year = [Query 8_11].Year
AND [Query 8_8].Quarters = [Query 8_11].Quarters
AND [Query 8_10].Year <= ([Forecast_Parameters].[forecast year] - 1);
```

Query 8_12a:

```
SELECT [Query 8_10].Year, [Query 8_8].[Quarter ID],
[Query 8_10].[Quarterly Average Sales], [Query 8_8].Sales,
[Query 8_11].[Seasonal Factor]
INTO [Table 8_6]
FROM Forecast_Parameters, [Query 8_10], [Query 8_8], [Query 8_11]
WHERE [Query 8_10].Year = [Query 8_8].Year
AND [Query 8_10].Year = [Query 8_11].Year
AND [Query 8_8].Quarters = [Query 8_11].Quarters
AND [Query 8_10].Year <= ([Forecast_Parameters].[forecast year] - 1);
```

Finally, the average seasonal factors are calculated by Query 8_13.

Query 8_13:

```
SELECT [Table 8_6].[Quarter ID],
Avg([Table 8_6].[Seasonal Factor]) AS [Average Seasonal Factor]
FROM [Table 8_6]
GROUP BY [Table 8_6].[Quarter ID];
```

Now we're ready to forecast the 1996 sales by quarter. We take the regression estimate for the total 1996 sales obtained above ($70,172.40) and divide it by 4, to give us a quarterly average of $17,543.10. We then multiply this quarterly average by each quarter's average seasonal factor, as shown in the following, to obtain the quarterly forecasts.

$$(17,543.10)(0.581576) = 10,202.65$$
$$(17,543.10)(1.036425) = 18,182.11$$
$$(17,543.10)(1.173279) = 20,582.95$$
$$(17,543.10)(1.208720) = 21,204.70$$

SQL/Query

The next set of queries performs the forecast for the 1996 sales by quarters.

Query 8_14 and Query 8_15 perform the linear regression over the first 5 years.

Query 8_14:

```
SELECT Count([Query 8_9].Year) AS n,
Sum([Query 8_9].Year) AS Sx,
Sum([Query 8_9].[Yearly Sales]) AS Sy,
Sum([Year] ^ 2) AS Sx2,
Sum([Yearly Sales] ^ 2) AS Sy2,
Sum([Year] * [Yearly Sales]) AS Sxy
FROM [Query 8_9], Forecast_Parameters
WHERE [Query 8_9].Year <= ([Forecast_Parameters].[forecast year] - 1);
```

Query 8_15:

```
SELECT ([Sy] * [Sx2] - [Sx] * [Sxy]) / ([n] * [Sx2] - [Sx] ^ 2) AS a,
([n] * [Sxy] - [Sx] * [Sy]) / ([n] * [Sx2] - [Ssx] ^ 2) AS b
FROM [Query 8_14];
```

Using the linear equation for the above regression, Query 8_16 forecasts the sales for year 1996, which is denoted in the table Forecast_Parameters as year 6.

Query 8_16:

```
SELECT [a] + [b] * ([Forecast_Parameters].[forecast year]) AS y,
([a] + [b] * ([Forecast_Parameters].[forecast year])) / 4 AS Q
FROM [Query 8_15], Forecast_Parameters;
```

Finally, Query 8_17 gives the new averages for the seasonal factors.

Query 8_17:

```
SELECT Forecast_Parameters.[forecast year] AS [Year],
[Query 8_13].[Quarter ID],
[Average Seasonal Factor] * [Q] AS [Quarter Forecast]
FROM [Query 8_16], [Query 8_13], Forecast_Parameters;
```

   Table 8-7 shows these forecasts, the actual 1996 sales figures, the 1996 seasonal factors, and the new average seasonal factors. These new averages are found by including the 1996 seasonal factors with the previous 5 years' seasonal factors, then finding the average for each quarter by dividing the total by 6. Recall that the 1996 seasonal factors are found by dividing each quarter's actual sales by the *average* quarterly sales (in this case, the average quarterly sales are equal to the total 1996 sales of 63,532.31 divided by 4, or 15,883.08).

Table 8-7. Forecast and Actual Sales for 1996, with Seasonal Factors

QUARTER	1996 FORECAST	ACTUAL 1996 SALES	1996 SEASONAL FACTORS	NEW AVERAGE SEASONAL FACTORS
1	10,202.65	9,484.70	0.597157	0.584173
2	18,182.11	16,979.71	1.069044	1.041862
3	20,582.95	15,217.08	0.958069	1.137411
4	21,204.70	21,850.82	1.375729	1.236555

SQL/Query

Query 8_18 gives the forecast and actual sales for 1996 along with the seasonal factors. Query 8_18a is the same as Query 8_18 except it appends the data value to Table 8_6. Once the new quarterly forecast values are appended to Table 8_6, we can run Query 8_13 again to get the new average seasonal factors.

Query 8_18:

```
SELECT [Query 8_17].Year, [Query 8_17].[Quarter ID],
[Query 8_17].[Quarter Forecast] AS Forecast,
[Query 8_8].Sales AS [Actual Sales],
[Sales] / [Quarterly Average Sales] AS [Seasonal Factors]
FROM [Query 8_8], [Query 8_17], [Query 8_10]
WHERE [Query 8_8].[Quarter ID] = [Query 8_17].[Quarter ID]
AND [Query 8_8].Year = [Query 8_17].Year
AND [Query 8_17].Year = [Query 8_10].Year
ORDER BY [Query 8_17].Year, [Query 8_17].[Quarter ID];
```

Query 8_18a:

```
INSERT INTO [Table 8_6]
([Year], [Quarter ID], [Quarterly Average Sales], Sales, [Seasonal Factor])
SELECT [Query 8_17].Year, [Query 8_17].[Quarter ID],
[Query 8_17].[Quarter Forecast] AS Forecast,
[Query 8_8].Sales AS [Actual Sales],
[Sales] / [Quarterly Average Sales] AS [Seasonal Factors]
FROM [Query 8_8], [Query 8_17], [Query 8_10]
WHERE [Query 8_8].[Quarter ID] = [Query 8_17].[Quarter ID]
AND [Query 8_8].Year = [Query 8_17].Year
AND [Query 8_17].Year = [Query 8_10].Year
ORDER BY [Query 8_17].Year, [Query 8_17].[Quarter ID];
```

If we change the forecast year in the Forecast_Parameters table from 6 to 7, we can run Query 8_18a again to get the forecast values for 1997 appended to Table 8_6. By changing the forecast year from 7 to 8 and running Query 8_18a again, we get the forecast for 1998 appended to Table 8_6. Continue changing the forecast year and running Query 8_18a to obtain the forecast values through year 2000.

When we are ready to forecast the 1997 sales, we assume we have the actual 1996 sales figures in hand. Therefore, we can find a new regression equation by fitting a line through the *six* annual sales figures from 1991 – 1996. This equation is different than the previous one because we have added another point on the graph. The new linear regression equation is

$$y = 49,641.92 + 2,842.06x$$

By substituting $x = 7$ into this equation, we find the projected annual sales for 1997 are $69,536.31. Dividing this by 4, we find the average quarterly sales projection to be $17,384.08. We multiply this figure by each of the new average seasonal factors from Table 8-7 to obtain the four quarterly sales forecasts for 1997, as follows.

$$(17,384.08)(0.584173) = 10,155.31$$
$$(17,384.08)(1.041862) = 18,111.81$$
$$(17,384.08)(1.137411) = 19,772.84$$
$$(17,384.08)(1.236555) = 21,496.37$$

Once the actual 1997 sales figures are known, we proceed in the same manner as before. We find the actual seasonal factors for 1997, add those to the previous 6 years' factors, and then find new average seasonal factors. These are then applied to the 1998 average quarterly sales, based on a new regression equation for the 1991–1997 annual sales. If we complete this seasonal forecasting methodology through the year 2000, we obtain the results shown in Table 8-8. Notice again that the forecast errors are given in the last column of the table. We'll use these shortly to get a handle on how well our various forecasting techniques are working.

Table 8-8. Seasonal (Quarterly) Forecasts Compared to Actual Sales

YEAR	QUARTER	ACTUAL SALES	FORECAST	ERROR
1996	1	9,484.70	10,202.65	-717.95
	2	16,979.71	18,182.11	-1,202.40
	3	15,217.08	20,582.95	-5,365.87
	4	21,850.82	21,204.70	646.12
1997	1	10,293.46	10,155.31	138.15
	2	18,475.83	18,111.81	364.02
	3	19,263.73	19,772.84	-509.11
	4	21,261.77	21,496.37	-234.60
1998	1	11,004.73	10,576.04	428.69
	2	18,817.88	18,879.71	-61.83
	3	18,029.76	20,476.15	-2,446.39
	4	22,721.15	22,308.47	412.68
1999	1	13,041.94	10,954.76	2,087.18
	2	19,606.93	19,447.03	159.90
	3	21,116.66	20,778.65	338.01
	4	31,449.09	23,042.68	8,406.41
2000	1	17,195.78	12,120.49	5,075.29
	2	25,578.97	21,138.38	4,440.59
	3	25,671.77	22,603.63	3,068.14
	4	27,083.57	25,923.00	1,160.57

SQL/Query

After the running of Query 8_18a repeatedly to append the forecast values to Table 8_6, we can run Query 8_19 to produce Table 8-8.

Query 8_19:

```
SELECT [Table 8_6].Year, [Table 8_6].[Quarter ID] AS Quarter,
[Table 8_6].Sales, [Table 8_6].[Quarterly Average Sales] AS Forecast,
[Sales] - [Quarterly Average Sales] AS Error
INTO [Table 8_8]
FROM [Table 8_6]
WHERE [Table 8_6].Year >= 6;
```

Criteria for Selecting
the Most Appropriate Forecasting Technique

There are several measures that are commonly used to determine the performance of forecasting methods. These measures are based on the forecast errors, or the differences between the actual and forecast values. The first of these measures is called the *CFE*, or cumulative sum of the forecast errors. To get the CFE, we add all the errors, keeping in mind that some are positive and some are negative. The closer the CFE is to zero, the better our forecasting method is performing. A positive value for the CFE means our forecasts have been too low on average; a negative CFE means the forecasts have been too high.

It is also helpful to know the standard deviation of forecast errors, as we'll see momentarily. It is just calculated from our old tried-and-true formula, first presented in Chapter 2. But we again need to keep track of those positive and negative signs.

If we don't care for the positive and negative signs, we may take the absolute value of each of the forecast errors. If we then add these absolute values and divide by the total number we have, we obtain what is called the *MAD*, or mean absolute deviation. It would be nice if this value would also be close to zero.

The *TS*, or tracking signal, is the ratio of the CFE to the MAD. It is desirable to have the TS between about −1.5 and +1.5, although it may range a little lower or higher (say from −3.0 to +3.0 or so) depending on the judgment of the data miner.

Since the forecast errors should usually be normally distributed, there is a rule of thumb that the ratio of the MAD to the standard deviation is about 0.8. This allows a quick check to see if the forecast technique is very accurate. We'll apply all these measures to our sales data example and the different forecasting methods we examined.

For the three-month simple moving average, we would have to add all 120 of the forecast errors to get the CFE. The result is 10,445.80, which is pretty high. To find the MAD, we add the absolute values of the forecast errors and divide the result by 120.

In this case, we obtain 1,621.47 for the MAD. The tracking signal is CFE/MAD, which is equal to 6.44. This is also higher than we want, so it appears that the simple 3-month moving average is not a very good predictor of actual sales. Incidentally, the standard deviation of forecast errors in this case is equal to 2,153.78, and the ratio of the MAD to the standard deviation is 0.75 (which of course is close to 0.8).

Measures are found for all the other forecasting techniques we presented in a similar way. Table 8-9 summarizes the results.

Table 8-9. Performance Measures for the Various Forecasting Techniques

MODEL	CFE	MAD	TS	STD. DEV	MAD/STD. DEV.
3-month Moving Average	10.445.80	1,621.47	6.44	2,153.78	0.75
3-month Weighted Moving Average	7,785.84	1,571.18	4.96	2,036.83	0.77
Single Exponential Smoothing, $\alpha = 0.1$	-31,432.22	1,615.89	-19.45	2,117.76	0.76
Single Exponential Smoothing, $\alpha = 0.2$	-17,708.52	1,592.11	-11.12	2,153.23	0.74
Single Exponential Smoothing, $\alpha = 0.5$	-5,941.89	1,636.75	-3.63	2,211.74	0.74
Double Exponential Smoothing, $\alpha = 0.3$, $\beta = 0.2$	-390.79	1,821.90	-0.21	2,379.89	0.77
Multiplicative Seasonal Adjustment	16,187.60	1,863.20	8.69	2,867.97	0.65

SQL/Query

The performance measures shown in Table 8-9 can be obtained by executing the following set of queries. Query 8_20 through Query 8_24 obtain the CFE, MAD, and standard deviation for the various models shown in Table 8-9. Query 8_25 combines the results of Query 8_20 through Query 8_24 and completes the calculations for the data shown in Table 8-9.

Query 8_20:

```
SELECT Sum([Table 8_2].Error) AS CFE,
Sum(Abs([Error])) / Count([Error]) AS MAD,
StDev([Table 8_2].Error) AS [Std Dev]
FROM [Table 8_2];
```

Query 8_21:

```
SELECT Sum([Table 8_3].Error) AS CFE,
Sum(Abs([Error])) / Count([Error]) AS MAD,
StDev([Table 8_3].Error) AS [Std Dev]
FROM [Table 8_3];
```

Query 8_22: ·

```
SELECT Sum([Table 8_4].[Forecast Error]) AS CFE,
Sum(Abs([Forecast Error])) / Count([Forecast Error]) AS MAD,
StDev([Table 8_4].[Forecast Error]) AS [Std Dev]
FROM [Table 8_4];
```

Query 8_23:

```
SELECT Sum([Table 8_5].[Forecast Error]) AS CFE,
Sum(Abs([Forecast Error])) / Count([Forecast Error]) AS MAD,
StDev([Table 8_5].[Forecast Error]) AS [Std Dev]
FROM [Table 8_5];
```

Query 8_24:

```
SELECT Sum([Table 8_8].Error) AS CFE,
Sum(Abs([Error])) / Count([Error]) AS MAD,
StDev([Table 8_8].Error) AS [Std Dev]
FROM [Table 8_8];
```

Query 8_25:

```
SELECT "3 month Moving Avg" AS Model,
CFE, MAD, [CFE] / [MAD] AS TS, [Std Dev], [MAD]/[Std Dev] AS [MAD/Std Dev]
FROM [Query 8_20]
UNION ALL
SELECT "3 month Weighted Moving Avg" AS Model,
CFE, MAD, [CFE] / [MAD] AS TS, [Std Dev], [MAD] / [Std Dev] AS [MAD/Std Dev]
FROM [Query 8_21]
UNION ALL
SELECT "Single Exponential" AS Model,
CFE, MAD, [CFE] / [MAD] AS TS, [Std Dev], [MAD] / [Std Dev] AS [MAD/Std Dev]
FROM [Query 8_22]
UNION ALL
```

```
SELECT "Double Exponential" AS Model,
CFE, MAD, [CFE] / [MAD] AS TS, [Std Dev], [MAD] / [Std Dev] AS [MAD/Std Dev]
FROM [Query 8_23]
UNION ALL
SELECT "Multiplicative Seasonal Adj" AS Mode,
CFE, MAD, [CFE] / [MAD] AS TS, [Std Dev], [MAD]  /[Std Dev] AS [MAD/Std Dev]
FROM [Query 8_24];
```

It is evident from the results in Table 8-9 that the double exponential smoothing method is the best of the various techniques we present. It has the CFE closest to zero and the best tracking signal. Notice also the relatively poor performance of the seasonal method. This is partially explained by the use of the linear regression equation to estimate the "next year's" sales. We might be able to improve this technique by combining it with the double exponential smoothing method as the annual sales predictor.

Conclusion

We have presented several of the better-known forecasting methods in use for time series analysis. Although the double exponential smoothing method proved the best for our example data, in other circumstances this may not be the case. Consequently, it is always wise to judge several methods based on their performance measures before settling on the preferred technique.

T-SQL Source Code

Following are five T-SQL source code listings for performing the various forecasting techniques discussed in this chapter. The first procedure is called Simple_Moving_Average and performs the simple moving average over a given time period. The second procedure is called Weighted_Moving_Average and performs the weighted moving average over a given time period. The third procedure is called Single_Exponential_Smoothing and uses the smoothing factor Alpha to smooth the forecasts. The fourth procedure is called Double_Exponential_Smoothing and it uses two smoothing factors in determining the forecast. The last procedure is called Seasonal_Adjustment and it uses yearly quarters for the adjustment. Following the procedure listings is a set of statements illustrating how to invoke each procedure.

Simple Moving Average

```
SET QUOTED_IDENTIFIER OFF
GO
SET ANSI_NULLS OFF
GO


ALTER PROCEDURE [Simple_Moving_Average]
@SrcTable Varchar(50) = 'Table 8_1',
@Time_Series Varchar(50) = 'Month',
@Data_Value Varchar(50) = 'Actual Gross Sales',
@Period_Size Int = 3,
@RsltTable Varchar(50) = 'Simple Moving Average'
AS


/*************************************************************/
/*                                                         */
/*                Simple_Moving_Average                    */
/*                                                         */
/*    This procedure performs the simple moving average on */
/* the data given in SrcTable. SrcTable must have at least */
/* two columns. One column gives the Time_Series and is    */
/* simply a time sequence numbering (1,2,3,...) of the data. */
/* The other column contains the actual data values. The   */
/* procedure calculates the average over a time period. The */
/* time period is the interval of data values to be used in */
/* the average or forecast. The size of the interval is    */
/* denoted by Period_Size. Once the forecasting is done,   */
/* the Time_Series, actual data value, forecast, and       */
/* forecasting error are recorded in a table denoted by    */
/* RsltTable.                                              */
/*                                                         */
/* INPUTS:                                                 */
/*   SrcTable - name of table containing sample data       */
/*   Time_Series - a column in ScrTable, and is the time   */
/*     sequencing number (1,2,3,...)                       */
/*   Data_Value - a column in ScrTable, and is the actual  */
/*     data values                                         */
/*   Period_Size - denotes the interval size of the period */
/*                                                         */
/* OUTPUTS:                                                */
/*   RsltTable - a name for the table that is created by   */
/*     this procedure to receive the forecasting information */
/*                                                         */
/*************************************************************/
```

```
/* Local Variables */
DECLARE @Q Varchar(500)        /* Query string */
DECLARE @Pbegin Int            /* Start of time period */
DECLARE @Pend Int              /* End of time period */
DECLARE @P_Avg Float           /* Average for time period */
DECLARE @P_Max Int             /* Last time period */

/* Create a work table for the moving average */
CREATE TABLE ##TempMovAvg
     (ID Int IDENTITY(1,1),
     Period Int,
     X Float,
     [Forecast] Float,
     Error Float)

/* Populate work table with time periods and data values */
SET @Q = 'INSERT INTO ##TempMovAvg(Period, X) ' +
     '(SELECT [' + @Time_Series + '], [' + @Data_Value + '] ' +
     'FROM [' + @SrcTable + '])'
EXEC(@Q)

/* Get upper limit for last period */
SELECT @P_max = (SELECT Max(ID) FROM ##TempMovAvg)

/* Initialize the beginning and end for first period */
SET @Pbegin = 1
SET @Pend = @Period_Size

/* For each period determine the average and forecast */
WHILE @Pend <= @P_max - 1
Begin

    /* Calculate forecast for this period */
    SELECT @P_Avg = (SELECT Sum(X) / Count (X) AS [A]
         FROM ##TempMovAvg
         WHERE ID Between @Pbegin And @Pend)

    /* Save forecast */
    UPDATE ##TempMovAvg
         SET [Forecast] = @P_Avg
         WHERE ID = @Pend + 1
```

```
            /* Prepare for next period */
            SET @Pbegin = @Pbegin +1
            SET @Pend = @Pend + 1

    End

    /* Calculate Error */
    UPDATE ##TempMovAvg
        SET [Error] = [X] - [Forecast]
        WHERE ID > @Period_Size

    /* If the result table exists, remove it */
    IF Exists (SELECT id FROM ..sysobjects
        WHERE name = @RsltTable)
    Begin
        SET @Q = 'DROP TABLE [' + @RsltTable + ']'
        EXEC(@Q)
    End

    /* Set up result table */
    SET @Q = 'CREATE TABLE [' + @RsltTable + '] ' +
        '([Time Period] Int, ' +
        '[Actual Data Value] Float, ' +
        '[Forecast] Float, ' +
        '[Error] Float)'
    EXEC(@Q)

    /* Fill result table */
    SET @Q = 'INSERT INTO [' + @RsltTable + '] ' +
        '([Time Period], [Actual Data Value], ' +
        '[Forecast], [Error]) ' +
        'SELECT Period, X, [Forecast], [Error] FROM ##TempMovAvg'
    EXEC(@Q)

GO
SET QUOTED_IDENTIFIER OFF
GO
SET ANSI_NULLS ON
GO
```

Weighted Moving Average

```
SET QUOTED_IDENTIFIER OFF
GO
SET ANSI_NULLS OFF
GO

ALTER PROCEDURE Weighted_Moving_Average
@SrcTable Varchar(50) = 'Table 8_1',
@Time_Series Varchar(50) = 'Month',
@Data_Value Varchar(50) = 'Actual Gross Sales',
@Period_Size Int = 3,
@Wt_Table Varchar(50) = 'Weights',
@Wt_ID Varchar(50) = 'Wt_ID',
@Wt_Value Varchar(50) = 'Wt_Value',
@RsltTable Varchar(50) = 'Weighted Moving Average'
AS

/************************************************************/
/*                                                        */
/*              Weighted_Moving_Average                   */
/*                                                        */
/*     This procedure performs the weighted moving average on */
/* the data given in SrcTable using the weights given in   */
/* Wt_Table. SrcTable must have at least two columns. One  */
/* column gives the Time_Series and is simply a time       */
/* sequence numbering (1,2,3,...) of the data. The other   */
/* column contains the actual data values. The Wt_Table also */
/* has two columns. One column is weight ID (1,2,3,...) and  */
/* the second column is the weight. The procedure calculates */
/* the weighted average over a time period. The time period  */
/* is the interval of data values to be used in computing a  */
/* weighted average or forecast. The size of the interval is */
/* denoted by Period_Size. The number of weights and the    */
/* Period_Size must be the same. Once the forecasting is    */
/* completed, the Time_Series, actual data value, forecast,  */
/* and forecasting error are recorded in a table denoted by  */
/* RsltTable.                                              */
/*                                                        */
/* INPUTS:                                                */
/*   SrcTable - name of table containing sample data      */
/*   Time_Series - a column in ScrTable, and is the time  */
/*     sequencing number (1,2,3,...)                      */
```

```
/*    Data_Value - a column in ScrTable, and is the actual    */
/*      data values                                            */
/*    Period_Size - denotes the interval size of the period    */
/*    Wt_Table - name of table containing the weights          */
/*    Wt_ID - a column in Wt_Table, and is the sequencing      */
/*      number (1,2,3,...)                                      */
/*    Wt_Value - a column in Wt_Table, and is the value of     */
/*      the weight                                             */
/*                                                             */
/* OUTPUTS:                                                    */
/*    RsltTable - a name for the table that is created by      */
/*      this procedure to receive the forecasting information */
/*                                                             */
/***************************************************************/


/* Local Variables */
DECLARE @Q Varchar(500)     /* Query string */
DECLARE @Pbegin Int         /* Start of time period */
DECLARE @Pend Int           /* End of time period */
DECLARE @P_Avg Float        /* Average for time period */
DECLARE @P_Max Int          /* Last time period */


/* Create a work table for the moving average */
CREATE TABLE ##TempWtMovAvg
    (ID Int IDENTITY(1,1),
    Period Int,
    X Float,
    [Forecast] Float,
    Error Float)


/* Populate work table with time periods and data values */
SET @Q = 'INSERT INTO ##TempWtMovAvg(Period, X) ' +
    '(SELECT [' + @Time_Series + '], [' + @Data_Value + '] ' +
    'FROM [' + @SrcTable + '])'
EXEC(@Q)


/* Create work table for weights */
CREATE TABLE ##TempWts
    (Wid Int,Wt Float)


/* Populate weights work table */
SET @Q = 'INSERT INTO ##TempWts(Wid, Wt) ' +
    '(SELECT [' + @Wt_ID + '], [' + @Wt_Value + '] ' +
    'FROM [' + @Wt_Table + '])'
EXEC(@Q)
```

```
/* Get upper limit for last period */
SELECT @P_max = (SELECT Max(ID) FROM ##TempWtMovAvg)

/* Initialize the beginning and end for first period */
SET @Pbegin = 1
SET @Pend = @Period_Size

/* For each period determine the average and forecast */
WHILE @Pend <= @P_max - 1
Begin

    /* Calculate forecast for this period */
    SELECT @P_Avg = (SELECT Sum(Wt*X) As [A]
        FROM ##TempWtMovAvg, ##TempWts
        WHERE ##TempWtMovAvg.[ID] = ##TempWts.Wid
        and ##TempWtMovAvg.[ID] Between @Pbegin and @Pend)

    /* Save forecast */
    UPDATE ##TempWtMovAvg
        SET [Forecast] = @P_Avg
        WHERE [ID] = @Pend + 1

    /* Increment weight ID so that it will match the next period */
    UPDATE ##TempWts
        SET Wid = Wid + 1

    /* Increment pointers to next period */
    SET @Pbegin = @Pbegin + 1
    SET @Pend = @Pend + 1

End

/* Calculate Error */
UPDATE ##TempWtMovAvg
    SET [Error] = [X] - [Forecast]
    WHERE [ID] > @Period_Size

/* If the result table exists, remove it */
IF Exists (SELECT id FROM ..sysobjects
    WHERE name = @RsltTable)
Begin
    SET @Q = 'DROP TABLE [' + @RsltTable + ']'
    EXEC(@Q)
End
```

```
/* Set up result table */
SET @Q = 'CREATE TABLE [' + @RsltTable + '] ' +
    '([Time Period] Int, ' +
    '[Actual Data Value] Float, ' +
    '[Forecast] Float, ' +
    '[Error] Float)'
EXEC(@Q)

/* Fill result table */
SET @Q = 'INSERT INTO [' + @RsltTable + '] ' +
    '([Time Period], [Actual Data Value], ' +
    '[Forecast], [Error]) ' +
    'SELECT Period, X, [Forecast], [Error] FROM ##TempWtMovAvg'
EXEC(@Q)

GO
SET QUOTED_IDENTIFIER OFF
GO
SET ANSI_NULLS ON
GO
```

Single Exponential Smoothing

```
SET QUOTED_IDENTIFIER OFF
GO
SET ANSI_NULLS OFF
GO

ALTER PROCEDURE Single_Exponential_Smoothing
@SrcTable Varchar(50) = 'Table 8_1',
@Time_Series Varchar(50) = 'Month',
@Data_Value Varchar(50) = 'Actual Gross Sales',
@First_Forecast_Period Int = 61,
@Alpha Float = 0.1,
@RsltTable Varchar(50) = 'Single Exponential Smoothing'
AS
```

```
/*************************************************************/
/*                                                         */
/*              Single_Exponential_Smoothing               */
/*                                                         */
/*     This procedure performs the single exponential      */
/* smoothing on the data given in SrcTable. SrcTable must  */
/* have at least two columns. One column gives the         */
/* Time_Series and is simply a time sequence numbering     */
/*(1,2,3,...) of the data. The other column contains the   */
/* actual data values. The procedure calculates the forecast */
/* and the forecast error. Once the forecasting is done,   */
/* the Time_Series, actual data value, forecast, and       */
/* forecasting error are recorded in a table denoted by    */
/* RsltTable.                                              */
/*                                                         */
/* INPUTS:                                                 */
/*    SrcTable - name of table containing sample data      */
/*    Time_Series - a column in ScrTable, and is the time  */
/*       sequencing number (1,2,3,...)                     */
/*    Data_Value - a column in ScrTable, and is the actual */
/*       data values                                       */
/*    First_Forecast_Period - the period to be forecasted  */
/*    Alpha - smoothing factor                             */
/*                                                         */
/* OUTPUTS:                                                */
/*    RsltTable - a name for the table that is created by  */
/*       this procedure to receive the forecasting information */
/*                                                         */
/*************************************************************/


/* Local Variables */
DECLARE @Q Varchar(500)          /* Query string */
DECLARE @Forecast Float          /* Forecast value */
DECLARE @Forecast_Error Float    /* Forecast error */
DECLARE @P Int                   /* A time period */
DECLARE @X_Data Float            /* Data value */

/* Create a work table for the exponential smoothing */
CREATE TABLE ##TempSES
    (ID Int IDENTITY(1,1),
    Time_Period Int,
    X Float,
    [Forecast] Float,
    [Error] Float)
```

```
/* Populate work table with time periods and data values */
SET @Q = 'INSERT INTO ##TempSES(Time_Period, X) ' +
    '(SELECT [' + @Time_Series + '], [' + @Data_Value + '] ' +
    'FROM [' + @SrcTable + '] '+
    'WHERE [' + @Time_Series + '] >=' +
    Convert(varchar(20), @First_Forecast_Period) + ')'
print @Q
EXEC(@Q)


/* Obtain the initial forecast */
SET @Q = 'SELECT Avg([' + @Data_Value + ']) AS [A] ' +
    'INTO ##TempIntAvg ' +
    'FROM [' + @SrcTable + '] ' +
    'WHERE [' + @Time_Series + '] < ' +
    Convert(varchar(20), @First_Forecast_Period)
print @Q
EXEC(@Q)
SELECT @Forecast = (SELECT A FROM ##TempIntAvg)


/* Obtain data value of the period just */
/* before the first forecast period */
SET @Q = 'SELECT [' + @Data_Value + '] AS Int_Val ' +
    'INTO ##TempInit ' +
    'FROM [' + @SrcTable + '] ' +
    'WHERE [' + @Time_Series + '] = ' +
    Convert(varchar(20), (@First_Forecast_Period -1))
print @Q
EXEC(@Q)
SELECT @X_Data = (SELECT Int_Val FROM ##TempInit)


/* Do first forecast */
SET @Forecast = @Alpha * @X_Data + (1.0 - @Alpha) * @Forecast
print 'First forecast is ' + convert(varchar(20), @Forecast)


/* Establish cursor pointing to record being forecasted */
DECLARE cr_F INSENSITIVE SCROLL CURSOR
    FOR SELECT Time_Period, X FROM ##TempSES


/* Open cursor and get the record for the first forecast */
OPEN cr_F
FETCH FIRST FROM cr_F INTO @P, @X_Data
```

```
/* Loop to calculate and save the forecast for each record */
WHILE @@FETCH_STATUS = 0
Begin

    SET @Forecast_Error = @Forecast - @X_Data

    /* Save forecast and error */
    UPDATE ##TempSES
        SET [Forecast] = @Forecast,
        [Error] = @Forecast_Error
        WHERE Time_Period = @p

    /* Calculate forecast for next period */
    SET @Forecast = @Alpha * @X_Data + (1.0 - @Alpha) * @Forecast

    /* Move to next record */
    FETCH NEXT FROM cr_F INTO @P, @X_Data

End

/* If the result table exists, remove it */
IF Exists (SELECT id FROM ..sysobjects
    WHERE name = @RsltTable)
Begin
    SET @Q = 'DROP TABLE [' + @RsltTable + ']'
    EXEC(@Q)
End

/* Set up result table */
SET @Q = 'CREATE TABLE [' + @RsltTable + '] ' +
    '([Time Period] Int, ' +
    '[Actual Data Value] Float, ' +
    '[Forecast] Float, ' +
    '[Error] Float)'
EXEC(@Q)

/* Fill result table */
SET @Q = 'INSERT INTO [' + @RsltTable + '] ' +
    '([Time Period], [Actual Data Value], ' +
    '[Forecast], [Error]) ' +
    'SELECT Time_Period, X, [Forecast], [Error] FROM ##TempSES'
EXEC(@Q)
```

```
GO
SET QUOTED_IDENTIFIER OFF
GO
SET ANSI_NULLS ON
GO
```

Double Exponential Smoothing

```
SET QUOTED_IDENTIFIER OFF
GO
SET ANSI_NULLS OFF
GO

ALTER PROCEDURE Double_Exponential_Smoothing
@SrcTable Varchar(50) = 'Table 8_1',
@Time_Series Varchar(50) = 'Month',
@Data_Value Varchar(50) = 'Actual Gross Sales',
@First_Forecast_Period Int = 61,
@Alpha Float = 0.3,
@Beta Float = 0.2,
@RsltTable Varchar(50) = 'Double Exponential Smoothing'
AS


/***************************************************************/
/*                                                           */
/*            Double_Exponential_Smoothing                   */
/*                                                           */
/*    This procedure performs the double exponential         */
/* smoothing on the data given in SrcTable. SrcTable must    */
/* have at least two columns. One column gives the           */
/* Time_Series and is simply a time sequence numbering       */
/*(1,2,3,...) of the data. The other column contains the     */
/* actual data values. The procedure calculates the forecast */
/* and the forecast error. Once the forecasting is done,     */
/* the Time_Series, actual data value, forecast, and         */
/* forecasting error are recorded in a table denoted by      */
/* RsltTable.                                                */
/*                                                           */
/* INPUTS:                                                   */
/*    SrcTable - name of table containing sample data        */
/*    Time_Series - a column in ScrTable, and is the time    */
/*       sequencing number (1,2,3,...)                       */
```

```
/*   Data_Value - a column in ScrTable, and is the actual    */
/*      data values                                          */
/*   First_Forecast_Period - the period to be forecasted     */
/*   Alpha - first smoothing factor                          */
/*   Beta - second smoothing factor                          */
/*                                                           */
/* OUTPUTS:                                                  */
/*   RsltTable - a name for the table that is created by     */
/*      this procedure to receive the forecasting information */
/*                                                           */
/*************************************************************/


/* Local Variables */
DECLARE @Q varchar(500)              /* Query string */
DECLARE @Initial_Forecast Float      /* Average of initial trend */
DECLARE @Forecast1 Float             /* Forecast from 1st exponential */
DECLARE @Forecast2 Float             /* Forecast from 2nd exponential */
DECLARE @Prev_Forecast1 Float        /* Previous 1st forecast */
DECLARE @Forecast_Error Float        /* Forecasting error */
DECLARE @Trend1 Float                /* Trend of 1st exponential */
DECLARE @Trend2 Float                /* Weighted trend of 2nd exponential */
DECLARE @Prev_Trend1 Float           /* Previous trend of 1st exponential */
DECLARE @P Int                       /* A time period */
DECLARE @X_Data Float                /* A data value */


/* Variables for linear regression */
DECLARE @Lin_Reg_Table Varchar(50)   /* Table of (x,y) for regression */
DECLARE @Lin_Reg_ColX Varchar(50     /* Column x for regression */
DECLARE @Lin_Reg_ColY Varchar(50     /* Column y for regression */
DECLARE @a Float                     /* Constant 'a' for line equation */
DECLARE @b Float                     /* Constant 'b' for line equation */
DECLARE @r Float                     /* Correlation Coef. for regression */


/*************************/
/*  DO LINEAR REGRESSION */
/*************************/


/* Create a work table of (x,y) values for linear regression */
CREATE Table ##TempLinReg
     (Lin_X Float, Lin_Y Float)
```

```
/* Populate linear regression work table with time */
/* periods and data values for the initial trend (slope) */
SET @Q = 'INSERT INTO ##TempLinReg(Lin_X, Lin_Y) ' +
    '(SELECT CAST([' + @Time_Series + '] AS Float), [' + @Data_Value + '] ' +
    'FROM [' + @SrcTable + '] ' +
    'WHERE [' + @Time_Series + '] < ' +
    convert(varchar(20), @First_Forecast_Period) + ')'
EXEC(@Q)

/* Set parameters for linear regression procedure */
SET @Lin_Reg_Table = '##TempLinReg'
SET @Lin_Reg_ColX = 'Lin_X'
SET @Lin_Reg_ColY = 'Lin_Y'
SET @a = 0
SET @b = 0
SET @r = 0

/* Invoke the linear regression procedure */
EXEC [CH5]..[Linear_Regression_2_Variables]
    @Lin_Reg_Table, @Lin_Reg_ColX, @Lin_Reg_ColY,
    @a OUTPUT , @b OUTPUT , @r OUTPUT


/**************************************************/
/*  PREPARE TO BEGIN DOUBLE EXPONENTIAL SMOOTHING  */
/**************************************************/

/* Use slope as initial trend */
SET @Prev_Trend1 = @b

/* Create a work table for the double exponential smoothing */
CREATE Table ##TempDES
    (ID Int IDENTITY(1,1),
    Time_Period Int,
    X Float,
    Smoothed Float,
    Trend Float,
    Weighted_Trend Float,
    Forecast Float,
    [Error] Float)
```

```
/* Populate work table with time periods and data values */
SET @Q = 'INSERT INTO ##TempDES(Time_Period, X) ' +
    '(SELECT [' + @Time_Series + '], [' + @Data_Value + '] ' +
    'FROM [' + @SrcTable + '] ' +
    'WHERE [' + @Time_Series + '] >= ' +
    convert(varchar(20), @First_Forecast_Period) + ')'
EXEC(@Q)

/* Obtain average of initial forecast */
SET @Q = 'SELECT Avg([' + @Data_Value + ']) AS [A] ' +
    'INTO ##TempIT ' +
    'FROM [' + @SrcTable + '] ' +
    'WHERE [' + @Time_Series + '] < ' +
    Convert(Varchar(20), @First_Forecast_Period)
EXEC(@Q)
SELECT @Initial_Forecast = (SELECT A FROM ##TempIT)

/* Get the data value just before the first forecast value */
SET @Q = 'SELECT [' + @Data_Value + '] AS Val ' +
    'INTO ##TempPD ' +
    'FROM [' + @SrcTable + '] ' +
    'WHERE [' + @Time_Series + '] = ' +
    convert(Varchar(20), (@First_Forecast_Period  - 1))
EXEC(@Q)
SELECT @X_data = (SELECT Val FROM ##TempPD)

/* Calculate first exponential smoothing value */
SET @Forecast1 = @Alpha * @X_Data +
    (1.0 - @Alpha) * (@Initial_Forecast + @Prev_Trend1)

/* Determine the trend and then calculate */
/* second exponential smoothing value */
SET @Trend1 = @Forecast1 - @Initial_Forecast
SET @Trend2 = @Beta * @Trend1 + (1.0 - @Beta) * @Prev_Trend1

/* Make forecast for next time period */
SET @Forecast2 = @Forecast1 + @Trend2

/* Establish cursor pointing to record being forecast */
DECLARE cr_F INSENSITIVE SCROLL CURSOR
    FOR SELECT Time_Period, X FROM ##TempDES
```

```
/* Open cursor and get the record for the first forecast */
OPEN cr_F
FETCH FIRST FROM cr_F INTO @P, @X_Data


/*****************************************/
/*  PERFORM DOUBLE EXPONENTIAL SMOOTHING  */
/*****************************************/

/* For each record determine the */
/* forecast and the error and save */
WHILE @@FETCH_STATUS = 0
Begin

    /* Determine Forecasting error */
    SET @Forecast_Error = @X_Data - @Forecast2

    /* Update table with calculated values */
    UPDATE ##TempDES
        SET Smoothed = @Forecast1,
        Trend = @Trend1,
        Weighted_Trend = @Trend2,
        Forecast = @Forecast2,
        [Error] = @Forecast_Error
        WHERE Time_Period = @P

    /* Calculate first smoothing */
    SET @Prev_Forecast1 = @Forecast1
    SET @Forecast1 = @Alpha * @X_Data + (1.0 - @Alpha) * @Forecast2

    /* Calculate trends */
    SET @Trend1 = @Forecast1 - @Prev_Forecast1
    SET @Trend2 = @Beta * @Trend1 + (1.0 - @Beta) * @Trend2

    /* Determine forecast for next time period */
    SET @Forecast2 = @Forecast1 + @Trend2

    /* Get next record */
    FETCH NEXT FROM cr_F INTO @P, @X_Data

End
```

```
/* If the result table exists, remove it */
IF Exists (SELECT id FROM ..sysobjects
     WHERE name = @RsltTable)
Begin
     SET @Q = 'DROP TABLE [' + @RsltTable + ']'
     EXEC(@Q)
End

/* Set up result table */
SET @Q = 'CREATE TABLE [' + @RsltTable + '] ' +
     '([Time Period] Int, ' +
     '[Actual Data Value] Float, ' +
     '[Smoothed Value] Float, ' +
     '[Trend] Float, ' +
     '[Weighted Trend] Float, ' +
     '[Forecast] Float, ' +
     '[Error] Float)'
EXEC(@Q)

/* Fill result table */
SET @Q = 'INSERT INTO [' + @RsltTable + '] ' +
     '([Time Period], [Actual Data Value], ' +
     '[Smoothed Value], [Trend], [Weighted Trend], ' +
     '[Forecast], [Error]) ' +
     'SELECT Time_Period, X, ' +
     'Smoothed, Trend, Weighted_Trend, ' +
     'Forecast, [Error] FROM ##TempDES'
EXEC(@Q)

GO
SET QUOTED_IDENTIFIER OFF
GO
SET ANSI_NULLS ON
GO
```

Seasonal Adjustment

```
SET QUOTED_IDENTIFIER OFF
GO
SET ANSI_NULLS OFF
GO
```

```
ALTER PROCEDURE Seasonal_Adjustment
@SrcTable varchar(50) = 'Table 8_1',
@Time_Series Varchar(50) = 'Month',
@SrcDate Varchar(50) = 'Date',
@Data_Value varchar(50) = 'Actual Gross Sales',
@Seasonal_Size Int = 3,
@ForecastYr Int = 6,
@RsltTable varchar(50) = 'Seasonal Forecast'
AS


/**************************************************************/
/*                                                          */
/*                  Seasonal_Adjustment                     */
/*                                                          */
/*    This procedure performs the seasonal forecast for the */
/* data given in SrcTable. SrcTable must contain three      */
/* columns. The first column is the Time_Series and is a    */
/* time sequencing number (1,2,3,...). The second column is */
/* the SrcDate and is the month/day/year (e.g., 02/01/98).  */
/* The third column is the actual data values. In this      */
/* procedure the season is defined to be a quarter or 3     */
/* months). The procedure does a seasonal forecast for the  */
/* next year. Once the forecasting is done, the Year,       */
/* seasonal ID, and forecast are recorded in a table denoted*/
/* by RsltTable.                                            */
/*                                                          */
/* INPUTS:                                                  */
/*   SrcTable - name of table containing sample data        */
/*   Time_Series - a column in ScrTable, and is the time    */
/*      sequencing number (1,2,3,...)                       */
/*   SrcDate - a column in ScrTable, and is the sample date */
/*      in the form of month/day/year (eg., 02/01/98)       */
/*   Data_Value - a column in ScrTable, and is the actual   */
/*      data values                                         */
/*   Seasonal_Size - number of months in a season           */
/*   ForecastYr - the year being forecast                   */
/*                                                          */
/* OUTPUTS:                                                 */
/*   RsltTable - a name for the table that is created by    */
/*      this procedure to receive the forecasting information*/
/*                                                          */
/**************************************************************/
```

```
/* Local Variables */
DECLARE @Q varchar(5000)      /* Query string */
DECLARE @SeasonsPerYr Int     /* Seasons per year */


/* Calculate number of seasons per year */
SET @SeasonsPerYr = 12 / @Seasonal_Size


/* Expand the date fields */
/* Query 8_7 */
SET @Q = 'SELECT Cast((((([' + @Time_Series + ']-1) / ' +
    convert(varchar(10), @Seasonal_Size) + ')+1) AS Int) AS Seasons, ' +
    'Datepart(q,[' + @SrcDate + ']) AS [Season_ID], ' +
    '[' + @Time_Series + '] AS [Time_Series], ' +
    'Cast(((([' + @Time_Series + ']-1)/12) AS Int)+1 AS [Year], ' +
    '[' + @Data_Value + '] AS [Data_Value] ' +
    'INTO ##TempExpDate ' +
    'FROM [' + @SrcTable + '] ' +
    'ORDER BY [' + @Time_Series + ']'
Exec(@Q)


/* Total data values for each season */
/* Query 8_8 */
SET @Q = 'SELECT Seasons, ' +
    'Avg([Season_ID]) AS [Season_ID], ' +
    'Avg([Year]) As [Year], ' +
    'Sum([Data_Value]) as [Seasonal_Value] ' +
    'INTO ##TempTotalBySeasons ' +
    'FROM ##TempExpDate ' +
    'GROUP BY Seasons'
EXEC(@Q)


/* Total data values for each year */
/* Query 8_9 */
SET @Q = 'SELECT Year, ' +
    'Sum([Data_Value]) AS [Year_Value] ' +
    'INTO ##TempTotalByYear ' +
    'FROM ##TempExpDate ' +
    'GROUP BY [Year]'
EXEC(@Q)
```

```
/* Find yearly average of seasonal totals */
/* Query 8_10 */
SET @Q = 'SELECT Year, ' +
    'Avg([Seasonal_Value]) AS [Seasonal_Average] ' +
    'INTO ##TempYrAvgBySeasons ' +
    'FROM ##TempTotalBySeasons ' +
    'GROUP BY Year'
EXEC(@Q)


/* Determine the seasonal factor as the ratio */
/* of seasonal value to seasonal average */
/* Query 8_11 */
SET @Q = 'SELECT Seasons, Season_ID, ##TempTotalBySeasons.[Year], ' +
    '[Seasonal_Value]/[Seasonal_Average] AS [Seasonal_Factor] ' +
    'INTO ##TempSeasonalFactors ' +
    'FROM ##TempTotalBySeasons, ##TempYrAvgBySeasons ' +
    'WHERE ##TempTotalBySeasons.[Year] = ' +
        '##TempYrAvgBySeasons.[Year]'
EXEC(@Q)


/* Combine all the calculations in a single table */
/* Query 8_12a */
SET @Q = 'SELECT ##TempYrAvgBySeasons.[Year] AS [Year], ' +
    '##TempTotalBySeasons.[Season_ID] AS [Season_ID], ' +
    '##TempYrAvgBySeasons.[Seasonal_Average] AS [Seasonal_Average], ' +
    '##TempTotalBySeasons.[Seasonal_Value] As [Seasonal_Value], ' +
    '##TempSeasonalFactors.[Seasonal_Factor] AS [Seasonal_Factor] ' +
    'INTO ##TempTable8_6 ' +
    'FROM ##TempYrAvgBySeasons, ##TempTotalBySeasons, ##TempSeasonalFactors ' +
    'WHERE ##TempYrAvgBySeasons.[Year] = ##TempTotalBySeasons.[Year] ' +
    'AND ##TempYrAvgBySeasons.[Year] = ##TempSeasonalFactors.[Year] ' +
    'AND ##TempTotalBySeasons.[Seasons] = ##TempSeasonalFactors.[Seasons] ' +
    'AND ##TempYrAvgBySeasons.[Year] <= ' +
    convert(varchar(20), (@ForecastYr - 1))
EXEC(@Q)


/* Calculate average seasonal factors */
/* Query 8_13 */
SET @Q = 'SELECT Season_ID, ' +
    'Avg([Seasonal_Factor]) AS [Average_Seasonal_Factor] ' +
    'INTO ##TempAvgSeasonalFactors ' +
    'FROM ##TempTable8_6 ' +
    'GROUP BY Season_ID'
EXEC(@Q)
```

```
/****************************************/
/*  FORECAST NEXT YEAR VALUES BY SEASONS  */
/****************************************/

/* Perform linear regression over all */
/* the years preceding the forecast year */
/* Query 8_14 */
SET @Q = 'SELECT Count([Year]) AS N, ' +
     'Sum([Year]) AS Sx, ' +
     'Sum([Year_Value]) AS Sy, ' +
     'Sum([Year]*[Year]) As Sx2, ' +
     'Sum([Year_Value]*[Year_Value]) AS Sy2, ' +
     'Sum([Year]*[Year_Value]) AS Sxy ' +
     'INTO ##TempSs ' +
     'FROM ##TempTotalByYear ' +
     'WHERE [Year] <= ' +
     convert(varchar(20), (@ForecastYr - 1))
EXEC(@Q)

/* Calculate the linear parameters 'a' and 'b' */
/* Query 8_15 */
SET @Q = 'SELECT (Sy*Sx2-Sx*Sxy) / (N*Sx2-Sx*Sx) as a, ' +
     '(N*Sxy-Sx*Sy) / (N*Sx2-Sx*Sx) As b ' +
     'INTO ##Tempab ' +
     'FROM ##TempSs'
EXEC(@Q)

/* Forecast the value for the forecast year */
/* Query 8_16 */
SET @Q = 'SELECT [a] + [b] * ' +
     convert(varchar(20), @ForecastYr) + ' AS y, ' +
     '([a] + [b] * ' +
     convert(varchar(20), @ForecastYr) + ')/' +
     convert(Varchar(20), @SeasonsPerYr) + ' AS Q ' +
     'INTO ##TempForecast ' +
     'FROM ##Tempab'
EXEC(@Q)

/* If the result table exists, remove it */
IF Exists (SELECT id FROM ..sysobjects
     WHERE name = @RsltTable)
Begin
     SET @Q = 'DROP TABLE [' + @RsltTable + ']'
     EXEC(@Q)
End
```

```
/* Calculate new averages for seasonal forecast */
/* Query 8_17 */
SET @Q = 'SELECT ' + convert(varchar(20), @ForecastYr) + ' AS [Year], ' +
    'Season_ID, ' +
    '[Average_Seasonal_Factor] * [Q] AS [Seasonal_Forecast] ' +
    'INTO [' + @RsltTable + '] ' +
    'FROM ##TempAvgSeasonalFactors, ##TempForecast'
EXEC(@Q)


GO
SET QUOTED_IDENTIFIER OFF
GO
SET ANSI_NULLS ON
GO
```

Procedure Calls

Next are examples of the call statements for invoking the Simple_Moving_Average, Weighted_Moving_Average, Single_Exponential_Smoothing, Double_Exponential_Smoothing, and Seasonal_Adjustment procedures.

```
DECLARE @SrcTable varchar(50)
DECLARE @Time_Series varchar(50)
DECLARE @Data_Value varchar(50)
DECLARE @Period_Size int
DECLARE @RsltTable varchar(50)

SET @SrcTable = 'Table 8_1'
SET @Time_Series = 'Month'
SET @Data_Value = 'Actual Gross Sales'
SET @Period_Size = 3
SET @RsltTable = 'Simple Moving Average'

EXEC [CH8]..[Simple_Moving_Average] @SrcTable, @Time_Series,
    @Data_Value, @Period_Size, @RsltTable
```

```
DECLARE @SrcTable varchar(50)
DECLARE @Time_Series varchar(50)
DECLARE @Data_Value varchar(50)
DECLARE @Period_Size int
DECLARE @Wt_Table varchar(50)
DECLARE @Wt_ID varchar(50)
DECLARE @Wt_Value varchar(50)
DECLARE @RsltTable varchar(50)

SET @SrcTable = 'Table 8_1'
SET @Time_Series = 'Month'
SET @Data_Value = 'Actual Gross Sales'
SET @Period_Size = 3
SET @Wt_Table = 'Weights'
SET @Wt_ID = 'Wt_ID'
SET @Wt_Value = 'Wt_Value'
SET @RsltTable = 'Weighted Moving Average'

EXEC [CH8]..[Weighted_Moving_Average] @SrcTable, @Time_Series, @Data_Value,
     @Period_Size, @Wt_Table, @Wt_ID, @Wt_Value, @RsltTable


DECLARE @SrcTable varchar(50)
DECLARE @Time_Series varchar(50)
DECLARE @Data_Value varchar(50)
DECLARE @First_Forecast_Period int
DECLARE @Alpha float
DECLARE @RsltTable varchar(50)

SET @SrcTable = 'Table 8_1'
SET @Time_Series = 'Month'
SET @Data_Value = 'Actual Gross Sales'
SET @First_Forecast_Period = 61
SET @Alpha = 0.1
SET @RsltTable = 'Single Exponential Smoothing'

EXEC [CH8]..[Single_Exponential_Smoothing] @SrcTable, @Time_Series, @Data_Value,
     @First_Forecast_Period, @Alpha, @RsltTable
```

```
DECLARE @SrcTable varchar(50)
DECLARE @Time_Series varchar(50)
DECLARE @Data_Value varchar(50)
DECLARE @First_Forecast_Period int
DECLARE @Alpha float
DECLARE @Beta float
DECLARE @RsltTable varchar(50)

SET @SrcTable = 'Table 8_1'
SET @Time_Series = 'Month'
SET @Data_Value = 'Actual Gross Sales'
SET @First_Forecast_Period = 61
SET @Alpha = 0.3
SET @Beta = 0.2
SET @RsltTable = 'Double Exponential Smoothing'

EXEC [CH8]..[Double_Exponential_Smoothing] @SrcTable, @Time_Series,
    @Data_Value, @First_Forecast_Period, @Alpha, @Beta, @RsltTable


DECLARE @SrcTable varchar(50)
DECLARE @Time_Series varchar(50)
DECLARE @SrcDate varchar(50)
DECLARE @Data_Value varchar(50)
DECLARE @Seasonal_Size int
DECLARE @ForecastYr int
DECLARE @RsltTable varchar(50)

SET @SrcTable = 'Table 8_1'
SET @Time_Series = 'Month'
SET @SrcDate = 'Date'
SET @Data_Value = 'Actual Gross Sales'
SET @Seasonal_Size = 3
SET @ForecastYr = 6
SET @RsltTable = 'Seasonal Forecast'

EXEC [CH8]..[Seasonal_Adjustment] @SrcTable, @Time_Series, @SrcDate,
    @Data_Value, @Seasonal_Size, @ForecastYr, @RsltTable
```

Success at Last

Overview of Relational Database Structure and SQL

THIS APPENDIX CONTAINS a brief overview of relational databases and the Structured Query Language (SQL). It provides the basic knowledge necessary to understand the SQL queries given throughout the chapters of this book. We highly encourage those who desire more information about relational databases and SQL to check the SQL sources listed at the end of this appendix.

Relational Databases

One of the most elegant breakthroughs in database management is the relational approach. The beauty of the relational database lies in its flexibility and simplicity. The relational approach allows us to view data collected into one or more tables, to relate the tables by sharing common column values, and to manage and manipulate the data with ease through the powerful, yet simple, query language known as SQL. In this section we examine the terminology, rules, and operations associated with relational databases.

A *relation* is a *table* that is composed of rows and columns. Table A-1 shows a table of e-commerce customers. The name of the table is Customers, and the table is composed of four columns and five rows. The columns are labeled CID (Customer ID), Name, Email, and Status. The rows of the table represent customers at some *instance* in time. That is, as time goes by the contents of the table change. New customers are added, inactive customers are removed, and customer information is updated. Sometimes we use the term *occurrence* instead of instance to refer to the contents of a table. The format of the table is referred to as the *relational structure*. We often write the relational structure as: Customers(CID, Name, Email, Status).

Table A-1. An Instance of the E-Commerce Table Customers

CID	NAME	EMAIL	STATUS
C010	Ali	Ali@my.com	10
C020	Baker	Baker@their.com	8
C030	Riley	Riley@his.com	10
C040	Tang	Tang@her.com	6
C050	Jones	Jones@our.com	8

Each table in a relational database must obey three important rules. Any table that violates any part of any rule is not a table in a relational database. The three rules are:

1. The entries in a table are single valued. That is, at the intersection of any column with any row there is only one value. Multiple values are not allowed. For example, a column called "Sports" cannot have the values "swimming," "soccer," and "skiing" in the same row-column entry. If you enter multiple values, the entry is treated as a single value. We resolve this problem by having multiple rows—one for each sporting activity.

2. All the values down a column are of the same type or kind. This means that the contents of a column can be those data values that are valid for only the characteristic identified by the column. For example, in Table A-1 the column named "Email" can contain only email addresses and not any other kinds of data. Each column of a table has a unique name that is used to identify the column. A column is referenced by its name, and the order of the columns in the table is not important. However, we humans like to arrange the columns in some meaningful order for our ease of use.

3. There are no duplicate rows. No two rows contain identical values for each and every column in the table. There must always be at least one column value that is different between any two rows in the table. The order of the rows is not important. However, we humans like to order (sort) data in some ascending or descending order for our ease of use.

One common problem that many folks have with relational databases is the terminology. Over many years different groups of people have worked on the development of relational database technology. Thus, many terms are synonymous. Table A-2 shows three different perspectives of relational database terminology. Those who are purists use the terms "relation," "attribute," and "tuple" (rhymes with couple) terminology. For more common everyday users, the terms "table," "field" or "column," and "record" or "row" are used. For example, Microsoft's Access and SQL Server use the terms table, field, and record.

Table A-2. Terminology

PURE RELATIONAL TERMS	PROGRAMMER TERMS	USER TERMS
Relation	File	Table
Attribute	Field	Column
Tuple	Record	Row

A relational database includes a collection of tables and a number of other objects, such as SQL procedures, Views, and so on. For example, suppose an e-commerce database contains the three tables Customers, Items, and Orders. An instance of the Customers table is given in Table A-1. Table A-3 shows an instance of the Items table, and Table A-4 shows an instance of the Orders table. The Items table has four columns. The first column is the item ID (IID), the second column is the name of the item, the third column is the unit price of the item, and the last column is the shipping weight of the item. The Orders table has three columns. The first column is the ID of the customer who placed the order (CID), the second column is the ID of the item ordered (IID), and the last column is the quantity ordered.

Table A-3. An Instance of the Items Table

IID	NAME	UNIT PRICE	WEIGHT
I010	Socks	4.75	0.4
I020	Hat	18.82	0.6
I030	Shirt	23.76	1.2
I040	Pants	29.99	2.3
I050	Shoes	72.39	3.3
I060	Shirt	26.49	1.8

Table A-4. An Instance of the Orders Table

CID	IID	QUANTITY
C010	I010	2
C010	I020	1
C010	I030	3
C010	I040	2
C010	I050	1
C020	I010	1
C020	I030	1
C020	I040	2
C020	I050	1
C030	I010	3
C030	I020	2
C030	I030	1
C040	I010	2
C040	I020	1
C040	I040	2
C040	I050	2
C050	I010	1
C050	I020	2
C050	I050	1

SQL Query Language

For relational databases, the most popular data access and data management language is SQL. SQL was introduced to the world in 1974 as SEQUEL (Structured English Query Language). Today SQL, or a dialogue of SQL, can be found in almost every relational database management system. As such, the term SQL does not necessarily refer to Microsoft's SQL Server relational database management system. However, in today's conversational terminology, SQL is becoming a popular short name for Microsoft's SQL Server system.

Looking at SQL, the language, as a whole, we can make several general observations. First, SQL queries can be viewed as an organized collection of clauses. Each clause begins with a keyword and is followed by a list of operands. The keywords identify query operations, and the operands identify those things to be used in the operation. Second, SQL can be logically divided into two major parts. One part deals with the definition and specification of tables and is referred to as *data definition*. Data definition is concerned with specifying table names, column names, column data types, as well as indexing for faster data access. The second part of SQL deals with the manipulation of data and is commonly called *data manipulation*. Data manipulation is concerned with extracting information from the tables, inserting new data into a table, updating existing data values, and deleting unwanted data. Each of these two major parts can be subdivided, as we see later. We first examine the SQL data manipulation features. Afterwards we take a look at the data definition features.

Data Manipulation

We can view data manipulation as being composed of four groups known as selection, insertion, update, and deletion. We examine each of these groups in the following sections.

Selection Queries

For selection, SQL provides us with seven clauses: SELECT, FROM, WHERE, GROUP BY, HAVING, ORDER BY, and UNION. The SELECT and FROM clauses are required in all selection queries. The other five clauses are optional. Now, let's take a look at some selection query examples.

Select Clause

Suppose we wish to see the names and emails of each customer from the Customers table shown in Table A-1. Our SQL query would look like this:

```
SELECT Name, Email
FROM Customers;
```

Table A-5 shows the result of the query.

Table A-5. Result of the Selection Query

NAME	EMAIL
Ali	Ali@my.com
Baker	Baker@their.com
Riley	Riley@his.com
Tang	Tang@her.com
Jones	Jones@our.com

The result of a selection query always appears as a table and is often referred to as a *recordset*. In the query we asked for Name and Email, and the results show only those two columns in the order they appear in the SELECT clause. The FROM clause specifies the table from which to take the information. In our example the information comes from the Customers table (see Table A-1).

If we wish, we can change the query result column names by using the following SELECT clause:

```
SELECT Name AS [Customer Name], Email AS [Customer Email]
```

The AS allows us to rename columns in the query result. Notice that we placed square brackets ([]) around the new column names. The square brackets are necessary whenever a name (table name or column name) is composed of more than one word. The reason is that SQL would detect the space between the two words and try to treat each word as a separate name.

Now, suppose we wish to see all the information in the Items table shown in Table A-3. The SQL query that follows accomplishes this:

```
SELECT *
FROM Items;
```

The result is the Items table shown in Table A-3. We could have listed each and every column name for the Items table in the SELECT clause; however, the shortcut notation "*" in SQL means all columns.

Suppose we wish to know the customer ID for all those customers who have placed an order (see Table A-4). The following query achieves just that:

```
SELECT DISTINCT CID
FROM Orders;
```

In the SELECT clause we used the DISTINCT keyword to eliminate duplicate CID values from the result table shown in Table A-6.

Table A-6. Results of the SELECT DISTINCT Query

CID
C010
C020
C030
C040
C050

SELECT clauses may contain expressions. For example, suppose we wish to calculate a sales tax of 8% for each item in the Items table (see Table A-3). In SQL we have:

```
SELECT Name, [Unit Price], [Unit Price] * 0.08 AS [Sales Tax]
FROM Items;
```

The query result is shown in Table A-7.

Table A-7. Result Query with an Expression

NAME	UNIT PRICE	SALES TAX
Socks	4.75	0.38
Hat	18.82	1.51
Shirt	23.76	1.90
Pants	29.99	2.40
Shoes	72.39	5.79
Shirt	26.49	2.12

WHERE Clause—Conditional Selections

Sometimes we wish to see only certain rows (or a subset of rows) from a table rather than all the rows. To choose a subset of rows we need to perform what is called a *conditional selection*. The WHERE clause in SQL is used to specify the condition. For example, suppose we wish to see only those items in Table A-3 with a shipping weight greater than or equal to 2. The following SQL query accomplishes this:

```
SELECT Name, [Unit Price]
FROM Items
WHERE Weight >= 2;
```

The query result is shown in Table A-8.

Table A-8. Result of the Selection and WHERE Query

NAME	UNIT PRICE
Pants	29.99
Shoes	72.39

Let's take a closer look at the WHERE clause condition. Notice that the condition is composed of three parts. First is the column name (Weight), next is the logical operator (>=), and finally is the value (2). In SQL, the selection condition always follows this pattern of column name, logical operator, and value. There are six logical operators as shown in Table A-9.

Table A-9. Logical Operators for the WHERE Cause

OPERATOR	MEANING
=	Equal
<> or !=	Not equal
<	Less than
>	Greater than
<=	Less than or equal to
>=	Greater than or equal to

Sometimes we may have more than one selection condition. For example, suppose we wish to know the price of all those shirts in the Items table shown in Table A-3 that are under $25.00. The following SQL query accomplishes this:

```
SELECT IID, Name, [Unit Price]
FROM Items
WHERE Name = "Shirt" AND [Unit Price] < 25.00;
```

The result is shown in Table A-10.

Table A-10. Result of Compound Conditional Selection Query

IID	NAME	UNIT PRICE
I030	Shirt	23.76

To combine the two selection conditions we use the logical AND connector. The AND selects those rows that satisfy both (or all) conditions. Notice the value "Shirt" in the first condition is placed between quotation marks. All string values in SQL must be between quotation marks. We can use single quotation marks rather than double quotation marks. SQL will accept either one as long as they are matching pairs. That is, if you begin with a single (double) quote mark, you must end with a single (double) quote.

Okay, let's consider a more complex selection condition example. Suppose we wish to find all those shirts in the Items table (see Table A-3) that are under $25.00 *or* have a shipping weight over 1.5. The SQL query below does just that:

```
SELECT IID, Name, [Unit Price], Weight
FROM Items
WHERE Name = "Shirt" AND ([Unit Price] < 25.00 OR Weight > 1.5);
```

The query result is shown in Table A-11.

Table A-11. Result of a Complex Conditional Selection Query

IID	NAME	UNIT PRICE	WEIGHT
I030	Shirt	23.76	1.2
I060	Shirt	26.49	1.8

The WHERE clause has three selection conditions. The first condition selects all the Shirts. The next two conditions are enclosed within parentheses and are combined by the logical OR. The logical OR selects those rows that meet either one or both (or any of the) conditions. So, in the example those rows that have a Unit Price under $25.00 are selected along with all those rows that have a shipping weight over 1.5. The Unit Price under $25.00 selects the first row of the query result, and the second row is selected by the "Weight over 1.5" selection condition. Whenever a WHERE clause contains a mixture of AND and OR connectors, SQL performs all the AND operators first and then performs the OR operators. To have SQL evaluate the OR before the AND, we must enclose those conditions connected by the OR in parentheses.

SQL offers an easier way of combining some compound conditions by using the BETWEEN and IN selection conditions. For example, suppose we wish to see the names of all customers in the Customers table (see Table A-1) who have a status between 8 and 10 inclusively. The following SQL query accomplishes this:

```
SELECT Name
From Customers
WHERE Status BETWEEN 8 AND 10;
```

Now, suppose we wish to see the Unit Price of all shirts, hats, and shoes from the Items table (see Table A-3). The query below does the selection.

```
SELECT [Unit Price]
FROM Items
WHERE Name IN ("Shirt", "Hat", "Shoes");
```

Looking at the Items table (see Table A-3), suppose we wanted to see all those items whose name begins with an "S." We can use the SQL LIKE operator as shown below.

```
SELECT Name
FROM Items
WHERE Name LIKE "S*";"
```

The quoted string "S*" after the LIKE operator is called the pattern. The asterisk (*) is a wildcard character denoting any number of characters. If we want the names of all those items whose name ends with an "s," we code the pattern as "*s." Similarly, we code the pattern as "*s*" to find all the item names that have an "s" somewhere in their names.

> **NOTE** *Different implementations of SQL may use different wildcard characters.*

Aggregate Functions

SQL provides five aggregate functions: COUNT, MIN, MAX, SUM, and AVG. They are called aggregate functions because they operate over all the rows in the query results (i.e., rows selected by the WHERE clause) and return a single value. The operations performed by these five functions are shown in Table A-12.

Table A-12. Aggregate Functions

FUNCTION	OPERATION PERFORMED
COUNT(*)	Returns the number of rows selected
MIN()	Returns the smallest value in a numeric column
MAX()	Returns the largest value in a numeric column
SUM()	Returns the total of all the values in a numeric column
AVG()	Returns the average of all the values in a numeric column

Consider the following query that gives the number of orders placed by customer C020 from the Orders table in Table A-4 and the total number of items ordered:

```
SELECT COUNT(*), SUM(Quantity)
FROM Orders
WHERE CID = "C020";
```

The result is shown in Table A-13.

Table A-13. Result of Summation Query

COUNT(*)	SUM(QUANTITY)
4	5

Microsoft Access 2000 and Microsoft SQL Server offer some additional functions. Those used throughout this book are listed in Table A-14. To get more detailed information about these functions, check out Help under Access and SQL Server. Also check Help under Visual Basic.

Table A-14. Additional Functions

FUNCTION	OPERATION PERFORMED
Abs()	Returns the positive value of a number
Ceiling()	Returns the smallest integer greater than or equal to a number
Floor()	Returns the largest integer less than or equal to a given number
Format()	Formats the values in a column as specified
Int()	Truncates the values in a column to whole numbers (integers)
Mid()	Extracts a substring from a text
Round()	Rounds the values to specified number of decimal places

Table A-14. Additional Functions (Continued)

FUNCTION	OPERATION PERFORMED
StDev	Returns the standard deviation of a column
Sqr()	Takes the square root of the values in a column
Var()	Returns the variance of a column

GROUP BY and HAVING

Suppose we wish to know how many items each customer has ordered by looking in the Orders table shown in Table A-4. The following query fulfills our wish:

```
SELECT CID, SUM(Quantity) AS [Total Orders]
FROM Orders
GROUP BY CID;
```

The result is shown in Table A-15.

Table A-15. Result of GROUP BY Query

CID	TOTAL ORDERS
C010	9
C020	5
C030	6
C040	7
C050	4

The GROUP BY clause forms groups of rows such that in each group the value of the CID (column specified in the GROUP BY clause) is the same value. GROUP BY associates an aggregate function that is applied to each group of rows. Notice that the CID is used in both the SELECT clause and the GROUP BY clause. We do this so that the result will show the identity of each group.

Now suppose we wish to see only those customers in the Orders table (see Table A-4) who have ordered more than 5 items:

```
SELECT CID, SUM(Quantity) AS [Total Orders]
FROM Orders
GROUP BY CID
HAVING SUM(Quantity) > 5;
```

The result is shown in Table A-16.

Table A-16. Result of GROUP BY *with* Having *Query*

CID	TOTAL ORDERS
C010	9
C030	6
C040	7

The HAVING clause is used to select those groups that meet a specified condition. Remember, the WHERE clause selects rows from the table, and the HAVING clause selects groups.

ORDER BY

Often we wish to order (sort or arrange) the rows of a query result table. The ORDER BY clause does that for us. Suppose we wish to see a listing of the items in the Items table (see Table A-3) in alphabetical order by item name. The following query accomplishes this:

```
SELECT Name, [Unit Price], IID
FROM Items
ORDER BY Name;
```

The result is shown in Table A-17.

Table A-17. Result of the ORDER BY *Query*

NAME	UNIT PRICE	IID
Hat	18.82	I020
Pants	29.99	I040
Shirt	23.76	I030
Shirt	26.49	I060
Shoes	72.39	I050
Socks	4.75	I010

The column name(s) used in the ORDER BY clause must also appear in the SELECT clause.

We can sort the query result on more than one column. For example, suppose we wish to sort the rows of the result table by Name and within each Name we wish to sort by the Unit Price in descending order. We can use a query like the one given below:

```
SELECT Name, [Unit Price], IID
FROM Items
ORDER BY Name, [Unit Price] DESC;
```

The result table is the same as the previous result table (see Table A-17) except that the shirt for $26.49 is listed before the shirt for $23.76.

Notice the DESC for descending after the Unit Price on the ORDER BY clause. We also can specify ASC for ascending, although ascending is the default order.

Since the ORDER BY clause only identifies those columns in the SELECT clause for sorting, SQL allows us to refer to the columns by position. Looking at the SELECT clause, Name is column position 1, Unit Price is column position 2, and IID is column position 3. We can use these position numbers in the ORDER BY clause as follows:

```
ORDER BY 1, 2 DESC
```

This ORDER BY clause is equivalent to our original ORDER BY clause. Using the positional number can be quite useful when the SELECT clause contains functions and/or expressions.

WHERE Clause – Join Conditions

One of the most powerful features of relational databases and SQL is the ability to join (combine) data in one table with data from another table by matching data values between the tables. For example, suppose we wish to list customer names with their orders rather than the CID. Looking at the Customers table (Table A-1) and the Orders table (Table A-4), we see there is a common column, CID. Thus, if we match up the CID in the Orders table with the CID in the Customers table, we can find the customers' names. We can do this with the following query.

```
SELECT Name, IID, Quantity
FROM Orders, Customers
WHERE Orders.CID = Customers.CID;
```

The result is shown in Table A-18.

Table A-18. Result of Joining Customers Table with Orders Table by CID

NAME	IID	QUANTITY
Ali	I010	2
Ali	I020	1
Ali	I030	3
Ali	I040	2
Ali	I050	1
Baker	I010	1
Baker	I030	1
Baker	I040	2
Baker	I050	1
Riley	I010	3
Riley	I020	2
Riley	I030	1
Tang	I010	2
Tang	I020	1
Tang	I040	2
Tang	I050	2
Jones	I010	1
Jones	I020	2
Jones	I050	1

Looking at the FROM clause, we see both tables are listed and that the table names are separated by a comma. The FROM clause always lists the names of all the tables that are participating in the query. The WHERE clause contains the join condition that specifies which columns are to be matched between the two tables. Since CID is used (i.e., spelled the same) in both tables we must distinguish the CIDs. In SQL we qualify the column name by prefixing the column name with the table name. Thus, the CID in the Orders table is uniquely identified as Orders.CID, and the CID in the Customers table is uniquely identified as Customers.CID. The period is used to connect, and yet separate, the table name with the column name. In the join condition we are matching Orders.CID to Customers.CID by specifying that the Orders.CID equals the Customers.CID. This join condition is called an "equal-join."

If we can join Customers with Orders, can we also join Items with Orders? Yes, we can. Let's see how we do it.

```
SELECT Customers.Name, Items.Name, Quantity
FROM Orders, Customers, Items
WHERE Orders.CID = Customers.CID
AND Orders.IID = Items.IID;
```

The result is shown in Table A-19.

Table A-19. Result of Joining Three Tables

CUSTOMERS.NAME	ITEMS.NAME	QUANTITY
Ali	Socks	2
Ali	Hat	1
Ali	Shirt	3
Ali	Pants	2
Ali	Shoes	1
Baker	Socks	1
Baker	Shirt	1
Baker	Pants	2
Baker	Shoes	1
Riley	Socks	3
Riley	Hat	2
Riley	Shirt	1
Tang	Socks	2
Tang	Hat	1
Tang	Pants	2
Tang	Shoes	2
Jones	Socks	1
Jones	Hat	2
Jones	Shoes	1

In the SELECT clause we had to qualify the Name column because it appears in both the Customers table and the Items table. Notice how the qualification carries over to the result table column headings. The WHERE clause has two join conditions that are connected with the logical AND operator. Join conditions are always connected by the logical AND. It does not make good sense to use a logical OR.

It is important to notice that the columns participating in a join condition must be of the same kind or type of data. In the examples, we joined CIDs to CIDs and IIDs to IIDs. We do not join things like Email addresses in the Customers table to Quantity in the Orders table because Email and Quantity are two different kinds of data.

Union Queries

There are occasions when we wish to combine the results of multiple queries into a single result table. The UNION operator allows us to do just that. Consider the following query:

```
SELECT CID, Status AS Rank, "Customer with high status" AS [Label]
FROM Customers
WHERE Status > 8
UNION ALL
SELECT CID, SUM(Quantity) AS Rank, "High order customer" AS [Label]
FROM Orders
GROUP BY CID
HAVING SUM(Quantity) > 6;
```

The result of the query following the UNION ALL is appended to the result of the query preceding the UNION ALL. The final query result appears in Table A-20.

Table A-20. Result of UNION *Query*

CID	RANK	LABEL
C010	10	Customer with high status
C030	10	Customer with high status
C010	9	High order customer
C040	7	High order customer

For each SELECT clause participating in the union, its columns must match in number and data type. Comparing the two SELECT clauses in the example we see there are three columns in the first SELECT clause and three columns in the second SELECT clause. The first column is CID in both SELECT clauses, and we know they are of the same data type.

The second column in the first SELECT clause is of type numeric and likewise the SUM function is of type numeric. The last columns match because they are both quoted text.

Insertion Queries

For insertion, SQL provides us with two ways: single row insertion and multiple row insertion. Single row insertion uses two clauses, the INSERT INTO and VALUES. Multiple row insertion uses the INSERT INTO clause along with a selection query. Let's take a look at some examples.

Single Row Insertion

Suppose we wish to add a new customer to the Customers table (Table A-1). To accomplish the insertion, we do the following:

```
INSERT INTO Customers
VALUES ("C060", "Sanchez", "sanchez@yours.com", 6);
```

Notice that the order of the values in the VALUES clause is the same as the columns in the Customers table.

Suppose we do not know Sanchez's email address. We can still insert the information that we do know as follows:

```
INSERT INTO Customers(Name, Status, CID)
VALUES ("Sanchez", 6, "C060");
```

In this insertion query we listed the column headings and we gave the values for each column as it appears in the list. We omitted Email since we have no value (or text) for it. Whenever we list the column headings (or names), the order in which they are listed does not necessarily have to match the order of the column headings in the Customers table.

Multiple Row Insertion

SQL allows us to extract data from existing tables and place the result in another table. Suppose for the moment that we have another table called Hot_Customers that has the same relational structure as the Customers table. We wish to place a copy of those customers with a Status greater than 8 in the Customers table in the Hot_Customers table. We can do this insertion as follows:

```
INSERT INTO Hot_Customers
(SELECT *
FROM Customers
WHERE Status > 8);
```

The selection query is enclosed in parentheses and retrieves the customer records for Ali and Riley from the Customers table shown in Table A-1. The records are appended to the Hot_Customers table.

Update Queries

We can change any value in a table by doing an update query. Suppose we wish to change Tang's status to 7 in the Customers table (see Table A-1). We can accomplish this with the following update query:

```
UPDATE Customers
SET Status = 7
WHERE Name = "Tang";
```

The query's action changes the Status to 7 for all the rows in the Customers table where Name has the value "Tang." Oh yes, if there is more than one Tang, all the rows are updated. As a precautionary measure, we often first do a selection query using the WHERE clause of the update query to see if the desired rows are selected by the condition specified in the WHERE clause.

Now suppose we need to change the Unit Price of every item in the Items table (Table A-3) because there is a 2% increase in price due to inflation. We can do this with the following update query:

```
UPDATE Items
SET [Unit Price] = [Unit Price] * 1.02;
```

In this situation we are setting the Unit Price to its existing value times 1.02. For example, if the Unit Price is $4.75, the new Unit Price is $4.75 * 1.02, or $4.85.

It is possible to update more than one field in a single update query. Suppose we have a different type of hat with a new Unit Price of $17.76 and a new weight of 0.7. We can do the update as follows:

```
UPDATE Items
SET [Unit Price] = 17.76,
Weight = 0.7
WHERE Name = "Hat";
```

With this update both column values are changed. Notice we only used SET once and that a comma is used to separate the two columns being updated.

Deletion Queries

At some point in time we need to remove unwanted rows from our tables. For example, suppose we need to remove Riley from the Customers table (Table A-1). We can accomplish this as follows:

```
DELETE *
FROM Customers
WHERE Name = "Riley";
```

This removes all rows in the Customers table where the Name is Riley. If there is more than one Riley in the Customers table, all Riley rows are removed. When doing deletions, it is best to use a column (or set of columns) that can uniquely identify the desired row(s) to be deleted. As a precautionary measure, we often first do a selection query with the deletion WHERE clause condition to see exactly which rows are targeted for removal.

Suppose we forgot to include the WHERE clause in our deletion query. Thus, the deletion query becomes:

```
DELETE *
FROM Customers;
```

In this case all of the rows in the Customers table are deleted. So, please be careful when doing deletions.

Data Definition

We can view data definition as being composed of four groups known as table creation, table removal, index creation, and index removal. We examine each of these groups in the following sections.

Table Creation

To create a table in SQL we use the CREATE TABLE command. Let's see how we can create a Customers table.

```
CREATE TABLE Customers
(CID Char(5) NOT NULL,
NAME Char(15) NOT NULL,
Email Char(30),
Status Integer);
```

After the CREATE TABLE command we give the name of the new table to be created. Following the table name in parentheses are the column names along with their data types. Some of the common data types are given in Table A-21. The NOT NULL means that the column must have a value in each and every row. In the Customers table only the CID and the Name are required to have values. Email and Status may be left unassigned, empty, or null in any given row.

Table A-21. Common Data Types

DATA TYPE	MEANING
Char(n)	A column of text where *n* is the maximum number of characters allowed.
Integer	A column of whole numbers, both positive and negative.
Single	A column of real numbers, both positive and negative. Each real number contains 7 significant digits.
Double	A column of real numbers, both positive and negative. Each real number contains 15 significant digits.
Date	A column of dates.
Currency	A column of monetary values.
Decimal(n, m)	A column of decimal numbers with a maximum of n digits with m digits after the decimal point.

Table Removal

When a table is no longer needed it can be removed or dropped from the database. SQL uses the DROP TABLE command. Suppose we wish to remove the Items table. The following single command does the job:

```
DROP TABLE Items;
```

Index Creation

To improve performance SQL allows us to create indexes on one or more columns. Suppose we wish to create an index on CID in the Customers table (Table A-1).

```
CREATE INDEX CID_Index
ON Customers (CID);
```

The name of the index is CID_Index. The Index is on the CID column in the Customers table.

To create an index on CID and IID in the Orders table (Table A-4) we do:

```
CREATE INDEX Orders_IDs
On Orders (CID, IID);
```

The index is created on pairs of values coming from CID and IID in each row of the Orders table.

Index Removal

When an index is no longer needed, it can be removed or dropped from the database. For example, to remove the index on the Orders table we do:

```
DROP INDEX Orders_IDs;
```

Other SQL Sources

Below is a list of Web sites that have information about SQL or books about SQL.
```
http://www.willcam.com/sql/default.htm
http://w3.one.net/~jhoffman/sqltut.htm
http://www.sjsu.edu/faculty/ddimov/fall2000/pt4sql/tsld001.htm
http://www.mkp.com/books_catalog/catalog.asp?ISBN=1-55860-245-3
http://www.geocities.com/SiliconValley/Way/6958/sql.htm
http://www.frick-cpa.com/ss7/Theory_SQL_Overview.asp
http://www.lnl.net/books/sql/
http://step.sdsc.edu/s96/sql/tutorial.html
```

Statistical Tables

THE TABLES IN this appendix are:

Table B-1. Area Under Standard Normal Distribution

Table B-2. Chi-Square Distribution (χ^2) Values

Table B-3. F Distribution at 1% Significance Level

Table B-4. F Distribution at 5% Significance Level

Table B-5. Student's t Distribution

Table B-6. Values of the Absolute Value of the Correlation Coefficient $|r|$

Table B-7. Factors for Determining Control Limits for $\bar{X}$ and R charts

Table B-1 gives the area under the standard normal distribution that has a mean of zero and a standard deviation of one. To use the table, we calculate what is known as the "normal variate" (z in the table). For example, suppose we have normally distributed data with mean $\bar{x} = 1.80$ and standard deviation $s = 0.44$. If we are interested in the chance of an observed value in the data falling below $x = 1.20$ (i.e., the area under the normal curve below 1.20), we first standardize the value by calculating

$$z = \frac{x - \bar{x}}{s} = \frac{1.20 - 1.80}{0.44} = -1.36$$

We then find -1.3 in the first column of the table and proceed across the column under 0.06 (the second decimal place of z). We read the area we need as 0.0869 (or 8.69%). We would expect 8.69% of the data observed to fall below 1.20.

Table B-2 gives the upper tail chi-square (χ^2) value corresponding to a given number of degrees of freedom (v) and significance level (α). For example, suppose we wish to find the χ^2 value for 9 degrees of freedom and 0.05 significance level (i.e., 5% in the upper tail). First we go down the column under v until we find 9. Then we proceed across until we are under the $\alpha = 0.05$ column. The result is $\chi^2 = 16.91896$.

Table B-3 and Table B-4 give the upper tail value of a random variable F that has an area in the upper tail equal to (or a significance level of) α for various combinations of the degrees of freedom v_1 and v_2. In Chapter 4 we used the F distribution to examine the ratio of two variances, with the larger variance in the numerator. The value of v_1 is the degrees of freedom associated with the numerator, and the value of v_2 is the degrees of freedom associated with the denominator. Table B-3 gives the values for $\alpha = 0.01$ (or 1%), and Table B-4 gives the values for $\alpha = 0.05$ (or 5%). For example, if $v_1 = 4$, $v_2 = 15$, and $\alpha = 0.05$ we use Table B-4 and obtain the value 3.056.

Table B-5 gives the total area in the upper and lower tails of the Student's t distribution for various values of the significance level α and degrees of freedom v. The table is set up for a two-tailed test, when the alternative hypothesis states that the two parameters being compared are not equal (see Chapter 4). For example, the value of t for 12 degrees of freedom with a significance level of $\alpha = 0.05$ (5%) is $t = 1.7823$. This means that 2.5% of the area under the t distribution curve is to the right of 1.7823, and 2.5% is to the left of -1.7823. If the table is used for a one-sided test, the significance level at the top of the table must be doubled. For example, if we are interested in whether a sample mean is greater than a population mean (rather than simply "not equal"), we look up the table t value for 2α.

Table B-6 gives the values of the correlation coefficient r for v degrees of freedom with a 5% and a 1% level of significance. The correlation coefficient indicates how well two or more variables relate to each other. Table B-6 gives the value of r for relating 2, 3, 4, or 5 variables. For example, the correlation coefficient of two variables at the 5% significance level with $v = 4$ degrees of freedom is $r = 0.811$.

Table B-7 gives the factors for determining control limits for $\bar{X}$ and R charts. The use of this table is demonstrated in Chapter 6.

For additional information, several sources are listed in the Bibliography. It might also be noted that the tables in this appendix were generated using Microsoft Excel.

Table B-1. Area Under Standard Normal Distribution to the Left of z

z	0.00	0.01	0.02	0.03	0.04	0.05	0.06	0.07	0.08	0.09
-3.40	0.0003	0.0003	0.0003	0.0003	0.0003	0.0003	0.0003	0.0003	0.0003	0.0002
-3.30	0.0005	0.0005	0.0005	0.0004	0.0004	0.0004	0.0004	0.0004	0.0004	0.0003
-3.20	0.0007	0.0007	0.0006	0.0006	0.0006	0.0006	0.0006	0.0005	0.0005	0.0005
-3.10	0.0010	0.0009	0.0009	0.0009	0.0008	0.0008	0.0008	0.0008	0.0007	0.0007
-3.00	0.0013	0.0013	0.0013	0.0012	0.0012	0.0011	0.0011	0.0011	0.0010	0.0010
-2.90	0.0019	0.0018	0.0018	0.0017	0.0016	0.0016	0.0015	0.0015	0.0014	0.0014
-2.80	0.0026	0.0025	0.0024	0.0023	0.0023	0.0022	0.0021	0.0021	0.0020	0.0019
-2.70	0.0035	0.0034	0.0033	0.0032	0.0031	0.0030	0.0029	0.0028	0.0027	0.0026
-2.60	0.0047	0.0045	0.0044	0.0043	0.0041	0.0040	0.0039	0.0038	0.0037	0.0036
-2.50	0.0062	0.0060	0.0059	0.0057	0.0055	0.0054	0.0052	0.0051	0.0049	0.0048
-2.40	0.0082	0.0080	0.0078	0.0075	0.0073	0.0071	0.0069	0.0068	0.0066	0.0064
-2.30	0.0107	0.0104	0.0102	0.0099	0.0096	0.0094	0.0091	0.0089	0.0087	0.0084
-2.20	0.0139	0.0136	0.0132	0.0129	0.0125	0.0122	0.0119	0.0116	0.0113	0.0110
-2.10	0.0179	0.0174	0.0170	0.0166	0.0162	0.0158	0.0154	0.0150	0.0146	0.0143
-2.00	0.0228	0.0222	0.0217	0.0212	0.0207	0.0202	0.0197	0.0192	0.0188	0.0183
-1.90	0.0287	0.0281	0.0274	0.0268	0.0262	0.0256	0.0250	0.0244	0.0239	0.0233
-1.80	0.0359	0.0351	0.0344	0.0336	0.0329	0.0322	0.0314	0.0307	0.0301	0.0294
-1.70	0.0446	0.0436	0.0427	0.0418	0.0409	0.0401	0.0392	0.0384	0.0375	0.0367
-1.60	0.0548	0.0537	0.0526	0.0516	0.0505	0.0495	0.0485	0.0475	0.0465	0.0455
-1.50	0.0668	0.0655	0.0643	0.0630	0.0618	0.0606	0.0594	0.0582	0.0571	0.0559
-1.40	0.0808	0.0793	0.0778	0.0764	0.0749	0.0735	0.0721	0.0708	0.0694	0.0681
-1.30	0.0968	0.0951	0.0934	0.0918	0.0901	0.0885	0.0869	0.0853	0.0838	0.0823
-1.20	0.1151	0.1131	0.1112	0.1093	0.1075	0.1056	0.1038	0.1020	0.1003	0.0985
-1.10	0.1357	0.1335	0.1314	0.1292	0.1271	0.1251	0.1230	0.1210	0.1190	0.1170
-1.00	0.1587	0.1562	0.1539	0.1515	0.1492	0.1469	0.1446	0.1423	0.1401	0.1379
-0.90	0.1841	0.1814	0.1788	0.1762	0.1736	0.1711	0.1685	0.1660	0.1635	0.1611
-0.80	0.2119	0.2090	0.2061	0.2033	0.2005	0.1977	0.1949	0.1922	0.1894	0.1867
-0.70	0.2420	0.2389	0.2358	0.2327	0.2296	0.2266	0.2236	0.2206	0.2177	0.2148
-0.60	0.2743	0.2709	0.2676	0.2643	0.2611	0.2578	0.2546	0.2514	0.2483	0.2451
-0.50	0.3085	0.3050	0.3015	0.2981	0.2946	0.2912	0.2877	0.2843	0.2810	0.2776
-0.40	0.3446	0.3409	0.3372	0.3336	0.3300	0.3264	0.3228	0.3192	0.3156	0.3121
-0.30	0.3821	0.3783	0.3745	0.3707	0.3669	0.3632	0.3594	0.3557	0.3520	0.3483
-0.20	0.4207	0.4168	0.4129	0.4090	0.4052	0.4013	0.3974	0.3936	0.3897	0.3859
-0.10	0.4602	0.4562	0.4522	0.4483	0.4443	0.4404	0.4364	0.4325	0.4286	0.4247
-0.00	0.5000	0.4960	0.4920	0.4880	0.4840	0.4801	0.4761	0.4721	0.4681	0.4641
0.00	0.5000	0.5040	0.5080	0.5120	0.5160	0.5199	0.5239	0.5279	0.5319	0.5359
0.10	0.5398	0.5438	0.5478	0.5517	0.5557	0.5596	0.5636	0.5675	0.5714	0.5753
0.20	0.5793	0.5832	0.5871	0.5910	0.5948	0.5987	0.6026	0.6064	0.6103	0.6141
0.30	0.6179	0.6217	0.6255	0.6293	0.6331	0.6368	0.6406	0.6443	0.6480	0.6517
0.40	0.6554	0.6591	0.6628	0.6664	0.6700	0.6736	0.6772	0.6808	0.6844	0.6879
0.50	0.6915	0.6950	0.6985	0.7019	0.7054	0.7088	0.7123	0.7157	0.7190	0.7224
0.60	0.7257	0.7291	0.7324	0.7357	0.7389	0.7422	0.7454	0.7486	0.7517	0.7549
0.70	0.7580	0.7611	0.7642	0.7673	0.7704	0.7734	0.7764	0.7794	0.7823	0.7852
0.80	0.7881	0.7910	0.7939	0.7967	0.7995	0.8023	0.8051	0.8078	0.8106	0.8133
0.90	0.8159	0.8186	0.8212	0.8238	0.8264	0.8289	0.8315	0.8340	0.8365	0.8389

Table B-1. Area Under Standard Normal Distribution to the Left of z (Continued)

z	0.00	0.01	0.02	0.03	0.04	0.05	0.06	0.07	0.08	0.09
1.00	0.8413	0.8438	0.8461	0.8485	0.8508	0.8531	0.8554	0.8577	0.8599	0.8621
1.10	0.8643	0.8665	0.8686	0.8708	0.8729	0.8749	0.8770	0.8790	0.8810	0.8830
1.20	0.8849	0.8869	0.8888	0.8907	0.8925	0.8944	0.8962	0.8980	0.8997	0.9015
1.30	0.9032	0.9049	0.9066	0.9082	0.9099	0.9115	0.9131	0.9147	0.9162	0.9177
1.40	0.9192	0.9207	0.9222	0.9236	0.9251	0.9265	0.9279	0.9292	0.9306	0.9319
1.50	0.9332	0.9345	0.9357	0.9370	0.9382	0.9394	0.9406	0.9418	0.9429	0.9441
1.60	0.9452	0.9463	0.9474	0.9484	0.9495	0.9505	0.9515	0.9525	0.9535	0.9545
1.70	0.9554	0.9564	0.9573	0.9582	0.9591	0.9599	0.9608	0.9616	0.9625	0.9633
1.80	0.9641	0.9649	0.9656	0.9664	0.9671	0.9678	0.9686	0.9693	0.9699	0.9706
1.90	0.9713	0.9719	0.9726	0.9732	0.9738	0.9744	0.9750	0.9756	0.9761	0.9767
2.00	0.9772	0.9778	0.9783	0.9788	0.9793	0.9798	0.9803	0.9808	0.9812	0.9817
2.10	0.9821	0.9826	0.9830	0.9834	0.9838	0.9842	0.9846	0.9850	0.9854	0.9857
2.20	0.9861	0.9864	0.9868	0.9871	0.9875	0.9878	0.9881	0.9884	0.9887	0.9890
2.30	0.9893	0.9896	0.9898	0.9901	0.9904	0.9906	0.9909	0.9911	0.9913	0.9916
2.40	0.9918	0.9920	0.9922	0.9925	0.9927	0.9929	0.9931	0.9932	0.9934	0.9936
2.50	0.9938	0.9940	0.9941	0.9943	0.9945	0.9946	0.9948	0.9949	0.9951	0.9952
2.60	0.9953	0.9955	0.9956	0.9957	0.9959	0.9960	0.9961	0.9962	0.9963	0.9964
2.70	0.9965	0.9966	0.9967	0.9968	0.9969	0.9970	0.9971	0.9972	0.9973	0.9974
2.80	0.9974	0.9975	0.9976	0.9977	0.9977	0.9978	0.9979	0.9979	0.9980	0.9981
2.90	0.9981	0.9982	0.9982	0.9983	0.9984	0.9984	0.9985	0.9985	0.9986	0.9986
3.00	0.9987	0.9987	0.9987	0.9988	0.9988	0.9989	0.9989	0.9989	0.9990	0.9990
3.10	0.9990	0.9991	0.9991	0.9991	0.9992	0.9992	0.9992	0.9992	0.9993	0.9993
3.20	0.9993	0.9993	0.9994	0.9994	0.9994	0.9994	0.9994	0.9995	0.9995	0.9995
3.30	0.9995	0.9995	0.9995	0.9996	0.9996	0.9996	0.9996	0.9996	0.9996	0.9997
3.40	0.9997	0.9997	0.9997	0.9997	0.9997	0.9997	0.9997	0.9997	0.9997	0.9998

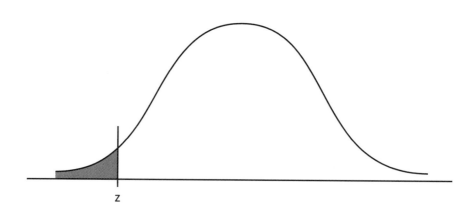

z

Table B-2. Chi-Square Distribution (χ^2) Values Corresponding to the Various Combinations of Degrees of Freedom (v) and Significance Levels (α)

v	α							
	0.995	**0.99**	**0.975**	**0.95**	**0.05**	**0.025**	**0.01**	**0.005**
1	0.0000393	0.000157	0.000982	0.00393	3.84146	5.02390	6.63489	7.87940
2	0.01002	0.02010	0.05064	0.10259	5.99148	7.37778	9.21035	10.59653
3	0.07172	0.11483	0.21579	0.35185	7.81472	9.34840	11.34488	12.83807
4	0.20698	0.29711	0.48442	0.71072	9.48773	11.14326	13.27670	14.86017
5	0.41175	0.55430	0.83121	1.14548	11.07048	12.83249	15.08632	16.74965
6	0.67573	0.87208	1.23734	1.63538	12.59158	14.44935	16.81187	18.54751
7	0.98925	1.23903	1.68986	2.16735	14.06713	16.01277	18.47532	20.27774
8	1.34440	1.64651	2.17972	2.73263	15.50731	17.53454	20.09016	21.95486
9	1.73491	2.08789	2.70039	3.32512	16.91896	19.02278	21.66605	23.58927
10	2.15585	2.55820	3.24696	3.94030	18.30703	20.48320	23.20929	25.18805
11	2.60320	3.05350	3.81574	4.57481	19.67515	21.92002	24.72502	26.75686
12	3.07379	3.57055	4.40378	5.22603	21.02606	23.33666	26.21696	28.29966
13	3.56504	4.10690	5.00874	5.89186	22.36203	24.73558	27.68818	29.81932
14	4.07466	4.66042	5.62872	6.57063	23.68478	26.11893	29.14116	31.31943
15	4.60087	5.22936	6.26212	7.26093	24.99580	27.48836	30.57795	32.80149
16	5.14216	5.81220	6.90766	7.96164	26.29622	28.84532	31.99986	34.26705
17	5.69727	6.40774	7.56418	8.67175	27.58710	30.19098	33.40872	35.71838
18	6.26477	7.01490	8.23074	9.39045	28.86932	31.52641	34.80524	37.15639
19	6.84392	7.63270	8.90651	10.11701	30.14351	32.85234	36.19077	38.58212
20	7.43381	8.26037	9.59077	10.85080	31.41042	34.16958	37.56627	39.99686
21	8.03360	8.89717	10.28291	11.59132	32.67056	35.47886	38.93223	41.40094
22	8.64268	9.54249	10.98233	12.33801	33.92446	36.78068	40.28945	42.79566
23	9.26038	10.19569	11.68853	13.09051	35.17246	38.07561	41.63833	44.18139
24	9.88620	10.85635	12.40115	13.84842	36.41503	39.36406	42.97978	45.55836
25	10.51965	11.52395	13.11971	14.61140	37.65249	40.64650	44.31401	46.92797
26	11.16022	12.19818	13.84388	15.37916	38.88513	41.92314	45.64164	48.28978
27	11.80765	12.87847	14.57337	16.15139	40.11327	43.19452	46.96284	49.64504
28	12.46128	13.56467	15.30785	16.92788	41.33715	44.46079	48.27817	50.99356
29	13.12107	14.25641	16.04705	17.70838	42.55695	45.72228	49.58783	52.33550
30	13.78668	14.95346	16.79076	18.49267	43.77295	46.97922	50.89218	53.67187

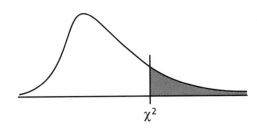

χ^2

Table B-3. The F Distribution at 1% Significance Level

v_2	v_1								
	1	2	3	4	5	6	7	8	9
1	4052.2	4999.3	5403.5	5624.3	5764.0	5859.0	5928.3	5981.0	6022.4
2	98.502	99.000	99.164	99.251	99.302	99.331	99.357	99.375	99.390
3	34.116	30.816	29.457	28.710	28.237	27.911	27.671	27.489	27.345
4	21.198	18.000	16.694	15.977	15.522	15.207	14.976	14.799	14.659
5	16.258	13.274	12.060	11.392	10.967	10.672	10.456	10.289	10.158
6	13.745	10.925	9.780	9.148	8.746	8.466	8.260	8.102	7.976
7	12.246	9.547	8.451	7.847	7.460	7.191	6.993	6.840	6.719
8	11.259	8.649	7.591	7.006	6.632	6.371	6.178	6.029	5.911
9	10.562	8.022	6.992	6.422	6.057	5.802	5.613	5.467	5.351
10	10.044	7.559	6.552	5.994	5.636	5.386	5.200	5.057	4.942
11	9.646	7.206	6.217	5.668	5.316	5.069	4.886	4.744	4.632
12	9.330	6.927	5.953	5.412	5.064	4.821	4.640	4.499	4.388
13	9.074	6.701	5.739	5.205	4.862	4.620	4.441	4.302	4.191
14	8.862	6.515	5.564	5.035	4.695	4.456	4.278	4.140	4.030
15	8.683	6.359	5.417	4.893	4.556	4.318	4.142	4.004	3.895
16	8.531	6.226	5.292	4.773	4.437	4.202	4.026	3.890	3.780
17	8.400	6.112	5.185	4.669	4.336	4.101	3.927	3.791	3.682
18	8.285	6.013	5.092	4.579	4.248	4.015	3.841	3.705	3.597
19	8.185	5.926	5.010	4.500	4.171	3.939	3.765	3.631	3.523
20	8.096	5.849	4.938	4.431	4.103	3.871	3.699	3.564	3.457
21	8.017	5.780	4.874	4.369	4.042	3.812	3.640	3.506	3.398
22	7.945	5.719	4.817	4.313	3.988	3.758	3.587	3.453	3.346
23	7.881	5.664	4.765	4.264	3.939	3.710	3.539	3.406	3.299
24	7.823	5.614	4.718	4.218	3.895	3.667	3.496	3.363	3.256
25	7.770	5.568	4.675	4.177	3.855	3.627	3.457	3.324	3.217
26	7.721	5.526	4.637	4.140	3.818	3.591	3.421	3.288	3.182
27	7.677	5.488	4.601	4.106	3.785	3.558	3.388	3.256	3.149
28	7.636	5.453	4.568	4.074	3.754	3.528	3.358	3.226	3.120
29	7.598	5.420	4.538	4.045	3.725	3.499	3.330	3.198	3.092
30	7.562	5.390	4.510	4.018	3.699	3.473	3.305	3.173	3.067
40	7.314	5.178	4.313	3.828	3.514	3.291	3.124	2.993	2.888
60	7.077	4.977	4.126	3.649	3.339	3.119	2.953	2.823	2.718
120	6.851	4.787	3.949	3.480	3.174	2.956	2.792	2.663	2.559
∞	6.635	4.605	3.782	3.319	3.017	2.802	2.639	2.511	2.407

Table B-3. The F Distribution at 1% Significance Level (Continued)

V_2	V_1									
	10	12	15	20	24	30	40	60	120	∞
1	6055.9	6106.7	6157.0	6208.7	6234.3	6260.4	6286.4	6313.0	6339.5	6365.6
2	99.397	99.419	99.433	99.448	99.455	99.466	99.477	99.484	99.491	99.499
3	27.228	27.052	26.872	26.690	26.597	26.504	26.411	26.316	26.221	26.125
4	14.546	14.374	14.198	14.019	13.929	13.838	13.745	13.652	13.558	13.463
5	10.051	9.888	9.722	9.553	9.466	9.379	9.291	9.202	9.112	9.020
6	7.874	7.718	7.559	7.396	7.313	7.229	7.143	7.057	6.969	6.880
7	6.620	6.469	6.314	6.155	6.074	5.992	5.908	5.824	5.737	5.650
8	5.814	5.667	5.515	5.359	5.279	5.198	5.116	5.032	4.946	4.859
9	5.257	5.111	4.962	4.808	4.729	4.649	4.567	4.483	4.398	4.311
10	4.849	4.706	4.558	4.405	4.327	4.247	4.165	4.082	3.996	3.909
11	4.539	4.397	4.251	4.099	4.021	3.941	3.860	3.776	3.690	3.602
12	4.296	4.155	4.010	3.858	3.780	3.701	3.619	3.535	3.449	3.361
13	4.100	3.960	3.815	3.665	3.587	3.507	3.425	3.341	3.255	3.165
14	3.939	3.800	3.656	3.505	3.427	3.348	3.266	3.181	3.094	3.004
15	3.805	3.666	3.522	3.372	3.294	3.214	3.132	3.047	2.959	2.868
16	3.691	3.553	3.409	3.259	3.181	3.101	3.018	2.933	2.845	2.753
17	3.593	3.455	3.312	3.162	3.083	3.003	2.920	2.835	2.746	2.653
18	3.508	3.371	3.227	3.077	2.999	2.919	2.835	2.749	2.660	2.566
19	3.434	3.297	3.153	3.003	2.925	2.844	2.761	2.674	2.584	2.489
20	3.368	3.231	3.088	2.938	2.859	2.778	2.695	2.608	2.517	2.421
21	3.310	3.173	3.030	2.880	2.801	2.720	2.636	2.548	2.457	2.360
22	3.258	3.121	2.978	2.827	2.749	2.667	2.583	2.495	2.403	2.305
23	3.211	3.074	2.931	2.780	2.702	2.620	2.536	2.447	2.354	2.256
24	3.168	3.032	2.889	2.738	2.659	2.577	2.492	2.403	2.310	2.211
25	3.129	2.993	2.850	2.699	2.620	2.538	2.453	2.364	2.270	2.169
26	3.094	2.958	2.815	2.664	2.585	2.503	2.417	2.327	2.233	2.131
27	3.062	2.926	2.783	2.632	2.552	2.470	2.384	2.294	2.198	2.097
28	3.032	2.896	2.753	2.602	2.522	2.440	2.354	2.263	2.167	2.064
29	3.005	2.868	2.726	2.574	2.495	2.412	2.325	2.234	2.138	2.034
30	2.979	2.843	2.700	2.549	2.469	2.386	2.299	2.208	2.111	2.006
40	2.801	2.665	2.522	2.369	2.288	2.203	2.114	2.019	1.917	1.805
60	2.632	2.496	2.352	2.198	2.115	2.028	1.936	1.836	1.726	1.601
120	2.472	2.336	2.191	2.035	1.950	1.860	1.763	1.656	1.533	1.381
∞	2.321	2.185	2.039	1.878	1.791	1.696	1.592	1.473	1.325	1.000

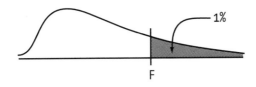

Table B-4. The F Distribution at 5% Significance Level

V_2	V_1								
	1	2	3	4	5	6	7	8	9
1	161.446	199.499	207.314	224.583	230.160	233.988	236.767	238.884	240.543
2	18.513	19.000	18.772	19.247	19.296	19.329	19.353	19.371	19.385
3	10.128	9.552	9.139	9.117	9.013	8.941	8.887	8.845	8.812
4	7.709	6.944	6.511	6.388	6.256	6.163	6.094	6.041	5.999
5	6.608	5.786	5.351	5.192	5.050	4.950	4.876	4.818	4.772
6	5.987	5.143	4.710	4.534	4.387	4.284	4.207	4.147	4.099
7	5.591	4.737	4.307	4.120	3.972	3.866	3.787	3.726	3.677
8	5.318	4.459	4.031	3.838	3.688	3.581	3.500	3.438	3.388
9	5.117	4.256	3.830	3.633	3.482	3.374	3.293	3.230	3.179
10	4.965	4.103	3.678	3.478	3.326	3.217	3.135	3.072	3.020
11	4.844	3.982	3.559	3.357	3.204	3.095	3.012	2.948	2.896
12	4.747	3.885	3.464	3.259	3.106	2.996	2.913	2.849	2.796
13	4.667	3.806	3.385	3.179	3.025	2.915	2.832	2.767	2.714
14	4.600	3.739	3.319	3.112	2.958	2.848	2.764	2.699	2.646
15	4.543	3.682	3.264	3.056	2.901	2.790	2.707	2.641	2.588
16	4.494	3.634	3.216	3.007	2.852	2.741	2.657	2.591	2.538
17	4.451	3.592	3.174	2.965	2.810	2.699	2.614	2.548	2.494
18	4.414	3.555	3.138	2.928	2.773	2.661	2.577	2.510	2.456
19	4.381	3.522	3.106	2.895	2.740	2.628	2.544	2.477	2.423
20	4.351	3.493	3.077	2.866	2.711	2.599	2.514	2.447	2.393
21	4.325	3.467	3.052	2.840	2.685	2.573	2.488	2.420	2.366
22	4.301	3.443	3.029	2.817	2.661	2.549	2.464	2.397	2.342
23	4.279	3.422	3.008	2.796	2.640	2.528	2.442	2.375	2.320
24	4.260	3.403	2.989	2.776	2.621	2.508	2.423	2.355	2.300
25	4.242	3.385	2.972	2.759	2.603	2.490	2.405	2.337	2.282
26	4.225	3.369	2.956	2.743	2.587	2.474	2.388	2.321	2.265
27	4.210	3.354	2.941	2.728	2.572	2.459	2.373	2.305	2.250
28	4.196	3.340	2.928	2.714	2.558	2.445	2.359	2.291	2.236
29	4.183	3.328	2.915	2.701	2.545	2.432	2.346	2.278	2.223
30	4.171	3.316	2.904	2.690	2.534	2.421	2.334	2.266	2.211
40	4.085	3.232	2.821	2.606	2.449	2.336	2.249	2.180	2.124
60	4.001	3.150	2.741	2.525	2.368	2.254	2.167	2.097	2.040
120	3.920	3.072	2.665	2.447	2.290	2.175	2.087	2.016	1.959
∞	3.841	2.996	2.605	2.372	2.214	2.099	2.010	1.938	1.880

Table B-4. *The F Distribution at 5% Significance Level (Continued)*

V_2	V_1									
	10	12	15	20	24	30	40	60	120	∞
1	241.882	243.905	245.949	248.016	249.052	250.096	251.144	252.196	253.254	254.317
2	19.396	19.412	19.429	19.446	19.454	19.463	19.471	19.479	19.487	19.496
3	8.785	8.745	8.703	8.660	8.638	8.617	8.594	8.572	8.549	8.527
4	5.964	5.912	5.858	5.803	5.774	5.746	5.717	5.688	5.658	5.628
5	4.735	4.678	4.619	4.558	4.527	4.496	4.464	4.431	4.398	4.365
6	4.060	4.000	3.938	3.874	3.841	3.808	3.774	3.740	3.705	3.669
7	3.637	3.575	3.511	3.445	3.410	3.376	3.340	3.304	3.267	3.230
8	3.347	3.284	3.218	3.150	3.115	3.079	3.043	3.005	2.967	2.928
9	3.137	3.073	3.006	2.936	2.900	2.864	2.826	2.787	2.748	2.707
10	2.978	2.913	2.845	2.774	2.737	2.700	2.661	2.621	2.580	2.538
11	2.854	2.788	2.719	2.646	2.609	2.570	2.531	2.490	2.448	2.404
12	2.753	2.687	2.617	2.544	2.505	2.466	2.426	2.384	2.341	2.296
13	2.671	2.604	2.533	2.459	2.420	2.380	2.339	2.297	2.252	2.206
14	2.602	2.534	2.463	2.388	2.349	2.308	2.266	2.223	2.178	2.131
15	2.544	2.475	2.403	2.328	2.288	2.247	2.204	2.160	2.114	2.066
16	2.494	2.425	2.352	2.276	2.235	2.194	2.151	2.106	2.059	2.010
17	2.450	2.381	2.308	2.230	2.190	2.148	2.104	2.058	2.011	1.960
18	2.412	2.342	2.269	2.191	2.150	2.107	2.063	2.017	1.968	1.917
19	2.378	2.308	2.234	2.155	2.114	2.071	2.026	1.980	1.930	1.878
20	2.348	2.278	2.203	2.124	2.082	2.039	1.994	1.946	1.896	1.843
21	2.321	2.250	2.176	2.096	2.054	2.010	1.965	1.916	1.866	1.812
22	2.297	2.226	2.151	2.071	2.028	1.984	1.938	1.889	1.838	1.783
23	2.275	2.204	2.128	2.048	2.005	1.961	1.914	1.865	1.813	1.757
24	2.255	2.183	2.108	2.027	1.984	1.939	1.892	1.842	1.790	1.733
25	2.236	2.165	2.089	2.007	1.964	1.919	1.872	1.822	1.768	1.711
26	2.220	2.148	2.072	1.990	1.946	1.901	1.853	1.803	1.749	1.691
27	2.204	2.132	2.056	1.974	1.930	1.884	1.836	1.785	1.731	1.672
28	2.190	2.118	2.041	1.959	1.915	1.869	1.820	1.769	1.714	1.654
29	2.177	2.104	2.027	1.945	1.901	1.854	1.806	1.754	1.698	1.638
30	2.165	2.092	2.015	1.932	1.887	1.841	1.792	1.740	1.683	1.622
40	2.077	2.003	1.924	1.839	1.793	1.744	1.693	1.637	1.577	1.509
60	1.993	1.917	1.836	1.748	1.700	1.649	1.594	1.534	1.467	1.389
120	1.910	1.834	1.750	1.659	1.608	1.554	1.495	1.429	1.352	1.254
∞	1.831	1.752	1.666	1.571	1.517	1.459	1.394	1.318	1.221	1.000

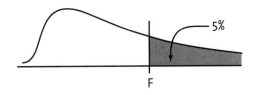

Table B-5. Student's t Distribution

ν	α				
	0.2	0.1	0.05	0.02	0.01
1	3.0777	6.3137	12.7062	31.8210	63.6559
2	1.8856	2.9200	4.3027	6.9645	9.9250
3	1.6377	2.3534	3.1824	4.5407	5.8408
4	1.5332	2.1318	2.7765	3.7469	4.6041
5	1.4759	2.0150	2.5706	3.3649	4.0321
6	1.4398	1.9432	2.4469	3.1427	3.7074
7	1.4149	1.8946	2.3646	2.9979	3.4995
8	1.3968	1.8595	2.3060	2.8965	3.3554
9	1.3830	1.8331	2.2622	2.8214	3.2498
10	1.3722	1.8125	2.2281	2.7638	3.1693
11	1.3634	1.7959	2.2010	2.7181	3.1058
12	1.3562	1.7823	2.1788	2.6810	3.0545
13	1.3502	1.7709	2.1604	2.6503	3.0123
14	1.3450	1.7613	2.1448	2.6245	2.9768
15	1.3406	1.7531	2.1315	2.6025	2.9467
16	1.3368	1.7459	2.1199	2.5835	2.9208
17	1.3334	1.7396	2.1098	2.5669	2.8982
18	1.3304	1.7341	2.1009	2.5524	2.8784
19	1.3277	1.7291	2.0930	2.5395	2.8609
20	1.3253	1.7247	2.0860	2.5280	2.8453
21	1.3232	1.7207	2.0796	2.5176	2.8314
22	1.3212	1.7171	2.0739	2.5083	2.8188
23	1.3195	1.7139	2.0687	2.4999	2.8073
24	1.3178	1.7109	2.0639	2.4922	2.7970
25	1.3163	1.7081	2.0595	2.4851	2.7874
26	1.3150	1.7056	2.0555	2.4786	2.7787
27	1.3137	1.7033	2.0518	2.4727	2.7707
28	1.3125	1.7011	2.0484	2.4671	2.7633
29	1.3114	1.6991	2.0452	2.4620	2.7564
30	1.3104	1.6973	2.0423	2.4573	2.7500
∞	1.2815	1.6448	1.9600	2.3264	2.5758

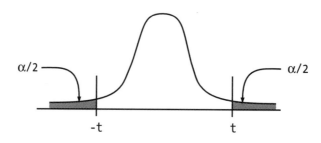

Table B-6. Values of the Absolute Value of the Correlation Coefficient |r|

	5% LEVEL OF SIGNIFICANCE (α) NUMBER OF VARIABLES					1% LEVEL OF SIGNIFICANCE (α) NUMBER OF VARIABLES			
ν	2	3	4	5	ν	2	3	4	5
1	0.997	0.999	0.999	0.999	1	1.000	1.000	1.000	1.000
2	0.950	0.975	0.983	0.987	2	0.990	0.995	0.997	0.998
3	0.878	0.930	0.950	0.961	3	0.959	0.976	0.983	0.987
4	0.811	0.881	0.912	0.930	4	0.917	0.949	0.962	0.970
5	0.754	0.836	0.874	0.898	5	0.874	0.917	0.937	0.949
6	0.707	0.795	0.839	0.867	6	0.834	0.886	0.911	0.927
7	0.666	0.758	0.807	0.838	7	0.798	0.855	0.885	0.904
8	0.632	0.726	0.777	0.811	8	0.765	0.827	0.860	0.882
9	0.602	0.697	0.750	0.786	9	0.735	0.800	0.836	0.861
10	0.576	0.671	0.726	0.763	10	0.708	0.776	0.814	0.840
11	0.553	0.648	0.703	0.741	11	0.684	0.753	0.793	0.821
12	0.532	0.627	0.683	0.722	12	0.661	0.732	0.773	0.802
13	0.514	0.608	0.664	0.703	13	0.641	0.712	0.755	0.785
14	0.497	0.590	0.646	0.686	14	0.623	0.694	0.737	0.768
15	0.482	0.574	0.630	0.670	15	0.606	0.677	0.721	0.752
16	0.468	0.559	0.615	0.655	16	0.590	0.662	0.706	0.738
17	0.456	0.545	0.601	0.641	17	0.575	0.647	0.691	0.724
18	0.444	0.532	0.587	0.628	18	0.561	0.633	0.678	0.710
19	0.433	0.520	0.575	0.615	19	0.549	0.620	0.665	0.698
20	0.423	0.509	0.563	0.604	20	0.537	0.608	0.652	0.685
21	0.413	0.498	0.552	0.592	21	0.526	0.596	0.641	0.674
22	0.404	0.488	0.542	0.582	22	0.515	0.585	0.630	0.663
23	0.396	0.479	0.532	0.572	23	0.505	0.574	0.619	0.652
24	0.388	0.470	0.523	0.562	24	0.496	0.565	0.609	0.642
25	0.381	0.462	0.514	0.553	25	0.487	0.555	0.600	0.633
26	0.374	0.454	0.506	0.545	26	0.478	0.546	0.590	0.624
27	0.367	0.446	0.498	0.536	27	0.470	0.538	0.582	0.615
28	0.361	0.439	0.490	0.529	28	0.463	0.530	0.573	0.606
29	0.355	0.432	0.482	0.521	29	0.456	0.522	0.565	0.598
30	0.349	0.426	0.476	0.514	30	0.449	0.514	0.558	0.591
35	0.325	0.397	0.445	0.482	35	0.418	0.481	0.523	0.556
40	0.304	0.373	0.419	0.455	40	0.393	0.454	0.494	0.526
45	0.288	0.353	0.397	0.432	45	0.372	0.430	0.470	0.501
50	0.273	0.336	0.379	0.412	50	0.354	0.410	0.449	0.479
60	0.250	0.308	0.348	0.380	60	0.325	0.377	0.414	0.442
70	0.232	0.286	0.324	0.354	70	0.302	0.351	0.386	0.413
80	0.217	0.269	0.304	0.332	80	0.283	0.330	0.362	0.389
90	0.205	0.254	0.288	0.315	90	0.267	0.312	0.343	0.368

Table B-6. Values of the Absolute Value of the Correlation Coefficient |r| (Continued)

ν	5% LEVEL OF SIGNIFICANCE (α) NUMBER OF VARIABLES				ν	1% LEVEL OF SIGNIFICANCE (α) NUMBER OF VARIABLES			
	2	3	4	5		2	3	4	5
100	0.195	0.241	0.274	0.300	100	0.254	0.297	0.327	0.351
125	0.174	0.216	0.246	0.269	125	0.228	0.266	0.294	0.316
150	0.159	0.198	0.225	0.247	150	0.208	0.244	0.270	0.290
200	0.138	0.172	0.196	0.215	200	0.181	0.212	0.234	0.253
300	0.113	0.141	0.160	0.176	300	0.148	0.174	0.192	0.208
400	0.098	0.122	0.139	0.153	400	0.128	0.151	0.167	0.180
500	0.088	0.109	0.124	0.137	500	0.115	0.135	0.150	0.162
1000	0.062	0.077	0.088	0.097	1000	0.081	0.096	0.106	0.116

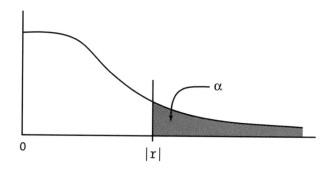

Table B-7. Factors for Determining Control Limits for $\bar{X}$ and R Charts

SAMPLE SIZE n	FACTOR FOR $\bar{X}$ CHART A_2	FACTORS FOR R CHART	
		LOWER CONTROL LIMIT D_3	UPPER CONTROL LIMIT D_4
2	1.88	0	3.27
3	1.02	0	2.57
4	0.73	0	2.28
5	0.58	0	2.11
6	0.48	0	2.00
7	0.42	0.08	1.92
8	0.37	0.14	1.86
9	0.34	0.18	1.82
10	0.31	0.22	1.78
11	0.29	0.26	1.74
12	0.27	0.28	1.72
13	0.25	0.31	1.69
14	0.24	0.33	1.67
15	0.22	0.35	1.65
16	0.21	0.36	1.64
17	0.20	0.38	1.62
18	0.19	0.39	1.61
19	0.19	0.40	1.60
20	0.18	0.41	1.59

Tables of Statistical Distributions and Their Characteristics

TABLES C-1 AND C-2 REPRESENT OUR EFFORT to condense a great deal of information about statistical distributions into one concise source. They will serve as a handy reference for goodness of fit tests (Chapter 3) and Tests of Hypothesis (Chapter 4), as well as other applications in which the shape and/or characteristics of a distribution are needed. The reader is referred to Chapter 3 for guidelines in using these tables.

Table C-1 contains five distributions that are discrete, or involve attributes data. The remaining distributions, shown in Table C-2, are continuous, or use variables data. The name of the distribution is given in the first column, followed by its mass or density function expression in the second column. The expression, you will recall, is termed a "mass" function if the distribution is discrete, or a "density" function if the distribution is continuous. The variable "x" in the expression represents the random variable that follows that statistical distribution. The third column of the tables defines the variables (parameters) and expressions used in the functional expression. The fourth and fifth columns include the expressions for the mean (μ) and variance (σ^2) of the statistical distribution, in terms of the parameters. (Recall that the standard deviation is the square root of the variance.) The next column of the tables illustrates typical shapes of the distribution. For discrete distributions, these are commonly shown as a series of vertical lines (i.e., bar charts), since the random variable may only assume discrete values (such as integers). For continuous distributions, the graphs are smooth curves. These shapes are by no means exhaustive, as distribution curves can vary widely depending on the values of the parameters. The seventh column of the tables is a guide to typical real-world phenomena that may follow a given distribution. Finally, the last column states whether cumulative statistical tables for the given distribution are readily available in statistical texts or handbooks, or whether the distribution values are easy to calculate.

Table C-1. Characteristics of Common Discrete Distributions

NAME OF DISTRIBUTION	MASS FUNCTION EXPRESSION	DEFINITION OF VARIABLES AND EXPRESSIONS	MEAN μ EQUAL TO
Binomial	$\binom{n}{x} p^x (1-p)^{n-x}$ $x = 0,1,2,\ldots,n$	p = probability of "success" (same for each trial) $1 - p$ = probability of "failure" ("success" and "failure" defined in terms of the phenomenon) n = number of independent trials of the phenomenon x = random variable that counts number of "successes" of the phenomenon $\binom{n}{x} = \dfrac{n!}{x!(n-x)!}$ where $n! = 1 \cdot 2 \cdot 3 \cdots n$	np
Poisson	$\dfrac{e^{-u} u^x}{x!}$ $x = 0,1,2,\ldots$	μ = mean of data = np (n and p defined as above, except a "failure" usually makes no sense). x defined as above. e = exponential constant = 2.71828	np
Geometric	$p(1-p)^{x-1}$ $x = 1,2,3,\ldots$	See Binomial above.	$\dfrac{1}{p}$
Negative Binomial	$\binom{x-1}{k-1} p^k (1-p)^{x-k}$ $x = k, k+1, k+2,\ldots$	See Binomial above. x = number of the trial on which the kth "success" occurs.	$\dfrac{k(1-p)}{p}$
Hypergeometric	$\dfrac{\binom{a}{x}\binom{N-a}{n-x}}{\binom{N}{n}}$ $x = 0,1,2,\ldots,n$	p (as defined above) = a/N N = lot (or population) size n = sample size drawn from N a = number of "successes" in N x counts the number of "successes" in n	np

Table C-1. Characteristics of Common Discrete Distributions (Continued)

VARIANCE σ^2 EQUAL TO	TYPICAL SHAPES	TYPICAL PHENOMENA	ARE CUMULATIVE TABLES EASY TO FIND?
$np(1-p)$		Classifying "defectives" and "non-defectives" in an industrial process.	Yes, but usually only through $n = 20$. May also be approximated under certain conditions by the Poisson or normal distributions.
		Classifying observations into one of two possible categories, such as black/white, male/female, pass/fail.	
		Sampling is considered "with replacement" since the likelihood of "success" is the same on every trial.	
np		Arrivals or departures. Objects distributed spatially. Defects appearing in products.	Yes, generally up to $\mu = 25$ and $x = 40$ or so.
$\dfrac{1-p}{p^2}$		Binomial situations, except that the interest is in the first "success" occurring on the xth trial. For example, what is the likelihood of inspecting 20 cars before the first one with a bad computer is found?	No
$\dfrac{k(1-p)}{p^2}$		Binomial situations, except that the interest is in the number of trials necessary to obtain a given number (k) of "successes."	No
$np(1-p)\left(\dfrac{N-n}{N-1}\right)$	(for $n = 5$)	Similar situations to Binomial, except lot or population is finite and sampling is done without replacement, thereby changing the likelihood of a "success" on each trial.	No, although occasionally tables are found up to $N = 20$.

Table C-2. Characteristics of Common Continuous Distributions

NAME OF DISTRIBUTION	DENSITY FUNCTION EXPRESSION	DEFINITION OF VARIABLES AND EXPRESSIONS	MEAN μ EQUAL TO
Uniform	$\dfrac{1}{b-a}$ x between a and b (0 elsewhere)	a and b are the limits over which x may range.	$\dfrac{a+b}{2}$
Normal	$\dfrac{1}{\sqrt{2\pi}\sigma}e^{\frac{-(x-u)^2}{2\sigma^2}}$ x may assume any value (positive or negative)	μ = mean (usually estimated from sample data) σ = standard deviation (usually estimated from sample data) σ^2 = variance e = exponential constant = 2.71828 π = 3.14159	μ
Gamma	$\dfrac{1}{\beta^{\alpha}\Gamma(\alpha)}x^{\alpha-1}e^{\frac{-x}{\beta}}$ for any $x > 0$ (also $\alpha, \beta > 0$)	α, β = parameters of the Gamma distribution, which may be derived from the data by using the mean and variance relationships. $\Gamma(\alpha)$ = the Gamma function $= (\alpha-1)! = 1 \cdot 2 \cdot 3 \cdots (\alpha-1)$ for α a positive integer. In general, $\Gamma(\alpha) = \int_0^{\infty} x^{\alpha-1}e^{-x}dx$ where $\infty =$ infinity and e = exponential constant = 2.71828	$\alpha\beta$
Exponential	$\dfrac{1}{\beta}e^{\frac{-x}{\beta}}$ for any $x > 0$ (also $B > 0$)	See Gamma distribution above.	β
Weibull	$\alpha\beta x^{\beta-1}e^{-\alpha x^{\beta}}$ for any $x > 0$ (also $\alpha, \beta > 0$)	See Gamma distribution above.	$\alpha^{\frac{-1}{\beta}}\Gamma\left(1+\dfrac{1}{\beta}\right)$ *Gamma function (Γ) as defined under Gamma distribution above.*

Table C-2. Characteristics of Common Continuous Distributions (Continued)

VARIANCE σ^2 EQUAL TO	TYPICAL SHAPES	TYPICAL PHENOMENA	ARE CUMULATIVE TABLES EASY TO FIND?
$\frac{1}{12}(b-a)^2$		All values of x are equally likely to occur within its range from a to b.	No, simple to calculate.
σ^2		Most frequently occurring of all continuous distributions. Covers wide range of data whose histograms appear bell-shaped.	Yes, in standard form (see Appendix B), usually up to 3.5 standard deviations either side of the mean.
$\alpha\beta^2$		Reliability and failure rate data. Other continuous data that is skewed rather than symmetric (i.e., non-normal)	No, although occasionally tables are found for limited values of α and β.
β^2		Special case of the Gamma distribution (see above) when $\alpha = 1$. Time between successive arrivals or departures.	No, relatively easy to calculate.
$\alpha^{\frac{-2}{\beta}}\left\{\Gamma\left(1+\frac{2}{\beta}\right)-\left[\Gamma\left(1+\frac{1}{\beta}\right)\right]^2\right\}$		Reliability, life, and failure rate data.	No

Table C-2. Characteristics of Common Continuous Distributions (Continued)

NAME OF DISTRIBUTION	DENSITY FUNCTION EXPRESSION	DEFINITION OF VARIABLES AND EXPRESSIONS	MEAN μ EQUAL TO
Beta	$$\frac{\Gamma(\alpha+\beta)}{\Gamma(\alpha)\cdot\Gamma(\beta)}x^{\alpha-1}(1-x)^{\beta-1}$$ x must be between 0 and 1 (also $\alpha, \beta > 0$)	See Gamma distribution.	$\dfrac{\alpha}{\alpha+\beta}$
Chi-square (χ^2)	$$\frac{x^{\frac{u-2}{2}}}{2^{\frac{\mu}{2}}\Gamma\left(\frac{\mu}{2}\right)}e^{\frac{-x}{2}}$$ for any $x > 0$	μ = mean Also see Gamma distribution.	μ
F	$$\frac{\Gamma\left(\frac{m+n}{2}\right)\left(\frac{m}{n}\right)^{\frac{m}{2}}}{\Gamma\left(\frac{m}{2}\right)\Gamma\left(\frac{n}{2}\right)\left(1+\frac{m}{n}\right)^{\frac{m+n}{2}}}$$ for the positive range	If X and Y are independent χ^2 distributions with m and n degrees of freedom, respectively, the quotient $$\frac{X/m}{Y/n}$$ is F distributed with m and n degrees of freedom. Also see Gamma distribution.	$\dfrac{n}{n-2}$ *for n > 2*
Student's t	$$\frac{\Gamma\left(\frac{n+1}{2}\right)}{\sqrt{n\pi}\,\Gamma\left(\frac{n}{2}\right)\left(1+\frac{x^2}{n}\right)^{\frac{1}{2}(n+1)}}$$ for any x	n = degrees of freedom Also see Gamma distribution.	0

Table C-2. Characteristics of Common Continuous Distributions (Continued)

VARIANCE σ^2 EQUAL TO	TYPICAL SHAPES	TYPICAL PHENOMENA	ARE CUMULATIVE TABLES EASY TO FIND?
$\dfrac{\alpha\beta}{(\alpha+\beta)^2(\alpha+\beta+1)}$		Commonly refers to data given in proportions from 0 to 100%, or 0 to 1.0.	No
2μ		Although a distribution in its own right, χ^2 is more often associated with contingency or goodness of fit tests.	Yes (see Appendix B)
$\dfrac{2n(m+n-2)}{m(n-2)^2(n-4)}$ for $n>4$		The F distribution is commonly associated with testing the ratio of variances from two (or more) sets of data to determine if they differ significantly.	Yes (see Appendix B)
$\dfrac{n}{n-2}$ for $n>2$		The t distribution looks much like the normal, and is often used to test distributions with unknown variances, or to compare means from two distributions.	Yes (see Appendix B)

Visual Basic Routines

THIS APPENDIX CONTAINS a set of Visual Basic routines (functions and procedures) that implement various formulas and methods used in statistical calculations. Table D-1 lists these Visual Basic functions and procedures alphabetically by name. Following Table D-1 is the actual Visual Basic code we used to implement these routines, and it is also in alphabetical order by the routine's name. These routines may be placed in an Access module for use with an Access database.

Table D-1. List of Visual Basic Routines

NAME	TYPE	DESCRIPTION
Chi_Square	Function	Links to Excel to obtain the Chi Square (CHIINV) Value
Close_Excel	Procedure	Closes an Excel workbook that is used to obtain statistical function values
cvNumToLetter	Function	Converts an Excel column number reference to the column letter reference
Database Gauss	Procedure	Handles database inputs and outputs for Gaussian elimination
F_Table	Function	Links to Excel to obtain the F statistic (FINV)
Factorial	Function	Calculates $n!$
Gauss Elimination	Procedure	Solves a set of linear equations using Gaussian elimination
Open_Excel	Procedure	Opens an Excel workbook that is used to obtain statistical function values
Poisson	Function	Poisson probability function
Standard_Normal	Function	Links to Excel to obtain the normal cumulative distribution (NORMALDIST)
t_Table	Function	Links to Excel to obtain the t statistic (tINV)
Zero_Or_Bigger	Function	Returns zero if argument is negative; otherwise, returns argument

Chi_Square: This function is used to obtain the Chi Square value for a given percentage (Prob) and degrees of freedom (DF). It calls the Open_Excel procedure to gain access to the temporary workbook that is used to do the calculations.

```
Public Function Chi_Sq(Prob As Double, DF As Double) As Double

    Dim R          As Integer    ' worksheet row index
    Dim C          As Integer    ' worksheet column index

    ' Default function to zero
    Chi_Sq = 0

    ' If not already open, Open the Excel
    ' workbook and clear first worksheet
    If XLOpen = False Then Call Open_Excel

    ' Initialize
    R = 1
    C = 1

    With XL.worksheets(1)

    ' Set probability parameter
    .cells(R, C).Value = Prob

    ' Set Degrees of Freedom parameter
    R = 2
    .cells(R, C).Value = DF

    ' Have Excel calculate Chi Square
    R = 3
    .Range("A3").Select
    .cells(R, C).Formula = "=CHIINV(R[-2]C[-0],R[-1]C[-0])"

    ' Return Chi Square
    Chi_Sq = .cells(R, 1)

    End With
End Function
```

Close_Excel: This procedure is used by those statistical functions that call Excel to obtain a statistical function value. This procedure closes both the worksheet and workbook. Before the procedure definition is a list of public variables that are used by the procedure.

```
Option Explicit

' EXCEL VARIABLES
Public XL              As Object    ' Excel object
Public XLFileName      As String    ' file name of Excel workbook
Public XLOpen          As Integer   ' Flag denoting an open worksheet

' MESSAGE BOX VARIABLES
Public msg  As String    ' message text
Public rsp  As Long       ' response from message box

Public Sub Close_Excel()

    ' CLOSE OUT EXCEL

    ' Trap close errors
    On Error GoTo HandleCloseError

    ' Save the Excel file
    If XLOpen = True Then
        ' Turn off the displaying of Excel messages
        XL.DisplayAlerts = False

        ' Close the workbook
        XL.Workbooks.Close
        DoEvents
        XLOpen = False

        ' Turn on the displaying of Excel messages
        XL.DisplayAlerts = True
    End If

    ' Reset error trap
    On Error GoTo 0

    ' close Excel
    XL.Application.Quit
    DoEvents
    Set XL = Nothing
    DoEvents
```

```
        ' All done, so exit sub
        Exit Sub


HandleCloseError:
    ' Display the errors
    Beep
    msg = "Error: " & Str(Err.Number) & Chr(13)
    msg = msg & Err.Description
    MsgBox msg
    Exit Sub


End Sub
```

cvNumToLetter: The following function converts an Excel numerical column reference (ColNum) to the equivalent Excel letter column reference. For example, if the column number is 4, the function returns the letter "D."

```
Public Function cvNumToLetter(ByVal ColNum As Integer) As String

    ' Convert an Excel column number to a column letter
    ' The range of column numbers is between 1 and 256.
    '

    ' For example, 1-->A, 2-->B, ..., 26-->Z,
    ' 27-->AA, ..., 53-->BA, ..., 256-->IV
    '

    ' To convert say, 3 to the letter "C," we add 64 to 3 which yields
    ' 67 which is the ASCII code number for the letter "C".

    Dim cn As Integer    ' column as a number [1,256]
    Dim n1 As Integer    ' ASCII code of first letter
    Dim n2 As Integer    ' ASCII for second letter
    Dim cc As String     ' Column reference as letter(s)

    ' Get work copy of column number
    cn = ColNum

    ' Check for a valid column number
    If cn < 1 Or cn > 256 Then
        ' Number is not a valid column number
        ' so default to null string
        cvNumToLetter = ""
        Exit Function
    End If
```

```
    ' Check for double letters
    If cn > 26 Then
        ' handle double letters
        n1 = 64 + (cn - 1) \ 26
        n2 = 65 + (cn - 1) Mod 26
        cc = Chr(n1) & Chr(n2)
    Else
        ' Handle single letter
        n2 = 64 + cn
        cc = Chr(n2)
    End If


    ' Return the letter as a column reference
    cvNumToLetter = cc
End Function
```

Database Gauss: The following procedure creates the matrix for applying Gaussian elimination, calls the Gauss_Elimination routine, and then records the result back in a database table. The two-dimensional matrix M and the vector X are public variables.

```
Public M()      As Single    ' the 2D matrix for the normal equations
Public X()      As Single    ' the unknowns of the equations


Public Sub Database_Gauss()

    ' This procedure performs the Gaussian Elimination
    ' on a set of normal equations whose coefficients
    ' are taken from the results of Query 5_11.  After
    ' Gaussian elimination has been performed, the
    ' results (coefficients of the regression equation)
    ' are saved in a Database table called "Polynomial
    ' Coefficients".

    Dim rs      As New ADODB.Recordset

    Dim fld     As Field        ' data field
    Dim Q       As String       ' A query

    Dim i       As Integer      ' loop index
    Dim r       As Integer      ' row in matrix
    Dim c       As Integer      ' column in matrix
    Dim N       As Integer      ' number of variables
```

```
' Build query to obtain number of variables
Q = "SELECT count(N) as N FROM [Query 5_11]; "


' Process the query
rs.Open Q, CurrentProject.Connection


' get the number of variables
rs.MoveFirst
N = rs("N")


' Define Matrix size
ReDim M(N, N + 1)
ReDim X(N)


' Close out query
rs.Close
Set rs = Nothing



' Build query to obtain normal equation coefficients
Q = "SELECT * "
Q = Q & "FROM [Query 5_11]; "


' Process the query
rs.Open Q, CurrentProject.Connection


' Build Matrix
rs.MoveFirst
r = 1


' Go through each record and populate the
' matrix for the Gaussian elimination
Do While Not rs.EOF
    c = 1
    ' For each record/row
    For Each fld In rs.Fields
        M(r, c) = fld.Value
        c = c + 1
    Next fld

    ' Go to next record
    rs.MoveNext
    r = r + 1
Loop
```

```vb
    ' Close out query
    rs.Close
    Set rs = Nothing

    ' Perform Gaussian elimination
    Call Gauss_Elimination(N, M)

    ' Turn off records modification confirmation message so that
    ' it does not appear while creating the coefficient table.
    DoCmd.SetWarnings False

    ' Execute the query to remove old table.  If table is
    ' not present, continue with next statement
    On Error Resume Next
    DoCmd.RunSQL "DROP TABLE [Polynomial Coefficients];"

    ' Recreate the table with the desired number of coefficients
    Q = "CREATE TABLE [Polynomial Coefficients] ("
    For i = 1 To N
        Q = Q & "Coef" & Format(i, "0") & " Single"
        If i < N Then Q = Q & ", "
    Next i
    Q = Q & ");"
    DoCmd.RunSQL Q

    ' Save the regression equation coefficients
    Q = "INSERT INTO [Polynomial Coefficients] VALUES ("
    For i = 1 To N
        Q = Q & X(i)
        If i < N Then Q = Q & ", "
    Next i
    Q = Q & ");"
    DoCmd.RunSQL Q

    ' Turn on the confirmation message
    DoCmd.SetWarnings True
End Sub
```

F_Table: This function is used to obtain the F statistic value for a given significance level (Alpha) and for a pair of degrees of freedom (v1, v2). It calls the Open_Excel procedure to gain access to the temporary workbook that is used to do the calculations.

```
Public Function F_Table(Alpha As Single, v2 As Long, v1 As Long) As Single

        Dim R           As Integer      ' worksheet row index
        Dim C           As Integer      ' worksheet column index

        ' Default function to zero
        F_Table = 0

        ' If not already open, open the Excel
        ' workbook and clear first worksheet
        If XLOpen = False Then Call Open_Excel

        ' Initialize
        R = 1
        C = 1

        With XL.worksheets(1)

        ' Set sample numerator degrees of freedom
        .cells(R, C).Value = v2

        ' Set sample denominator degrees of freedom
        R = 2
        .cells(R, C).Value = v1

        ' Set level of confidence
        R = 3
        .cells(R, C).Value = Alpha

        ' Have excel calculate F table value
        R = 4
        .Range("A4").Select
        .cells(R, C).Formula = "=FINV(R[-1]C[-0],R[-2]C[-0],R[-3]C[-0])"

        ' Return F table value
        F_Table = CSng(.cells(R, C))

        End With
End Function
```

Factorial: This function calculates the factorial of an integer *n* greater than or equal to zero.

```
Public Function Fact(n As Long) As Double

    ' This function calculates the factorial of n

    Dim i As Integer ' loop index
    Dim p As Double  ' cumulative product

    ' Handle special case of zero and one
    If n = 0 Or n = 1 Then
        Fact = 1
        Exit Function
    End If

    ' Initialize
    p = 1

    ' Accumulate the products
    For i = n To 2 Step -1
        p = p * i
    Next i

    ' Return results
    Fact = p
End Function
```

Gauss Elimination: This procedure performs the Gaussian elimination method for solving a system of linear equations. The two dimensional matrix M and the vector X are public variables.

```
Public M()    As Single   ' the 2D matrix for the normal equations
Public X()    As Single   ' the unknowns of the equations

Public Sub Gauss_Elimination(Nv As Integer)

    ' Perform the Gaussian Elimination

    ' Parameters: nv -- number of variables
    '             m  -- a matrix of NR rows and NC columns

    Dim NC  As Integer ' Number of columns in matrix
    Dim NR  As Integer ' Number of rows in matrix
```

```
Dim i   As Integer  ' Matrix row pointer
Dim j   As Integer  ' Matrix column pointer
Dim jj  As Integer  ' Pointer to pivot row
Dim k   As Integer  ' loop index

Dim Big As Single  ' largest value
Dim ab  As Single  ' absolute value
Dim q   As Single  ' quotient of m(i,k) / m(k,k)
Dim s   As Single  ' sum
Dim Tmp As Single  ' temporary variable for swap

' Initialize
NR = Nv
NC = Nv + 1

' Loop through each equation
For k = 1 To NR - 1

    ' Assume first value is largest
    jj = k
    Big = Abs(M(k, k))

    ' Search for largest possible pivot element
    For i = k + 1 To NR
        ab = Abs(M(i, k))
        If Big < ab Then
            jj = i
            Big = ab
        End If
    Next i

    ' Decide if we need to do row interchange
    If jj <> k Then
        ' swap rows
        For j = k To NC
            Tmp = M(jj, j)
            M(jj, j) = M(k, j)
            M(k, j) = Tmp
        Next j
    End If
```

```
        ' Calculate elements of new matrix
        For i = k + 1 To NR
            q = M(i, k) / M(k, k)
            For j = k + 1 To NC
                M(i, j) = M(i, j) - q * M(k, j)
            Next j
        Next i

        ' Clear column below this (i-th) variable
        For i = k + 1 To NR
            M(i, k) = 0
        Next i

    Next k

    ' First step in back substitution
    X(NR) = M(NR, NC) / M(NR, NR)

    ' Remainder of back substitutions
    For k = 1 To NR - 1
        s = 0
        i = NR - k
        For j = i + 1 To NR
            s = s + M(i, j) * X(j)
        Next j
        X(i) = (M(i, NC) - s) / M(i, i)
    Next k
End Sub
```

Open_Excel: This procedure is used by those functions that are extracting statistical values from Excel. This procedure only opens a temporary Excel workbook and clears the first worksheet. Before the procedure definition is a list of public variables that are used by the procedure.

```
Option Explicit

' EXCEL VARIABLES
Public XL            As Object    ' Excel object
Public XLFileName    As String    ' file name of Excel workbook
Public XLOpen        As Integer   ' Flag denoting an open worksheet
```

```
' MESSAGE BOX VARIABLES
Public msg  As String   ' message text
Public rsp  As Long     ' response from message box

Public Sub Open_Excel()

    ' Open a temporary Excel workbook
    Dim RowCount   As Long      ' worksheet number of rows
    Dim ColCount   As Long      ' worksheet number of columns
    Dim RangeOver  As String    ' worksheet range

    ' Set up an Excel object
    Set XL = CreateObject("EXCEL.APPLICATION")
    XL.Visible = False

    ' Set up temp Excel file
    XLFileName = "C:\Temp\Temp_Excel.xls"

    ' Set error trap
    On Error GoTo XLopenError

    ' Open the Excel file
    XL.Workbooks.Open FileName:=XLFileName

    ' Turn off error trap
    On Error GoTo 0

    ' Set flag to denote workbook has been opened
    XLOpen = True

With XL.worksheets(1)

    ' Activate the first worksheet
    .Activate

    ' Get number of rows
    RowCount = .Rows.Count
    ColCount = .Columns.Count

    ' Set overall range
    RangeOver = "A1" & ":" & cvNumToLetter(ColCount) & Trim(Str(RowCount))
```

```
        ' Clear the worksheet
        .Range(RangeOver).Clear

    End With

        ' All done, exit
    Exit Sub


XLopenError:
    ' Handle errors
    If Err.Number = 1004 Then
        ' File not found
        Beep
        msg = "Do you wish to create a new Excel workbook?"
        rsp = MsgBox(msg, vbYesNo + vbQuestion, "File Creation Confirmation")
        If rsp = vbNo Then
            ' Do not create new file
            Exit Sub
        Else
            ' Create new Excel File
            ' first close current workbook
            If XLOpen = True Then XL.Quit
            ' Create a new workbook
            XL.Workbooks.Add
            DoEvents
            Resume Next
        End If
    End If

        ' Dispay all other errors
    Beep
    msg = "Error number: " & Str(Err.Number)
    msg = msg & Chr(13) & Chr(13)
    msg = msg & Err.Description
    MsgBox msg

End Sub
```

Poisson: This function calculates the Poisson statistic for a given variable *x* and mean. It uses the function Fact that is also given in this appendix.

```
Public Function Poisson(x As Long, mean As Double) As Double
    ' Calculate the Poisson statistic for x.
    Poisson = (Exp(-(mean)) * (mean) ^ x) / Fact(x)
End Function
```

Standard_Normal: This function links to Excel for determining the normal cumulative distribution value for x. The Excel statistical function NORMALDIST is used to perform the calculations. Two additional procedures called by this function are Open_Excel and Close_Excel. Before the function definition is a list of public variables that are needed by the function or the procedure it calls.

```
Option Explicit

' EXCEL VARIABLES
Public XL            As Object    ' Excel object
Public XLFileName    As String    ' file name of Excel workbook
Public XLOpen        As Integer   ' Flag denoting an open worksheet

' MESSAGE BOX VARIABLES
Public msg As String   ' message text
Public rsp As Long     ' response from message box

Public rsp As Long       ' response from message box Function Standard_Normal(x As
Double) As Double

    Dim R        As Integer   ' worksheet row index
    Dim C        As Integer   ' worksheet column index

    ' Default function to zero
    Standard_Normal = 0

    ' If not already open, Open the Excel
    ' workbook and clear first worksheet
    If XLOpen = False Then Call Open_Excel

    ' Initialize
    R = 1
    C = 1

    With XL.worksheets(1)
```

```
   ' Set parameter values
   .cells(R, C).Value = x

   ' Have Excel calculate standard normal
   R = 2
   .Range("A2").Select
   .cells(R, C).Formula = "=NORMDIST(R[-1]C[-0],0,1,TRUE)"

   ' Return standard normal
   Standard_Normal = .cells(R, 1)

   End With
End Function
```

T_Table: This function is used to obtain the *t* statistic value for a given significance level (Alpha) and degrees of freedom (v1). It calls the Open_Excel procedure to gain access to the temporary workbook that is used to do the calculations.

```
Public Function t_Table(Alpha As Single, v1 As Long) As Single

   Dim R          As Integer    ' worksheet row index
   Dim C          As Integer    ' worksheet column index

   ' Default function to zero
   t_Table = 0

   ' If not already open, open the Excel
   ' workbook and clear first worksheet
   If XLOpen = False Then Call Open_Excel

   ' Initialize
   R = 1
   C = 1

   With XL.worksheets(1)

   ' Set sample degrees of freedom
   .cells(R, C).Value = v1

   ' Set level of confidence
   R = 2
   .cells(R, C).Value = Alpha
```

```
' Have Excel calculate t table value
R = 3
.Range("A4").Select
.cells(R, C).Formula = "=tINV(R[-1]C[-0],R[-2]C[-0])"

' Return t table value
t_Table = CSng(.cells(R, C))

    End With
End Function
```

Zero_Or_Bigger: This function returns zero if the value of the argument (x) is negative; otherwise, the function returns the value of the argument.

```
Public Function Zero_Or_Bigger(x As Variant) As Variant

    ' Returns zero if x is negative; otherwise, returns the value of x.

    If x< 0 Then
        Zero_Or_Bigger = 0
    Else
        Zero_Or_Bigger = x
    End If
End Function
```

Bibliography

ALTHOUGH MOST OF THE MATERIAL presented in this book has been assimilated from our experience and course notes, we would like to list several texts to which we referred during our writing. The editions listed below are the most current to our knowledge. These are recommended to any readers who would like additional, or in-depth, treatments of the subjects we covered.

Colburn, Rafe, *Special Edition Using SQL*, 2000. Que (a division of Macmillan USA), Indianapolis, Indiana.

Deming, W. Edwards, *Out of the Crisis*, 2000. Massachusetts Institute of Technology (MIT) Press, Cambridge, Massachusetts.

Devore, Jay L., *Probability and Statistics for Engineering and the Sciences*, Fifth Edition, 1999. PWS Publishing Company, Boston, Massachusetts.

Duncan, Acheson J., *Quality Control and Industrial Statistics*, Fifth Edition, 1994. McGraw-Hill Professional Publishing, New York.

Grant, Eugene L., and Richard S. Leavenworth, *Statistical Quality Control*, Seventh Edition, 1996. McGraw-Hill Higher Education, New York.

Kennedy, John B., and Adam M. Neville, *Basic Statistical Methods for Engineers and Scientists*, Third Edition, 1986. Harper & Row Publishers, New York.

Krajewski, Lee J., and Larry P. Ritzman, *Operations Management: Strategy and Analysis*, Fifth Edition, 1998. Longman, Inc., New York.

Scheaffer, Richard L., and James T. McClave, *Probability and Statistics for Engineers*, Fourth Edition, 1994. Wadsworth Publishing Company, Stamford, Connecticut.

Sinclair, Russell, *From Access to SQL Server*, 2000. Apress, Berkeley, California.

Vaughn, William R., *ADO Examples and Best Practices*, 2000. Apress, Berkeley, California.

Walpole, Ronald E., and Raymond H. Myers, *Probability and Statistics for Engineers and Scientists*, Sixth Edition, 1997. Prentice Hall Inc., Upper Saddle River, New Jersey.

Wells, Garth, *Code Centric: T-SQL Programming with Stored Procedures and Triggers,* 2001. Apress, Berkeley, California.

Index

Numbers and Symbols

$\bar{x}$ (mean symbol)
    in normal density function, 48

s (standard deviation symbol)
    in normal density function, 48

Σ (summation symbol)
    using, 124

100 percent inspection
    versus sampling inspection, 2

A

Abs() function
    operation performed by, 347

aggregate functions
    in SQL, 346–349

α (Greek alpha character)
    notation in hypothesis testing, 85

analysis of variance (ANOVA)
    concept of, 229–275
    involving three factors for life expectancy example, 245–260
    one-way, 230
    two-way, 230

ANOVA. *See* analysis of variance (ANOVA)

ANOVA life expectancy table
    SQL/Query for generating, 253–259

ANOVA table
    for life expectancy example source of variance, 251–252
    of results for the mileage data example, 238

Area Under Standard Normal
    statistical table, 361–362

arithmetic mean
    as measure of central tendency, 10–12

Array_2D
    T-SQL source code for, 158–160

asterisk (*)
    use of as wildcard in SQL, 346

attributes control charts
    for fraction nonconforming, 206–212
    for "yes/no" type decisions, 194

attributes data
    defined, 2

automobile sales example
    for p chart, 207–212

averages
    as data measures, 10

AVG() function
    operation performed by, 347

B

bell curve, 47. *See also* normal statistical distribution

beta distribution
    characteristics of, 378

between-samples mean square
    SQL/Query for calculating, 237–238

between-sample variance
    calculating the sum of squares for, 236
    SQL/Query for determining, 233–235

BETWEEN selection condition
    using, 346

bibliography, 393

bimodal
    defined, 16

Binomial distribution
    characteristics of, 374

C

Calculate_T_Statistic
    T-SQL source code, 103–104

Calculate_Z_Statistic
    T-SQL source code, 105–114

c chart
    for plotting nonconforming numbers, 213–215

C_Chart
    procedure calls for, 227–228
    T-SQL source code for, 224–226

Ceil function
    using in place of Int function, 15

Ceiling() function
    operation performed by, 347

Central Limit Theorem, 184

central line (CL), 185

central line for $\bar{X}$
    notation used for, 200
central tendency
    measure of, 10
central tendency and dispersion
    measures of, 9–40
CFE (cumulative sum of the forecast errors)
    determining forecasting methods performance
        with, 308–311
Char(n) data type, 357
Chart Wizard
    creating a histogram with, 30–31
Chi_Square
    Visual Basic routine, 382
Chi-square distribution X^2
    characteristics of, 378
Chi-Square Distribution (X^2)
    statistical table, 363
chi-square statistics table
    SQL/Query to calculate the X^2 value, 60–61
chi-square test (X^2)
    for goodness of fit, 42
circuit board example
    stabilized p chart and values for the data, 211–212
    table of results over 15-day period, 210
Close_Excel
    Visual Basic routine, 383–384
columnar format
    for three-variable linear regression model, 142
column names
    format of in book, 8
Combine_Intervals
    procedure calls for, 82–83
    T-SQL source code listing, 74–80
common continuous distributions
    types and characteristics of, 376–378
common discrete distributions
    types and characteristics of, 374–375
Compare_Means_2_Samples
    T-SQL source code, 106–111
Compare_Observed_And_Expected
    procedure calls for, 82–83
    T-SQL source code listing, 80–82
complex conditional selection query
    results of, 345
compound conditional selection query
    result of, 345
Contingency_Test
    T-SQL source code, 111–114

contingency test
    for comparing more than two samples, 99–102
continuous data. *See* variables data
continuous statistical distribution
    representation of by a density function, 47
control charts
    common indicators of lack of statistical control, 194
    dissecting, 193–194
    for fraction nonconforming, 206–212
    importance of in business, 181–228
    for sample range and mean values, 195–206
    special causes of variations in, 194
control limits
    on control chart, 185
corn
    forecasting needs for grits, 286
corn drying hypothesis
    testing, 99
correction term
    calculating, 235
    SQL/Query for finding, 235
correlation
    in statistical data analysis, 121–127
correlation coefficient
    alternative way to calculate, 128–129
    calculating for the curve fit, 127–128
    software for calculating, 135
COUNT(*) function
    operation performed by, 347
Counter (or autonumber) data type, 13
CREATE INDEX command
    for creating indexes on one or more columns,
        357–358
CREATE TABLE command
    in SQL, 356–357
Currency data type, 357
curve fit
    calculating the correlation coefficient for, 127–128
curve fitting, 119–180
    additional nonlinear regression models available
        for, 136–141
cvNumToLetter
    Visual Basic routine, 384–385

D

data
    categories of, 2
Database Gauss
    Visual Basic routine, 385–387

data definition
    groups composed of, 356–357
data miner
    importance of data classification by, 2
data organization and measures of central tendency and dispersion
    example of, 35–39
data sampling, 3–4
data types
    common in SQL, 357
Date data type, 357
Decimal(n, m) data type, 357
degree of significance
    establishing before testing a hypothesis, 43–46
degrees of freedom
    determining for our variance estimates test, 234–235
    explanation of, 45–46
    SQL/Query for calculating for two-way ANOVA table, 242
deletion queries
    removing unwanted rows from a table with, 356
Deming, Dr. W. Edwards
    promotion of control chart usage by, 182
    use of control charts by, 186
density function
    for a Poisson distribution, 64–67
dependent variables
    versus independent variables, 120–121
diagnostic tree, 5–6
diagnostic tree and statistical principals
    basic, 1–8
discrete data. *See also* attributes data
    attributes data as, 2
discrete distributions
    characteristics of common, 374–375
Double_Exponential_Smoothing
    procedure calls for, 332–334
    T-SQL source code for, 322–327
Double data type, 357
double exponential smoothing
    general formula for forecast calculations, 294–295
    graph of results versus actual sales, 299
    incorporating a trend with, 292–299
    linear regression performed for, 292–294
    Visual Basic function for calculating results of, 295–299
DROP INDEX command
    removing indexes from databases with, 358

DROP TABLE command
    using to drop or remove tables from databases, 357

E

empty table
    SQL/Query for creating, 13
equal-join, 351
experimental designs
    analysis of, 229–275
Exponential_Model
    call statements for calling, 179–180
    T-SQL source code for, 169–172
exponential curve
    typical with a negative portion of its range, 137
exponential distribution
    characteristics of, 376
    fitting to observed data, 68–72
exponential smoothing
    time series forecasting method, 278

F

F_Table
    Visual Basic routine, 388
factor
    in ANOVA, 230
Factorial
    Visual Basic routine, 389
factorial design, 246
Factors for Determining Control Limits for $\overline{X}$ and R charts
    statistical table, 371
Falls Mill & Country Store
    tracking sales figures for, 281
F distribution
    characteristics of, 378
    using to compare means and variance of two samples, 92–99
F Distribution at 1% Confidence Level
    statistical table, 364–365
F Distribution at 5% Confidence Level
    statistical table, 366–367
Fisher, Sir Ronald Aylmer
    ANOVA concept by, 244
Floor function
    using in place of Int function, 15
Floor() function
    operation performed by, 347
forecasting, 277–334. *See also* time series analysis

forecasting methods

measures used to determine performance of, 308–311

forecasting techniques

criteria for selecting most appropriate, 308–311

performance measures for various, 309

Format() function

operation performed by, 347

fraction nonconforming

control chart for, 206–212

F ratio test

for comparing basketball and football teams, 94

FROM clause

for SQL selection queries, 342

G

Galton, Sir Francis

theory of correlation developed by, 121

gamma distribution

characteristics of, 376

gamma statistical distribution

graph, 62

gas milaege example. *See also* truck mileage study

calculations for the straight line fit in, 124

Gauss Elimination

Visual Basic routine, 389–391

Gaussian_Elimination

call statements for calling, 179–180

manual method for, 133–134

T-SQL source code for, 150–157

Gen_Histogram_Data() function

code listing for, 28–30

Geometric distribution

characteristics of, 374

geometric mean ($\bar{x}_g$)

as measure of central tendency, 17–20

SQL/Query for calculating, 18–20

goodness of fit

test, 46–47

testing for, 41–83

Gosset, W. S.

The Probable Error of a Mean by, 90

gross sales

using simple moving average for predicting, 281–286

GROUP BY clause

using, 348

H

H_0 (null hypothesis), 84

HAVING clause

using, 348–349

H function distibution

using to derive known distributions, 72

histogram

with bar height curve added, 44

constructing to illustrate measures of dispersion, 22–32

options for constructing, 23–25

representing data as, 10

SQL/Query for constructing, 25–32

horse kick analysis

by Simeon Denis Poisson, 64

Hypergeometric distribution

characteristics of, 374

hypothesis testing

additional methods, 84–115

defined, 43–46

for differences among sample means in vehicle mileage versus payloads, 232

notation for significance level or rejection region, 85

possible outcomes for two sales figures, 85

I

independent variables

versus dependent variable, 120–121

indexes

creating on one or more columns in database tables, 357–358

removing from databases, 358

IN selection condition

using, 346

insertion queries, 354–355

instance

of the e-commerce table customers, 338

of the items table, 339

of the orders table, 340

Integer data type, 357

interaction sum of squares values

calculating for life expectancy example, 250–251

Internet usage

comparing a single mean to a specified value for, 86–91

SQL/Query for creating the histogram, 87–89

Internet usage histogram

intervals required for adequate X^2 testing, 87

Internet Web site
  tracking elapsed time between hits on, 68–72
Int() function
  operation performed by, 347
inventory quantities
  determining daily stock levels needed, 66–68

J

John's Jewels
  The Chance of an Accident Given the
    Opportunity, 62
  The Chaos of Boring a Hole, 86
  The Danger of Enthusiasm, 231
  The Degrees of Freedom of Flightless Birds, 46
  Forecasting Grits, 286
  The Godfather of Statistical Distributions, 72
  If Only More Than Half Were Above the Median, 22
  It Got Here Just in Time, 300
  Let Statistics Work for You, 6–7
  The Man Who Advocated Honest Coat Hangers,
    182
  Mean or Median?, 16
  Testing the Hypothesis That the Corn Dries the
    Same, 99
  Would You Want to Accept Bad Quality?, 213
  Yesterday's Data, 193
join conditions
  using with WHERE clause, 350–353
Juran, Dr. Joseph, 186
just-in-time inventory control
  importance of accurate forecasting for, 300

K

kurtosis
  a moment of, 35

L

LCL (lower control limit). *See* lower control limit (LCL)
life expectancy example
  analysis of variance, 250
  analysis of variance involving three factors,
    245–260
  calculating factor C for, 250
  calculating the sum of squares for factor B in, 250
  data table with totals shown, 247
  pulling sums for three main factors from table, 250
  three-factor data table for, 246
LIKE operator
  using, 346

limousine service
  example, 35–39
  testing data to see if it fits a normal distribution,
    48–61
Linear_Regression_2_Variables
  call statements for calling, 179–180
  T-SQL source code for, 148–150
linear correlation
  in two variables, 127–130
linear regression
  in more than two variables, 141–147
  in two variables, 121–127
linear regression model
  normal equations for, 123
logical AND connector
  combining two selection conditions with, 345
logical operators
  for the WHERE clause, 344
logical OR connector
  use of, 345
lower control limit (LCL)
  commonly set distance of, 194
  on control chart, 185

M

MAD (mean absolute deviation)
  determining forecasting methods performance
    with, 308–311
Make_Intervals
  procedure calls for, 82–83
  T-SQL source code listing, 73–74
mass function, 64
MAX() function
  operation performed by, 347
mean ($\bar{x}$)
  comparing a single to a specified value, 86–91
  as measure of central tendency, 10–12
mean absolute deviation (MAD)
  determining forecasting methods performance
    with, 308–311
means and variances
  comparing two samples, 92–98
  general procedure for comparing two samples, 93
  illustrating use of general procedure for comparing
    samples, 94–99
measures of central tendency, 10–22
  mean, 10–12
  median, 12–16
  mode as, 16–17

measures of central tendency and dispersion
  example illustrating data organization and, 35–39
measures of dispersion
  illustrating by constructing a histogram, 22–32
  range, 32–33
  standard deviation, 33–35
median
  as measure of central tendency, 12–16
MedianTable
  SQL/Query for populating, 13–14
median value $\tilde{x}$
  calculating from an even number of records, 13–15
  determining, 12–16
  using an SQL query to find, 13–15
Microsoft Access 2000
  SQL functions offered by, 347
Microsoft SQL Server
  SQL functions offered by, 347
Mid() function
  operation performed by, 347
mileage data example
  ANOVA table of results for, 238
MIN() function
  operation performed by, 347
mode
  as measure of central tendency, 16–17
  SQL/Query for obtaining, 16–17
moment-generating function, 35
moving averages. *See also* simple moving averages
  general formula for calculating weighted, 284
  simple, 278–286
  time series forecasting method, 278
  using to predict gross sales, 281–286
MS values
  SQL/Query for obtaining for two-way ANOVA table, 242–243
Multiple_Linear_Regression
  call statements for calling, 179–180
  T-SQL source code for, 172–179
multiple row insertion
  using, 354–355
multiplicative seasonal adjustment
  incorporating into time series analysis, 300–308

N

Negative Binomial distribution
  characteristics of, 374

nonlinear regression models
  additional available for curve fitting, 136–141
  transforming into linear forms, 136–137
normal distribution. *See also* normal statistical distribution
  characteristics of, 376
  showing area under the curve to the right of 40 miles, 61
  theoretical of limousine customer pick-ups, 50
normal equations
  for linear regression model, 123
normal statistical distribution
  fitting to observed data, 47–62
  rule to follow for testing, 48
  testing to see if limousine data fit, 48–61
normal variate
  using to find the desired area in a normal table, 50
null hypothesis (H_0), 84
number nonconforming, 206–207
number of nonconformities
  control chart for, 213–215

O

observed data
  fitting an exponential distribution to, 68–72
  fitting a Poisson distribution to, 62–68
one-tailed test, 43–44
one-way ANOVA, 231–238
Open_Excel
  Visual Basic routine, 391–393
ORDER BY clause
  using, 349–350
outlier, 12
  effect of on figuring mean or median, 16

P

parabola
  general equation that characterizes, 130–131
parabolic shape
  of gas mileage graph, 130–136
parameter design
  by Dr. Genichi Taguchi of Japan, 260
parcel delivery service
  calculating sample means and variances for, 97–98
  determining the values of t' and v, 98
payload sum of squares
  for between-sample variance, 236

p chart
  for analyzing fraction nonconforming data, 207–212
  calculating the control limits of, 207–208
  modifying to accommodate varying sample sizes, 210–212
$\bar{p}$
  for central line of auto sales example chart, 207–212
Poisson
  Visual Basic routine, 394
Poisson distribution, 63
  characteristics of, 374
  density function for, 64–67
  fitting to observed data, 62–68
  using for control charting plotting of nonconforming numbers, 213–215
Poisson, Simeon Denis
  horse kick analysis by, 64
Polynomial_Regression
  call statements for calling, 179–180
  T-SQL source code for, 160–168
polynomial regression
  in two variables, 130–136
population or universe
  collecting samples from, 2–4
principle of least squares
  regression based on, 123
procedure calls, 227–228, 332–334
  for Make_Intervals, Combine_Intervals, and Compare_Observed_And_Expected, 82–83
  T-SQL source code for, 115

Q

query
  to calculate the sample mean and standard deviation, 52
  creating a seven column temporary work table with, 52
query names
  format of in book, 8

R

random sampling
  importance of, 2
range
  SQL/Query for calculating, 32–33
range and mean
  SQL/Query for finding for training times sample, 197
range and mean charts, 195–206
  determining appropriate sample size for, 195

ranges and sample means
  SQL/Query for determining for training times, 205–206
R chart
  calculating UCL and LCL for, 199
  SQL/Query for eliminating a sample in, 200–201
  SQL/Query for getting data for, 198–199
  for the training data, 199
recordset
  of SELECT clause query results, 342
registered voters example
  using nonlinear regresstion models, 138–141
regression
  basis for, 123
  in statistical data analysis, 121–127
regression curve
  plotted on a graph of data points, 134
regression plane
  fitting to the mileage data, 143–147
relation
  in relational databases, 337
relational database query language. *See* SQL (Structured Query Language)
relational databases
  rules that apply to, 338
  terminology, rules, and operations associated with, 337–340
relational database structure and SQL
  overview of, 337–358
relational structure
  of a table, 337
$\bar{R}$
  SQL/Query for accomplishing the calculations for, 198
Robert's Rubies
  The Creeping Menace of Round-Off Errors, 130
  Density Contours, 40
  The Father of ANOVA, 244
robust design
  by Dr. Genichi Taguchi of Japan, 260
Round() function
  operation performed by, 347
round-off errors
  Robert's Rubies about, 130
rules
  for relational database tables, 338
run chart
  of daily stock index values, 183
  showing mean or central line, 183

S

sales data
  fitting a Poisson distribution to, 62–68
sales figures
  graphing, 280–281
  tracking for a small business, 279–286
Sample_Range_and_Mean_Charts
  procedure calls for, 227–228
  T-SQL source code for, 216–219
sample data
  commonly used approaches for collecting, 2–4
sample range chart (R)
  setting up, 195–206
samples
  comparison of more than two, 99–102
sample sizes
  determining appropriate for range and mean
    charts, 195
sampling inspection, 2
sampling methods, 2–4
Sampling Paradox
  example of, 4
Seasonal_Adjustment
  procedure calls for, 332–334
  T-SQL source code for, 327–332
seasonal factors
  for making adjustments to sales forecasts, 300–308
seasonal sales
  forecasting, 300–308
  tracking variations for a small business, 279–286
SELECT clause
  calculating a sales tax with, 343
  query results, 342
  for SQL selection queries, 341–343
SELECT DISTINCT query
  results from, 343
selection queries
  in SQL, 341–354
SEQUEL (Structured English Query Language)
  introduction of, 340
Shewhart, Dr. Walter
  control chart originator, 186
Simple_Moving_Average
  procedure calls for, 332–334
  T-SQL source code for, 312–314
simple moving averages
  general formula for calculating, 282
  graph of average forecasts versus actual sales, 283

sample table for three month, 282
  versus single exponential smoothing, 286
Single_Exponential_Smoothing
  procedure calls for, 332–334
  T-SQL source code for, 318–322
Single data type, 357
single exponential smoothing, 286–292
  general formula for calculating, 287
  graph of results versus actual sales, 291
  versus moving average methods, 286
  running a Visual Basic function for calculating,
    288–291
single mean
  comparing to a specified value, 86–91
single row insertion
  using, 354
smoothing factor
  use of for forecasting sales, 287
software training course example. *See* training course
  example
software training times
  data table, 196
  data table for the new sampling period, 204
SPC. *See* Statistical Process Control (SPC)
specified value
  comparing a single mean to, 86–91
spreads
  as data measures, 10
S & P's weekly closing index averages for 2000
  table of, 188–189
  Web site address for, 187
SQL query
  for joining three tables, 352–353
SQL/Query
  for accomplishing the calculations for $\bar{R}$, 198
  to calculate the X^2 value, 60–61
  for calculating average fraction defective (p bar)
    for circuit boards, 212
  for calculating LCL and UCL for training hours,
    201
  for calculating mean for number of retirees leaving
    monthly, 11–12
  for calculating mean, standard deviation, and
    estimated standard deviation, 190
  for calculating r, 128–129
  for calculating $\bar{R}$ central line (CL), 201
  for calculating the correction term in two-way
    ANOVA, 240
  for calculating the geometric mean, 18–20

for calculating the range, 33

for calculating the sume of squares total in two-way ANOVA, 240

for calculating the sum of squares for payload in two-way ANOVA, 241

for calculating the sum of squares for speed in two-way ANOVA, 241

calculating the t statistic of the Internet usage example, 90–91

for calculating the t value for the teams, 96

for calculating the value for three standard deviations, 191

for calculating the weighted mean, 21–22

for calculating the z statistic, 89–90

for calculation the within-sample variance estimate, 232–233

for clearing the LinePoints table of old data, 126–127

for comparing basketball and football teams, 95–96

for computing a standard deviation, 34–35

for computing the value for F calculated, 243

for constructing a histogram, 25–32

for creating a histogram with date grouped into intervals, 31–32

for creating a MedianTable, 13

for creating an initial version of a table, 49–51

for creating a two-way ANOVA table, 242–244

for creating data for three month simple moving average table, 283–284

for creating the LinePoints table, 126

for the data for making the $\overline{X}$ chart for training hours, 201–202

for determing the within-sample sum of squares, 237

for determining |r|, 135–136

for determining a second degree polynomial curve, 135–136

for determining number of speeds and degrees of freedom, 240

for determining ranges and sample means for training times, 205–206

for determining seasonal factors, 302–304

for determining the average sales per 100 prospects, 209–210

for determining the between-sample variance, 233–235

determining the coefficients of a regression equation with, 134–135

for determining the degrees of freedom for the between-samples mean square, 237–238

for determining the number of payloads and degrees of freedom, 240

for determining the sum of squares values for two-way ANOVA, 239–241

for eliminating a sample in an R chart, 200–201

for eliminating samples in the $\overline{X}$ chart, 203

for finding a correction term, 235

for finding minimum and maximum x values, 126

for finding range and mean of training times samples, 197

for finding weekly S & P closing index average, 187–190

for forecasting seasonal sales by quarters, 304–305

for generating data for weighted moving average table, 285

for generating the ANOVA life expectancy table, 253–259

for getting data to make the R chart, 198–199

for getting the forecast and sales with seasonal factors, 305–308

for getting the mean and standard deviation for hourly Web site hits, 214–215

for giving the count, total, and degrees of freedom in two-way ANOVA, 240

for giving the Totals column in two-way ANOVA example, 239

for giving the Totals row in two-way ANOVA, 240

for looking up the F values from the F table, 244

to obtain average increase in registered voters, 139–141

for obtaining expected frequencies from vehicle mileage tables, 101–102

for obtaining graph data for weekly S & P closing index averages control chart, 192–193

for obtaining mileage values, 129–130

for obtaining sums shown in life expectancy table, 248–249

for obtaining the average sample size, 190–191

for obtaining the central line and control limits on a control chart, 192

for obtaining the CFE, MAD, and standard deviation, 309–311

for obtaining the mode, 16–17

for obtaining the sums needed for a regression equation, 132–133

for obtaining values for curve fitting, 125–126

to obtain the answer to the "more than 40 miles" question, 61

for performing a linear regression, 292–299

for populating a MedianTable, 13–14

using to find the median value, 13–15

using with a VB procedure, 134–135

SQL query language, 340–358

SQL Servers

versus SQL (Structured Query Language), 7

SQL sources

Web site addresses of, 358

SQL (Structured Query Language)

data extraction examples, 7–8

versus SQL Servers, 7

using to find the median value, 13–15

Sqr() function

operation performed by, 348

SS values

SQL/Query for getting for two-way ANOVA table, 243

Stabilized_P_Chart

procedure calls for, 227–228

T-SQL source code for, 222–224

Standard_Normal

Visual Basic routine, 394–395

Standard_P_Chart

procedure calls for, 227–228

T-SQL source code for, 219–222

standard deviation (*s*)

as measure of dispersion, 33–35

Standard & Poor's (S & P's) Index

charting values to measure mean and standard deviation, 183–193

statistical control chart. *See* control chart

statistical distributions

steps necessary to fit to a set of observed data, 42

tables of, 373–379

testing for goodness of fit, 42

statistical principles and diagnostic tree

basic, 1–8

Statistical Process Control (SPC)

importance of, 6–7

statistical tables, 359–371

statistics

letting them work for you, 6–7

StDev function

operation performed by, 348

Student's *t* distribution

characteristics of, 378

statistical table, 368

using instead of standard deviation test, 89–90

SUM() function

operation performed by, 347

summation symbol (Σ)

using, 124

sum of squares (SS)

calculating correction term for estimating, 235

calculating for factor B in life expectancy table, 250

calculations for two-way analysis of variance, 239

SQL/Query for finding, 236

symbols

geometric mean, 17

mean, 10–12

median value, 12

for sum of a group of data values, 33

summation, 124

for variance estimate, 34

weighted mean, 20

symmetric data, 17

T

T_Table

Visual Basic routine, 395–396

table creation group, 356–357

table names

format of in book, 8

tables

SQL command for removing from databases, 357

tabular format

for three-variable linear regression model, 141

Taguchi, Dr. Genichi

design approach by, 260

tests

of hypothesis, 43–46

The Probable Error of a Mean

publication of, 90

three-variable linear regression model

columnar format for, 142

tabular format for, 141

time series analysis, 277–334

incorporating seasonal influences into, 300–308

Total Quality Management (TQM), 6

Total VB Statistics (FMS, Corp.)

software for calculating correlation coefficients by, 135

TQM. *See* Total Quality Management (TQM)

training course example, 195–206

training times for new sampling period, 204

treatment

in ANOVA, 230

treatment sum of squares

for between-sample variance, 236

truck mileage study. *See also* gas mileage example
    comparison of observed to predicted mileage values, 129
    fitting a regression plane to the mileage data, 143–147
    graph showing best guess line, 123
    graph showing degradation with increasing payload, 122
    graph showing regression line, 126
    observed and predicted milaeage values, with residuals, 144
    table showing degradation with increasing payload, 120
T-SQL source code
    ANOVA, 261–274
    ANOVA procedure calls, 275
    for Array_2D, 158–160
    for C_Chart, 224–226
    Calculate_T_Statistic, 103–104
    Calculate_Z_Statistic, 105–114
    Compare_Means_2_Samples, 106–111
    Contingency_Test, 111–114
    Double_Exponential_Smoothing, 322–327
    for Exponential_Model, 169–172
    for Gaussian_Elimination, 150–157
    for Linear_Regression_2_Variables, 148–150
    for Multiple_Linear_Regression, 172–179
    for performing chi-square goodness of fit tests, 72
    for performing regressions, 147–180
    for Polynomial_Regression, 160–168
    for procedure calls, 115
    for Sample_Range_and_Mean_Charts, 216–219
    for Seasonal_Adjustment, 327–332
    Simple_Moving_Average, 312–314
    Single_Exponential_Smoothing, 318–322
    for Stabilized_P_Chart, 222–224
    for Standard_P_Chart, 219–222
    Weighted_Moving_Average, 314–318
T-SQL source code listing
    Combine_Intervals, 74–80
    Compare_Observed_And_Expected, 80–82
    Make_Intervals, 73–74
t statistic
    SQL/Query for calculating for Internet usage example, 90–91
    T-SQL source code for calculating, 103–104
TS (tracking signal), 308–309
two-tailed test, 43
    typical, 45

two-way analysis of variance
    sum of squares calculations for, 239
    with two treatments, 238–244
two-way ANOVA
    table for mileage example source of variance, 241
    with two treatments, 238–244
Type I error, 85
    example, 86
Type II error, 84–85

U

UCL (upper control limit). *See* upper control limit (UCL)
uniform distribution
    characteristics of, 376
UNION ALL operator
    using, 353–354
UNION operator
    using, 353–354
universe. *See* population or universe
update queries
    changing values in a table with, 355
upper control limit (UCL)
    commonly set distance of, 194
    on control chart, 185

V

Values of Correlation Coefficient *r*
    statistical table, 369–370
Var() function
    operation performed by, 348
variables control charts
    for analyzing the process generating the data, 194
variables data
    defined, 2
    how one variable may be influnced by others, 120–121
variance estimates
    determining degrees of freedom for the test, 234–235
variate value (*z*)
    for Internet usage histogram, 87–88
variation
    common and special causes of, 183–193
    special causes of in control charts, 194
vehicle mileage data tables
    for comparing fuel efficiency, 100–101
    for testing vehicle mileage for various payload levels, 231
Visual Basic routines, 381–396

W

Web site
  graph of average number of hits per hour, 132
  recording hits per hour for, 131
Web site address
  for S & P's weekly closing index averages for
    2000, 187
Weibull distribution
  characteristics of, 376
Weighted_Moving_Average
  procedure calls for, 332–334
  T-SQL source code for, 314–318
weighted mean ($\bar{x}_w$)
  calculating, 20–22
weighted moving average
  general formula for calculating, 284
  graph of forecasts versus actual sales, 285
  table of results for first nine months sales, 284
WHERE clause
  conditional selections, 343–344
  join conditions, 350–353
  logical operators for, 344
within-samples mean square
  SQL/Query for calculating, 237–238

within-sample sum of squares
  SQL/Query for calculating, 237
within-sample variance
  SQL/Query for calculating, 232–233

X

x
  variable in normal density function, 48
X^2 statistical distribution, 42
$\bar{X}$
  notation for central line for $\bar{X}$ chart, 200
$\bar{X}$ chart (sample mean chart)
  constructing for software training example,
    200–201
  setting up, 195–206
  for software training hours, 201
$\tilde{x}$ (median value)
  notation for median value, 12
$\bar{x}_g$ (geometric mean)
  as measure of central tendency, 17–20
  SQL query for calculating, 18–20

Z

Zero_Or_Bigger
  Visual Basic routine, 396

books for professionals by professionals™

apress™

About Apress

Apress, located in Berkeley, CA, is an innovative publishing company devoted to meeting the needs of existing and potential programming professionals. Simply put, the "A" in Apress stands for the "Author's Press™." Apress' unique author-centric approach to publishing grew from conversations between Dan Appleman and Gary Cornell, authors of best-selling, highly regarded computer books. In 1998, they set out to create a publishing company that emphasized quality above all else, a company with books that would be considered the best in their market. Dan and Gary's vision has resulted in over 30 widely acclaimed titles by some of the industry's leading software professionals.

Do You Have What It Takes to Write for Apress?

Apress is rapidly expanding its publishing program. If you can write and refuse to compromise on the quality of your work, if you believe in doing more then rehashing existing documentation, and if you're looking for opportunities and rewards that go far beyond those offered by traditional publishing houses, we want to hear from you!

Consider these innovations that we offer all of our authors:

- **Top royalties with *no* hidden switch statements**
  Authors typically only receive half of their normal royalty rate on foreign sales. In contrast, Apress' royalty rate remains the same for both foreign and domestic sales.

- **A mechanism for authors to obtain equity in Apress**
  Unlike the software industry, where stock options are essential to motivate and retain software professionals, the publishing industry has adhered to an outdated compensation model based on royalties alone. In the spirit of most software companies, Apress reserves a significant portion of its equity for authors.

- **Serious treatment of the technical review process**
  Each Apress book has a technical reviewing team whose remuneration depends in part on the success of the book since they too receive royalties.

Moreover, through a partnership with Springer-Verlag, one of the world's major publishing houses, Apress has significant venture capital behind it. Thus, we have the resources to produce the highest quality books *and* market them aggressively.

If you fit the model of the Apress author who can write a book that gives the "professional what he or she needs to know™," then please contact one of our Editorial Directors, Gary Cornell (gary_cornell@apress.com), Dan Appleman (dan_appleman@apress.com), Karen Watterson (karen_watterson@apress.com) or Jason Gilmore (jason_gilmore@apress.com) for more information.

Apress Titles

ISBN	LIST PRICE	AUTHOR	TITLE
1-893115-01-1	$39.95	Appleman	Appleman's Win32 API Puzzle Book and Tutorial for Visual Basic Programmers
1-893115-23-2	$29.95	Appleman	How Computer Programming Works
1-893115-97-6	$39.95	Appleman	Moving to VB.NET: Strategies, Concepts and Code
1-893115-09-7	$29.95	Baum	Dave Baum's Definitive Guide to LEGO MINDSTORMS
1-893115-84-4	$29.95	Baum, Gasperi, Hempel, and Villa	Extreme MINDSTORMS
1-893115-82-8	$59.95	Ben-Gan/Moreau	Advanced Transact-SQL for SQL Server 2000
1-893115-85-2	$34.95	Gilmore	A Programmer's Introduction to PHP 4.0
1-893115-17-8	$59.95	Gross	A Programmer's Introduction to Windows DNA
1-893115-62-3	$39.95	Gunnerson	A Programmer's Introduction to C#, Second Edition
1-893115-10-0	$34.95	Holub	Taming Java Threads
1-893115-04-6	$34.95	Hyman/Vaddadi	Mike and Phani's Essential C++ Techniques
1-893115-50-X	$34.95	Knudsen	Wireless Java: Developing with Java 2, Micro Edition
1-893115-79-8	$49.95	Kofler	Definitive Guide to Excel VBA
1-893115-56-9	$39.95	Kofler/Kramer	MySQL
1-893115-75-5	$44.95	Kurniawan	Internet Programming with VB
1-893115-19-4	$49.95	Macdonald	Serious ADO: Universal Data Access with Visual Basic
1-893115-06-2	$39.95	Marquis/Smith	A Visual Basic 6.0 Programmer's Toolkit
1-893115-22-4	$27.95	McCarter	David McCarter's VB Tips and Techniques
1-893115-76-3	$49.95	Morrison	C++ For VB Programmers
1-893115-80-1	$39.95	Newmarch	A Programmer's Guide to Jini Technology

ISBN	LIST PRICE	AUTHOR	TITLE
1-893115-81-X	$39.95	Pike	SQL Server: Common Problems, Tested Solutions
1-893115-20-8	$34.95	Rischpater	Wireless Web Development
1-893115-93-3	$34.95	Rischpater	Wireless Web Development with PHP and WAP
1-893115-24-0	$49.95	Sinclair	From Access to SQL Server
1-893115-94-1	$29.95	Spolsky	User Interface Design for Programmers
1-893115-53-4	$39.95	Sweeney	Visual Basic for Testers
1-893115-65-8	$39.95	Tiffany	Pocket PC Database Development with eMbedded Visual Basic
1-893115-59-3	$59.95	Troelsen	C# and the .NET Platform
1-893115-54-2	$49.95	Trueblood/Lovett	Data Mining and Statistical Analysis Using SQL
1-893115-16-X	$49.95	Vaughn	ADO Examples and Best Practices
1-893115-83-6	$44.95	Wells	Code Centric: T-SQL Programming with Stored Procedures and Triggers
1-893115-95-X	$49.95	Welschenbach	Cryptography in C and C++
1-893115-05-4	$39.95	Williamson	Writing Cross-Browser Dynamic HTML
1-893115-78-X	$49.95	Zukowski	Definitive Guide to Swing for Java 2, Second Edition
1-893115-92-5	$49.95	Zukowski	Java Collections

Available at bookstores nationwide or from Springer Verlag New York, Inc. at 1-800-777-4643; fax 1-212-533-3503. Contact us for more information at sales@apress.com.

Apress Titles Publishing SOON!

ISBN	AUTHOR	TITLE
1-893115-99-2	Cornell/Morrison	Programming VB.NET: A Guide for Experienced Programmers
1-893115-72-0	Curtin	Building Trust: Online Security for Developers
1-893115-55-0	Frenz	Visual Basic for Scientists
1-893115-96-8	Jorelid	J2EE FrontEnd Technologies: A Programmer's Guide to Servlets, JavaServer Pages, and Enterprise
1-893115-87-9	Kurata	Doing Web Development: Client-Side Techniques
1-893115-58-5	Oellerman	Fundamental Web Services with XML
1-893115-89-5	Shemitz	Kylix: The Professional Developer's Guide and Reference
1-893115-29-1	Thomsen	Database Programming with VB.NET

Available at bookstores nationwide or from Springer Verlag New York, Inc. at 1-800-777-4643; fax 1-212-533-3503. Contact us for more information at sales@apress.com.